Springer Handbook of Auditory Research

Volume 63

More information about this series at http://www.springer.com/series/2506

The ASA Press

The ASA Press imprint represents a collaboration between the Acoustical Society of America and Springer dedicated to encouraging the publication of important new books in acoustics. Published titles are intended to reflect the full range of research in acoustics. ASA Press books can include all types of books published by Springer and may appear in any appropriate Springer book series.

ASA Press

Rolf M. Quam • Marissa A. Ramsier
Richard R. Fay • Arthur N. Popper
Editors

Primate Hearing and Communication

ASA Press

Springer

Editors
Rolf M. Quam
Department of Anthropology
Binghamton University
Binghamton, NY, USA

Marissa A. Ramsier
Department of Anthropology
Humboldt State University
Arcata, CA, USA

Richard R. Fay
Loyola University Chicago
Chicago, IL, USA

Arthur N. Popper
Department of Biology
University of Maryland
College Park, MD, USA

ISSN 0947-2657 ISSN 2197-1897 (electronic)
Springer Handbook of Auditory Research
ISBN 978-3-319-59476-7 ISBN 978-3-319-59478-1 (eBook)
DOI 10.1007/978-3-319-59478-1

Library of Congress Control Number: 2017947697

Printed on acid-free paper

This Springer imprint is published by Springer Nature
The registered company is Springer International Publishing AG
The registered company address is: Gewerbestrasse 11, 6330 Cham, Switzerland

Acoustical Society of America

The purpose of the Acoustical Society of America (www.acousticalsociety.org) is to generate, disseminate, and promote the knowledge of acoustics. The Acoustical Society of America (ASA) is recognized as the world's premier international scientific society in acoustics, and counts among its more than 7,000 members, professionals in the fields of bioacoustics, engineering, architecture, speech, music, oceanography, signal processing, sound and vibration, and noise control.

Since its first meeting in 1929, the ASA has enjoyed a healthy growth in membership and in stature. The present membership of approximately 7,000 includes leaders in acoustics in the United States of America and around the world. The ASA has attracted members from various fields related to sound including engineering, physics, oceanography, life sciences, noise and noise control, architectural acoustics; psychological and physiological acoustics; applied acoustics; music and musical instruments; speech communication; ultrasonics, radiation, and scattering; mechanical vibrations and shock; underwater sound; aeroacoustics; macrosonics; acoustical signal processing; bioacoustics; and many more topics.

To assure adequate attention to these separate fields and to new ones that may develop, the Society establishes technical committees and technical groups charged with keeping abreast of developments and needs of the membership in their specialized fields. This diversity and the opportunity it provides for interchange of knowledge and points of view has become one of the strengths of the Society.

The ASA's publishing program has historically included the *The Journal of the Acoustical Society of America, JASA-Express Letters, Proceedings of Meetings on Acoustics,* the magazine *Acoustics Today*, and various books authored by its members across the many topical areas of acoustics. In addition, ASA members are involved in the development of acoustical standards concerned with terminology, measurement procedures, and criteria for determining the effects of noise and vibration.

Series Preface

The following preface is the one that we published in volume 1 of the Springer Handbook of Auditory Research back in 1992. As anyone reading the original preface, or the many users of the series, will note, we have far exceeded our original expectation of eight volumes. Indeed, with books published to date and those in the pipeline, we are now set for over 70 volumes in SHAR, and we are still open to new and exciting ideas for additional books.

We are very proud that there seems to be consensus, at least among our friends and colleagues, that SHAR has become an important and influential part of the auditory literature. While we have worked hard to develop and maintain the quality and value of SHAR, the real value of the books is very much because of the numerous authors who have given their time to write outstanding chapters and to our many co-editors who have provided the intellectual leadership to the individual volumes. We have worked with a remarkable and wonderful group of people, many of whom have become great personal friends of both of us. We also continue to work with a spectacular group of editors at Springer. Indeed, several of our past editors have moved on in the publishing world to become senior executives. To our delight, this includes the current president of Springer US, Dr. William Curtis.

But the truth is that the series would and could not be possible without the support of our families, and we want to take this opportunity to dedicate all of the SHAR books, past and future, to them. Our wives, Catherine Fay and Helen Popper, and our children, Michelle Popper Levit, Melissa Popper Levinsohn, Christian Fay, and Amanda Fay Sierra, have been immensely patient as we developed and worked on this series. We thank them and state, without doubt, that this series could not have happened without them. We also dedicate the future of SHAR to our next generation of (potential) auditory researchers—our grandchildren—Ethan and Sophie Levinsohn, Emma Levit, and Nathaniel, Evan, and Stella Fay.

Preface 1992

The Springer Handbook of Auditory Research presents a series of comprehensive and synthetic reviews of the fundamental topics in modern auditory research. The volumes are aimed at all individuals with interests in hearing research including advanced graduate students, postdoctoral researchers, and clinical investigators. The volumes are intended to introduce new investigators to important aspects of hearing science and to help established investigators to better understand the fundamental theories and data in fields of hearing that they may not normally follow closely.

Each volume presents a particular topic comprehensively, and each serves as a synthetic overview and guide to the literature. As such, the chapters present neither exhaustive data reviews nor original research that has not yet appeared in peer-reviewed journals. The volumes focus on topics that have developed a solid data and conceptual foundation rather than on those for which a literature is only beginning to develop. New research areas will be covered on a timely basis in the series as they begin to mature.

Each volume in the series consists of a few substantial chapters on a particular topic. In some cases, the topics will be ones of traditional interest for which there is a substantial body of data and theory, such as auditory neuroanatomy (Vol. 1) and neurophysiology (Vol. 2). Other volumes in the series deal with topics that have begun to mature more recently, such as development, plasticity, and computational models of neural processing. In many cases, the series editors are joined by a co-editor having special expertise in the topic of the volume.

Richard R. Fay, Chicago, IL, USA
Arthur N. Popper, College Park, MD, USA

SHAR logo by Mark B. Weinberg, Potomac, Maryland, used with permission.

Volume Preface

The order Primates comprises over 300 species that constitute our closest living evolutionary relatives. The large degree of diversity in body size, habitat use, diet, and social organization across primates makes them particularly useful for understanding the evolution of hearing and communication in mammals, including humans.

The hearing abilities of primates have been tested experimentally in a large number of species across the primate order, revealing both consistent patterns and considerable variation within and between taxonomic groups. These differences reflect some combination of evolutionary history, anatomical changes, and social and environmental selective pressures—the relative influence of each of these factors is the subject of considerable debate.

At the same time, ongoing studies of primate acoustic communication continue to expand our understanding of the acoustics and social and environmental contexts of primate calls. Historically, primate calls have been evaluated largely by the qualitative assessment of the researcher, but new methods and approaches now enable a greater appreciation for how signals are used and perceived by the primates under consideration. The frequency range and complexity of known primate vocalizations has also expanded substantially with the development of more sophisticated and accessible acoustic technologies. Of growing interest is how selective pressures such as ambient environmental acoustics may have influenced both the vocalizations produced and their propagation through the habitat. New data also expand our understanding of relationships between vocal acoustics and anatomy.

The confluence of factors influencing vocal behavior and the accumulating evidence for the sophistication and complexity in both the signal and its interpretation were the motivations for the present volume. We anticipate that the comprehensive approach to the volume will provide new insights into the intimately related topics of primate hearing and communication and will represent an important contribution to the literature.

The first section of the book involves a discussion of functional anatomy and physiology of sound production, reception, and perception in primates, as well as the acoustic properties of their natural habitats. In Chap. 1, Marissa A. Ramsier and

Rolf M. Quam provide an overview of the volume. This is followed by Chap. 2, wherein Sirpa Nummela discusses the anatomy of the peripheral auditory system in primates, and Chap. 3, wherein Marissa A. Ramsier and Josef Rauschecker consider hearing and sound perception/processing. In Chap. 4, the final chapter in section 1, Charles Brown and Peter Waser examine the acoustic properties of the natural habitats occupied by primates, including both background noise and sound transmission characteristics.

The second section of the book focuses on vocal communication in extant primates. In Chap. 5, Elke Zimmermann examines vocal communication in the most primitive living primates as models for the origin of primate communication. This is followed by a consideration of vocal communication in pair-bonded primates in Chap. 6 by Charles Snowdon. In the last two chapters (Chaps. 7 and 8), the relationship between primate communication and human language is examined by Klaus Zuberbühler, and the evolution of hearing and language in the human lineage is discussed by Rolf M. Quam, Ignacio Martínez, Manuel Rosa, and Juan Luis Arsuaga.

Rolf M. Quam, Binghamton, NY, USA
Marissa A. Ramsier, Arcata, CA, USA
Richard R. Fay, Chicago, IL, USA
Arthur N. Popper, College Park, MD, USA

Contents

1 **Introduction to Primate Hearing and Communication** 1
Marissa A. Ramsier and Rolf M. Quam

2 **The Primate Peripheral Auditory System and the Evolution of Primate Hearing** 13
Sirpa Nummela

3 **Primate Audition: Reception, Perception, and Ecology** 47
Marissa A. Ramsier and Josef P. Rauschecker

4 **Primate Habitat Acoustics** 79
Charles H. Brown and Peter M. Waser

5 **Evolutionary Origins of Primate Vocal Communication: Diversity, Flexibility, and Complexity of Vocalizations in Basal Primates** 109
Elke Zimmermann

6 **Vocal Communication in Family-Living and Pair-Bonded Primates** 141
Charles T. Snowdon

7 **The Primate Roots of Human Language** 175
Klaus Zuberbühler

8 **Evolution of Hearing and Language in Fossil Hominins** 201
Rolf M. Quam, Ignacio Martínez, Manuel Rosa, and Juan Luis Arsuaga

Contributors

Juan Luis Arsuaga Centro de Investigación (UCM-ISCIII) sobre la Evolución y Comportamiento Humanos, Madrid, Spain

Charles H. Brown Department of Psychology, University of South Alabama, Mobile, AL, USA

Ignacio Martinez Departamento de Ciencias de la Vida, Universidad de Alcalá, Alcalá de Henares, Spain

Sirpa Nummela Department of Biosciences, University of Helsinki, Helsinki, Finland

Rolf M. Quam Department of Anthropology, Binghamton University (SUNY), Binghamton, NY, USA

Marissa A. Ramsier Department of Anthropology, Humboldt State University, Arcata, CA, USA

Josef P. Rauschecker Department of Neurosciences, Georgetown University, Washington, DC, USA

Institute for Advanced Study, TUM, Munich-Garching, Munich, Germany

Manuel Rosa Departamento de Teoría de la Señal y Comunicaciones, Universidad de Alcalá, Escuela Politécnica, Alcalá de Henares, Spain

Charles T. Snowdon Department of Psychology, University of Wisconsin, Madison, WI, USA

Peter M. Waser Department of Biological Sciences, Purdue University, West Lafayette, IN, USA

Elke Zimmermann Institute of Zoology, University of Veterinary Medicine, Hannover, Germany

Klaus Zuberbühler Department of Comparative Cognition, Institute of Biology, University of Neuchatel, Neuchatel, Switzerland

School of Psychology and Neuroscience, University of St Andrews, Fife, Scotland, UK

Chapter 1
Introduction to Primate Hearing and Communication

Marissa A. Ramsier and Rolf M. Quam

Abstract The diverse and well-studied order Primates serves as an excellent model for understanding the evolution of acoustic communication among mammals. Over the past 60 million years, primates have evolved into more than 300 extant species that range from nocturnal to diurnal, arboreal to terrestrial, and solitary to groups of thousands, and they range in body mass from the 30-g pygmy mouse lemur (*Microcebus myoxinus*) to the 175-kg eastern lowland gorilla (*Gorilla beringei graueri*). Nonhuman primates vary in their auditory sensitivity and perceptual capabilities and emit a wide range of often complex vocalizations. Some aspects of primate audition and vocalizations have been related to each other and/or phylogeny, anatomy, and ecology, but many aspects have yet to be fully understood. The integration of anatomical and behavioral data on acoustic communication, and the correlates thereof, have significant potential for reconstructing behavior in the fossil record, including that of humans. This volume presents a comprehensive review of nonhuman primate audition and vocal communication to bridge these closely related topics that are often addressed separately. The first section of the book is a discussion of primate sound production, reception, and perception, as well as habitat acoustics in the environmental settings occupied by primates in the wild. The second section focuses on vocal communication in extant primates, including consideration of spectral analyses of primate calls and the evolutionary relationships among hearing, vocal communication, and human language. The goal for this comprehensive approach is to provide new insights into these related topics.

Keywords Acoustic communication • Audition • Bioacoustics • Psychoacoustics • Sensory ecology • Vocalization

M.A. Ramsier (✉)
Department of Anthropology, Humboldt State University,
1 Harpst Street, Arcata, CA 95521, USA
e-mail: marissa.ramsier@humboldt.edu

R.M. Quam
Department of Anthropology, Binghamton University (SUNY),
Binghamton, NY 13902-6000, USA
e-mail: rquam@binghamton.edu

R.M. Quam et al. (eds.), *Primate Hearing and Communication*, Springer Handbook of Auditory Research 63, DOI 10.1007/978-3-319-59478-1_1

1.1 Introduction to the Primates

The order Primates is diverse and well studied and thus serves as an excellent model for understanding the evolution of acoustic communication among mammals. The first primates evolved along with the proliferation of mammals that occurred following the Cretaceous-Paleogene extinction approximately 65 million years ago (Ma) (Ravosa and Dagosto 2007). The closest extant relatives of the order Primates include colugos (order Dermoptera) and tree shrews (order Scandentia) (Godinot 2007; Perelman et al. 2011).

The earliest primates likely evolved from a small (less than about 200 g) insectivorous mammal, perhaps of the order Plesiadapiformes (Ravosa and Dagosto 2007), and they were likely nocturnal (Ross et al. 2007), although many primates have since moved into diurnal niches and evolved color vision (Kawamura et al. 2012). Early primates occupied Africa, Asia, Europe, and North America but became extinct in North America in the Eocene (56–34 Ma). Traits characteristic of the primate order are largely related to precision grip and arboreal mobility (e.g., opposable thumbs, nails instead of claws) and increased reliance on vision (fully enclosed bony eye orbits, forward-facing eyes with increased depth perception and visual acuity) (Fleagle 2013). There are multiple hypotheses for the evolution of these traits, the most popular being tied to an arboreal adaptive niche: increased grasp-leaping locomotion (Szalay and Dagosto 1980); visual predation of small prey within a complex arboreal environment (Cartmill 1970); and the exploitation of angiosperm (flowering, fruiting) plants at the ends of branches where dexterity is advantageous (Sussman 1991).

Over the past 60 million years, primates have proliferated into what are now over 300 diverse extant species separated into two major taxonomic semiorders (Fig. 1.1) (Fleagle 2013). The semiorder Strepsirrhini is the most basal primate group, and, as a whole, its members retain many primitive mammalian features. They range in body size from less than 100 g to approximately 6.5 kg, are primarily arboreal quadrupeds and clingers and leapers, and are mostly nocturnal. Compared to other primates, they have a relatively small brain size and relatively enhanced olfaction, and they display mainly solitary behavior or live in small social groups (Campbell et al. 2010).

Within the strepsirrhines, the infraorder Lorisiformes includes nine genera of nocturnal, insectivorous species that occupy Asia or Africa, such as the slow loris (*Nycticebus coucang*) and the galago (*Galago* sp.). The infraorder Lemuriformes includes fifteen genera that have been isolated on the island of Madagascar for approximately 60 million years and have proliferated into many niches. The Lemuriformes are more variable in their behavior and ecology than the Lorisformes, including species such as the aye-aye (*Daubentonia madagascariensis*) that is nocturnal and solitary and the ring-tailed lemur (*Lemur catta*) that is diurnal and has social groups containing an average of approximately fifteen individuals (Mittermeier et al. 2010).

The second primate semiorder, Haplorhini, includes two suborders. The suborder Tarsiiformes is made up of three genera of small (about 50–150 g), arboreal,

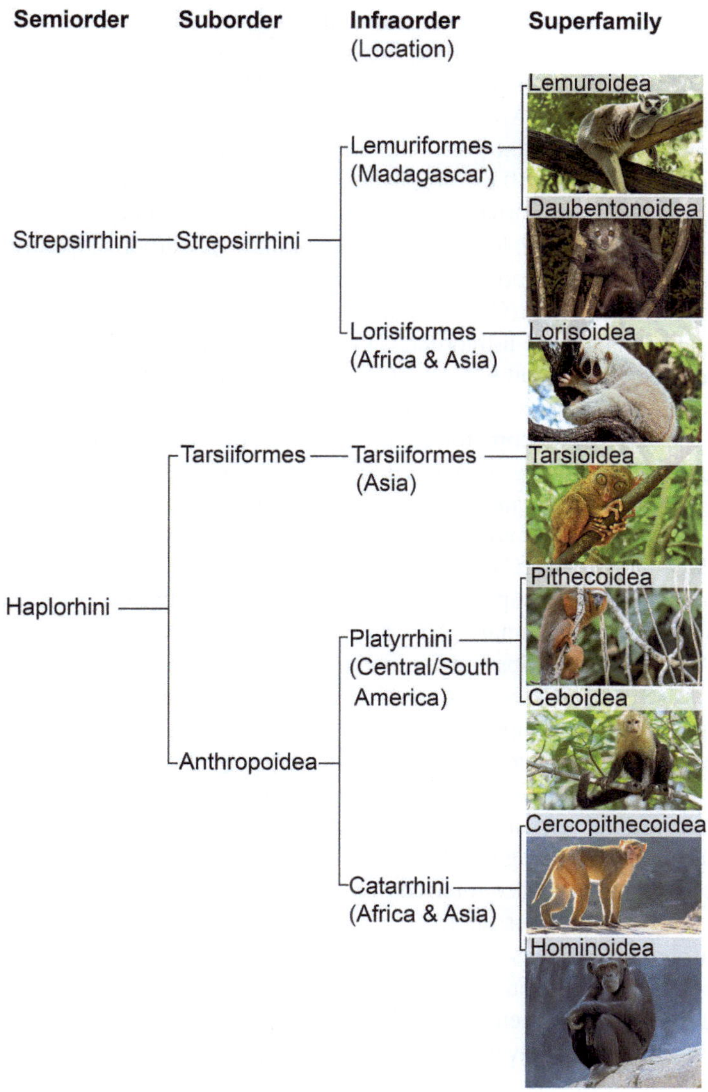

Fig. 1.1 Taxonomy and geographic location of the major primate groups based on Fleagle (2013). Photos from top to bottom: Ring-tailed lemur (*Lemur catta*) (iStock.com/voraorn); Aye-aye (*Daubentonia madagascariensis*) (iStock.com/javarman3); Slow loris (*Nycticebus coucang*) (iStock.com/GreyCarnation); Tarsier (*Carlito* sp.) (iStock.com/LaserLens); Brown titi monkey (*Callicebus brunneus*) (iStock.com/alexakriesphotography); White-faced capuchin (*Cebus capucinus*) (iStock.com/DamianPEvans); Rhesus macaque (*Macaca mulatta*) (iStock.com/Donyanedomam); and Chimpanzee (*Pan troglodytes*) (iStock.com/Fotoamator)

leaping, nocturnal, faunivorous tarsiers (*Tarsius* sp.) from Asia. Although the tarsiers resemble strepsirrhines more closely in their behavior and appearance, they are grouped with the haplorhines on the basis of several evolutionarily derived features not found in strepsirrhines (Wright et al. 2003).

The haplorhine suborder Anthropoidea includes two infraorders. The infraorder Platyrrhini is composed of the New World monkeys that currently live in Central and South America. They likely arrived around 40 Ma after drifting on debris across the Atlantic Ocean from Africa and have since been isolated from other primates (Poux et al. 2006). The New World monkeys include eighteen genera that range in body size from 100 g to 10 kg, are highly arboreal, and consume mostly fruit and plant materials. All are diurnal except for the owl monkey (*Aotus* sp.), which is the only nocturnal anthropoid. Most New World monkeys are arboreal quadrupeds, but some, such as the spider monkey (*Ateles* sp.), practice brachiation, and four genera have fully prehensile tails (Garber et al. 2009).

The second haplorhine infraorder, Catarrhini, includes two major superfamilies. The superfamily Cercopithecoidea (Old World monkeys) includes twenty-three genera from Africa and Asia, approximately 2–31 kg, all of which are diurnal. They show a wide variety of locomotor, social, and dietary habits, ranging from the arboreal, quadrupedal colobus monkeys (e.g., *Colobus* sp.), which have specialized stomachs for processing plant materials, to the more terrestrial geladas (*Theropithecus gelada*), which move across the highlands of Ethiopia in hordes of hundreds to thousands of individuals (Campbell et al. 2010).

The other haplorhine superfamily, the Hominoidea, comprises eight extant genera, including apes and humans. Apes are diurnal primates that lack tails and evince complex social relationships and intelligence. The "lesser" apes (about 5–12 kg) are the gibbons (e.g., *Hylobates* sp.), which are monogamous arboreal brachiators that inhabit Asia. The "great" apes are larger in body size (about 34–175 kg) and include the largely solitary, slow-moving, arboreal clambering orangutans (*Pongo*) of Asia and the gregarious, omnivorous chimpanzees (*Pan* sp.) and folivorous gorillas (*Gorilla gorilla*) of Africa that practice terrestrial knuckle walking and arboreal suspensory behavior (Fleagle 2013).

The earliest humans, also referred to as hominins, diverged from their common ancestor with chimpanzees approximately 6-7 Ma in or near Africa. The earliest traits that defined humans include changes in the dentition (reduction in size of canine teeth) and the adoption of habitual bipedal locomotion. Since the late 1900s, the general thought has been that the driving force in hominin evolution was exploitation of a more open savanna habitat in the face of reduced rainforest habitat, although there is evidence to suggest that the forest still played an important role in early hominin behavior (Wood and Harrison 2011). The earliest hominins, dating to 4–6 Ma, were transitional bipeds but still retained the approximate body and brain size of chimpanzees; there is no evidence of tool use. The Australopithecines (2–4 Ma) were well adapted to bipedalism and had further canine reduction but still retained small body and brain size and limited tool use (Wood and Richmond 2000).

The evolution of the genus *Homo* around 2.5 Ma saw the adoption of consistent tool use, an expansion of brain size, and, eventually, an increase in stature and leg length and a reduction in arm length (Collard and Wood 2007). It was the genus *Homo*, specifically *Homo erectus*, that first left Africa approximately 1.8 Ma. The ancestors of the Neandertals (*Homo* sp.) first emerged in Europe around 500 thousand years ago (Ka) (Arsuaga et al. 2014), and the Neandertals themselves (*Homo neanderthalensis*) evolved in Europe around 200 Ka, roughly at the same time that modern humans, *Homo sapiens*, evolved in Africa (McDougall et al. 2005).

1.2 Primate Hearing and Communication

The hearing abilities of primates have been tested experimentally in nearly 10% of species across the order, and these studies have revealed consistent patterns as well as interesting variations (Ramsier and Rauschecker, Chap. 3). Recent studies have shed light on how variation in anatomical structures along the auditory pathway relates to variations in auditory sensitivity (Coleman and Colbert 2010; Nummela, Chap. 2; Quam, Martínez, Rosa, and Arsuaga, Chap. 8). The work of Brown and Waser (1984) represents one of the rare cases of a study that focused on evolutionary relationships between vocal acoustics and audition in primates. There remains much variation in audition within the order that is not fully understood (Ramsier et al. 2012a; Ramsier and Rauschecker, Chap. 3).

Regarding acoustic communication in general, primates are varied and interesting. For early primates, the dense and discontinuous substrates of an arboreal niche may have decreased the utility of olfactory and visual cues for all but close-range communication, which may have been part of the initial pressure that led to the complex nature of acoustic communication among extant primates (Zimmermann, Chap. 5). Within the larger realm of bioacoustics, anthropologists have long been particularly interested in the vocalizations of nonhuman primates as a model for understanding the evolution of and unique aspects of language in humans (Fedurek and Slocombe 2011). The vocalizations of primates are often species specific and function widely, from communicating with group members and potential mates to warning off competitors and predators (Zuberbühler et al. 1997).

Early studies that included consideration of vocalizations mainly focused on qualitative analyses and descriptions. It was not until the middle of the 20th century that researchers began to include recordings and analyses of the acoustic structure of various calls more routinely. By the turn of the millennium, technological advances and new methodological approaches enabled a greater appreciation for how signals are perceived and used by primates. The range of known primate vocalizations has increased dramatically, and it is now possible to readily obtain high-fidelity, broadband recordings (Maciej et al. 2011) that include very high frequency (i.e., ultrasonic) vocalizations (Ramsier et al. 2012b). New data are also emerging that provide evidence for how variations in vocal acoustics are related to anatomy (Fitch 2006).

Studies of vocal communication in wild primate populations continue to reveal new insights into the social and environmental contexts of many primate calls. Across the order, there is evidence that the number and complexity of vocalizations are tied to social complexity, with more social species producing more complex calls of various types (Semple and McComb 2000; Zimmermann, Chap. 5; Snowdon, Chap. 6; Zuberbühler, Chap. 7). Some primates also communicate with other primate species and emit functionally referential vocalizations that can alert other individuals to specific information, such as the type of predator (e.g., aerial versus terrestrial, leopard versus snake) (Seyfarth et al. 1980; Snowdon, Chap. 6; Zuberbühler, Chap. 7). Nevertheless, there is lingering debate as to the degree to which primate vocalizations have informational content or merely reflect the emotional state of the caller. Most primate vocalizations are considered innate with subtle variations due to voluntary changes in the shape of the vocal tract; however, babbling behavior in some primate species does imply vocal learning (Snowdon, Chap. 6). It is largely agreed that nonhuman primates, while sharing similarities with humans in terms of the function of vocalizations, do not evince the level of complexity and plasticity displayed by humans (Zuberbühler, Chap. 7; Quam, Martínez, Rosa, and Arsuaga, Chap. 8). Phylogenetically, neural control of complex movements of the vocal folds is considered important for the emergence of human speech and language (Fitch et al. 2016).

Several studies have explored how selective pressures, such as habitat structure and ambient environmental acoustics, have influenced the structure of vocalizations produced by primates and the auditory sensitivity of the receivers (Maciej et al. 2011; Brown and Waser, Chap. 4). An analysis of the strepsirrhine semiorder supports a sensory drive hypothesis for primate vocalizations: vocalizations are shaped by external factors such as the sounds produced by predators and habitat acoustics (Zimmermann, Chap. 5). Furthermore, Brown et al. (1995) demonstrated that the broader trend of use of lower frequencies in more forested niches applies to nonhuman primates, agreeing with findings for other taxonomic groups.

1.3 Volume Overview

This volume presents a comprehensive review of nonhuman primate audition and vocal communication and bridges these closely related topics that are often addressed separately. The first section of the book (Chapters 2–4) involves a discussion of functional anatomy and physiology of sound production, reception, and perception in primates, as well as the acoustic properties of their natural habitats.

In Chap. 2, Sirpa Nummela discusses the functional anatomy of the peripheral auditory system of primates, including the outer, middle, and inner ears. Topics include the basic function of the auditory system, auditory anatomy, differences among primate groups, and the evolutionary development of variation in the order. Relationships between audition and the other senses are also discussed.

Marissa Ramsier and Joseph Rauschecker (Chap. 3) follow with a discussion of the auditory sensitivity of primates, including the underlying auditory physiology and the perception of vocalizations, sound localization, and differences in sensitivity across frequencies. They consider what makes nonhuman primates unique among mammals and how nonhuman primate audition is both similar and different from that of humans. Evolutionary explanations for the variations are explored.

In Chap. 4, Charles Brown and Peter Waser discuss environmental constraints on the emission and perception of vocal signals. They particularly focus on how the acoustics of primate habitats affect the propagation of sound, including differences between forested and more open environments, and the elevation of the signaler and receiver high in the canopy versus on the ground. They discuss the corresponding utility of high and low frequency, tonal, and modulated calls and of the acoustic modality in general.

The second section of the book (Chaps. 5–8) focuses on vocal communication in extant primates, including some consideration of spectral analyses of primate calls and the evolutionary relationships between hearing, vocal communication, and human language. In Chap. 5, Elke Zimmermann provides an overview of the origins of primate acoustic communication, using those primates that typically are solitary or form small groups—the strepsirrhines and tarsiers—as models. More specifically, Zimmermann explores the relationship between vocal communication, social complexity, body size, and ecology. This is followed by Charles Snowdon's exploration of vocal communication in primates that form pair-bonded and family groups (Chap. 6). Snowdon focuses on how primates use vocal communication to recognize mates, locate food sources, strengthen social relationships, and defend against competitors and predators. Also included is a discussion of how these factors have led to larger vocal repertoires, flexibility in the structure and use of vocalizations, and social learning.

In Chap. 7, Klaus Zuberbühler continues along this trajectory by exploring the primate roots of human language. Topics include comparison of the complexities of human language and nonhuman primate vocal communication and consideration of processes such as coding, possible syntax, inferential capacities and semantics, and social cognition.

Finally, Rolf Quam, Ignacio Martínez, Manuel Rosa, and Juan Luis Arsuaga explore the evolution of hearing and language in humans in Chap. 8. The focus of the chapter is on the estimation of hearing abilities of fossil hominins based on functional and physiological models of the outer and middle ear in extant apes and humans. The evolution of acoustic communication in primates, including human language, is discussed.

1.4 Open Questions and Future Directions

The chapters in the present volume provide an overview of the present state of knowledge of primate audition and vocal communication. Despite a relative abundance of data on the topic compared to many other mammalian orders, there are still many questions left to be answered, and a more complete understanding of these topics will require an integrated approach. Previous studies have largely focused on primate hearing or vocal communication in isolation, the former often taking a more biomedical approach and the latter a more ecological approach. Both hearing and vocal communication are also topics of interest in the field of psychoacoustics but are largely treated independently. Only through an integrated consideration of auditory reception and perception, vocal acoustics and meaning, and habitat conditions, and the continued exploration of multiple methods of data collection and analysis, will the complexities of primate acoustic communication be more fully understood.

One topic of current debate is how variable nonhuman primates are in their auditory sensitivities. Do they simply follow a general mammalian trend, with variation tied to interaural distance, or do other factors, such as the complexities of social relationships and ecology, exert selective pressures strong enough to be reflected in the audiograms of different species? To address this question, additional data are needed. However, studies aiming to produce traditional behaviorally based audiograms for primates have slowed to a halt; since 1990, only five behavioral audiograms have been reported and only one since 1999 (Osmanski and Wang 2011). This dearth of studies can partially be attributed to a general notion that researchers had adequately defined primate hearing patterns and were unlikely to find anything novel in studying additional species. In addition, due to the logistical constraints of behaviorally based methods, most of the existing data are for species common to laboratory settings and amenable to behavioral training. Whereas this provides useful baseline data for biomedical studies and the primate order in general, it has resulted in a skewed dataset that omits numerous groups that differ in anatomy and behavior and may contain auditory specialists.

Researchers interested in bioacoustic communication in primates have often been forced to assume that the species in question are generalized in terms of auditory sensitivity, likely having capabilities similar to the most closely related species for which data exist or similar to that of the human observers. Thus, there is a clear need to test additional primate species from diverse taxonomic groups. Such an undertaking may best be accomplished by further developing minimally invasive physiologically based techniques, such as the auditory brainstem response, which would enable construction of a more comprehensive dataset of species both in captive facilities and in the field. Such methods have already revealed data for multiple additional primate species (Ramsier and Rauschecker, Chap. 3). Researchers have made progress in understanding how such data compare to behaviorally based audiograms (e.g., Ramsier and Dominy 2010), but there is more to be done, particularly in terms of estimating absolute behavioral thresholds at lower frequencies.

In addition, there is still considerable debate regarding the degree to which the vocalizations and perceptual abilities of nonhuman primates are innate or plastic, simple or complex and, consequently, the extent to which they serve as models for the evolution of acoustic communication and language in humans. Since a seminal study more than 30 years ago suggested the use of external referents by vervet monkeys (*Chlorocebus aethiops*) (Seyfarth et al. 1980), debate has continued as to the extent to which the vocalizations of these and other primates refer to specific external objects in the environment. Related to this is the issue of whether most primate calls contain a semantic element or merely reflect the emotional state of the caller. In both cases, there seems to be a growing consensus that primate calls do contain information relevant to both caller and receiver and that this information content can, but does not always, involve the use of external referents. The fact that some primate alarm calls can be understood by other species, such as the calls of Campbell's monkeys (*Cercopithecus campbelli*) being understood by Diana monkeys (*Cercopithecus diana*) (Zuberbühler 2000), would suggest a role for both social learning and imply the use of external referents. Both aspects are considered crucial for human language since learning a large vocabulary is necessary to communicate increasingly complex ideas, and external referents are functionally similar to words and therefore suggest symbolic communication.

Perhaps related to that, Fitch et al. (2016) have demonstrated that the vocal tract in the long-tailed macaque (*Macaca fascicularis*) can produce five clearly distinguishable vowels, implying that their vocal tract is "speech ready." The fact that they do not produce speech is due in part to a lack of neural control of the vocal structures, mainly of the vocal chords, and also to a lack of fine coordination of respiration and phonation rather than any inherent limitations of their vocal tract anatomy. This finding has the potential to dramatically shift the study of language evolution away from a focus on the anatomy and proportions of the vocal tract and place it more squarely in the cognitive realm. This might help to explain why previous attempts to infer hominin speech capabilities from reconstructions of their vocal tracts (Lieberman et al. 1972) or the shape of their hyoid bones (Arensburg et al. 1990) have been largely inconclusive. This leaves the study of audition as one of the few remaining approaches to studying language evolution in fossil hominins.

In addition, future research on primate hearing and communication would benefit from a more thorough consideration of the influences of habitat acoustics. In addition to generalizations for forest (closed) versus savanna (open) habitats, such studies should consider the acoustics of specific habitats occupied by the species of interest, including deciduous forested environments, different types of savanna landscapes, and the forest margins long considered to be important during the earliest stages of hominin evolution.

Acknowledgments The authors wish to thank Arthur Popper and Richard Fay for their invitation to edit this volume, as well as for their patience and invaluable editorial efforts. We also wish to thank the numerous colleagues with whom discussions of the topics contained in this volume were instrumental to its inspiration and completion.

Compliance with Ethics Requirements

Marissa A. Ramsier declares that she has no conflicts of interest.

Rolf M. Quam declares that he has no conflicts of interest.

References

Arensburg, B., Schepartz, L., Tillier, A., Vandermeersch, B., & Rak, Y. (1990). A reappraisal of the anatomical basis for speech in Middle Paleolithic hominids. *American Journal of Physical Anthropology*, 83, 137–146.

Arsuaga, J. L., Martínez, I., Arnold, L., Aranburu, A., et al. (2014). Neandertal roots: Cranial and chronological evidence from Sima de los Huesos. *Science*, 344(6190), 1358–1363.

Brown, C. H., & Waser, P. M. (1984). Hearing and communication in blue monkeys (*Cercopithecus mitis*). *Animal Behaviour*, 32, 66–75.
Brown, C., Gomez, R., & Waser, P. (1995). Old world monkey vocalizations: Adaptation to the local environment? *Animal Behaviour*, 50, 945–961.
Campbell, C. J., Fuentes, A., Mac Kinnon, K. C., Bearder, S., & Stumpf, R. (2010). *Primates in perspective* (2nd ed.). New York: Oxford University Press.
Cartmill, M. (1970). *The orbits of arboreal mammals: A reassessment of the arboreal theory of primate evolution*. Ph.D. dissertation, University of Chicago. Available from the University of Chicago Library.
Coleman, M. N., & Colbert, M. W. (2010). Correlations between auditory structures and hearing sensitivity in non-human primates. *Journal of Morphology*, 271, 511–532.
Collard, M., & Wood, B. (2007). Defining the genus *Homo*. In W. Henke & I. Tattersall (Eds.), *Handbook of paleoanthropology, Phylogeny of hominids* (Vol. 3, pp. 157–161). Berlin: Springer Science+Business Media.
Fedurek, P., & Slocombe, K. E. (2011). Primate vocal communication: A useful tool for understanding human speech and language evolution? *Human Biology*, 83(2), 153–173.
Fitch, W. (2006). Production of vocalizations in mammals. *Visual Communication*, 3, 115–121.
Fitch, W. T., de Boer, B., Mathur, N., & Ghazanfar, A. A. (2016). Monkey vocal tracts are speech-ready. *Science Advances*, 2(12), e1600723.
Fleagle, J. G. (2013). *Primate adaptation and evolution* (3rd ed.). San Diego, CA: Academic Press.
Garber, P. A., Estrada, A., Bicca-Marques, J. C., Heymann, E. W., & Strier, K. B. (2009). *South American primates*. Chicago, IL: Springer Science+Business Media.
Godinot, M. (2007). Primate origins: A reappraisal of historical data favoring tupaiid affinities. In M. J. Ravosa & M. Dagosto (Eds.), *Primate origins: Adaptations and evolution* (pp. 83–142). New York: Springer Science+Business Media.
Kawamura, S., Hiramatsu, C., Melin, A. D., Schaffner, C. M., et al. (2012). Polymorphic color vision in primates: Evolutionary considerations. In H. Hirai, H. Imai, & Y. Go (Eds.), *Post-genome biology of primates* (pp. 93–120). Tokyo: Springer Tokyo.
Lieberman, P., Crelin, E., & Klatt, D. (1972). Phonetic ability and related anatomy of the newborn and adult human, Neandertal man, and the chimpanzee. *American Anthropologist*, 74, 287–307.
Maciej, P., Fischer, J., & Hammerschmidt, K. (2011). Transmission characteristics of primate vocalizations: Implications for acoustic analyses. *PLoS One*, 6(8), e23015.
McDougall, I., Brown, F., & Fleagle, J. (2005). Stratigraphic placement and age of modern humans from Kibish, Ethiopia. *Nature*, 433, 733–736.
Mittermeier, R., Louis Jr., E., Richardson, M., Schwitzer, C., et al. (2010). *Lemurs of Madagascar* (3rd ed.). Arlington, VA: Conservation International.
Osmanski, M. S., & Wang, X. (2011). Measurement of absolute auditory thresholds in the common marmoset (*Callithrix jacchus*). *Hearing Research*, 227, 127–133.
Perelman, P., Johnson, W. E., Ross, C., Seuánez, H. N., et al. (2011). A molecular phylogeny of living primates. *PLoS Genetics*, 7, e1001342.
Poux, C., Chevret, P., Huchon, D., de Jong, W. W., & Douzery, E. J. P. (2006). Arrival and diversification of caviomorph rodents and platyrrhine primates in South America. *Systematic Biology*, 55(2), 228–244.
Ramsier, M. A., & Dominy, N. J. (2010). A comparison of auditory brainstem responses and behavioral estimates of hearing sensitivity in *Lemur catta* and *Nycticebus coucang*. *American Journal of Primatology*, 72, 217–233.
Ramsier, M. A., Cunningham, A. J., Finneran, J. J., & Dominy, N. J. (2012a). Social drive and the evolution of primate hearing. *Philosophical Transactions of the Royal Society of London B: Biological Sciences*, 367, 1860–1868.
Ramsier, M. A., Cunningham, A. J., Moritz, G. L., Finneran, J. J., et al. (2012b). Primate communication in the pure ultrasound. *Biology Letters*, 8, 508–511.
Ravosa, M. J., & Dagosto, M. (Eds.). (2007). *Primate origins: Adaptations and evolution*. New York: Springer Science+Business Media.

Ross, C. F., Hall, I. M., & Heesy, C. P. (2007). Were basal primates nocturnal? Evidence from eye and orbit shape. In M. Ravosa & M. Dagosto (Eds.), *Primate origins: Adaptations and evolution* (pp. 233–256). New York: Springer Science+Business Media.
Semple, S., & McComb, K. (2000). Perception of female reproductive state from vocal cues in a mammal species. *Philosophical Transactions of the Royal Society of London B: Biological Sciences*, 267, 707–712.
Seyfarth, R. M., Cheney, D. L., & Marler, P. (1980). Monkey responses to three different alarm calls: Evidence of predator classification and semantic communication. *Science*, 210(4471), 801–803.
Sussman, R. W. (1991). Primate origins and the evolution of angiosperms. *American Journal of Primatology*, 23, 209–233.
Szalay, F. S., & Dagosto, M. (1980). Locomotor adaptations as reflected in the humerus of Paleogene primates. *Folia Primatologica*, 34, 1–45.
Wood, B., & Harrison, T. (2011). The evolutionary context of the first hominins. *Nature*, 470(7334), 347–352.
Wood, B., & Richmond, B. (2000). Human evolution: Taxonomy and paleobiology. *Journal of Anatomy*, 196, 19–60.
Wright, P. C., Simons, E. L., & Gursky, S. (2003). *Tarsiers: Past, present, and future*. New Brunswick, NJ: Rutgers University Press.
Zuberbühler, K. (2000). Interspecies semantic communication in two forest primates. *Proceedings of the Royal Society of London B: Biological Sciences*, 267, 713–718.
Zuberbühler, K., Noë, R., & Seyfarth, R. M. (1997). Diana monkey long-distance calls: messages for conspecifics and predators. *Animal Behaviour*, 53(3), 589–604.

Chapter 2
The Primate Peripheral Auditory System and the Evolution of Primate Hearing

Sirpa Nummela

Abstract The primate peripheral auditory organ closely resembles that of other terrestrial mammals. Acoustic communication has an important role in primate communities, and hearing characteristics are well-known for several species. In this chapter, morphological variation of the primate outer, middle, and inner ears is reviewed and is related to auditory data known from experimental work. Differences can be discerned among various primate groups, the greatest differences being between small, mainly nocturnal strepsirrhines and larger, mainly diurnal haplorhines. The evolutionary history of primate hearing is discussed in relation to different hypotheses of primate origins with the view that, as in the earliest mammals, the earliest primates were nocturnal and had good high-frequency hearing. Increased sensitivity to lower frequencies evolved later, although relatively early in the history of primates. This was made possible by an elongation of the cochlea and the disappearance of the secondary spiral lamina. The body size and ecology of primates is related to their ear size, and the role of hearing together with other sensory modalities, mainly vision and olfaction, is discussed.

Keywords Acoustics • Allometry • Auditory sensitivity • Cochlea • Eocene • Fossil primates • Haplorhines • Inner ear • Middle ear • Outer ear • Pinna • Primate origins • Sensory ecology • Sound localization • Strepsirrhines

2.1 Introduction

The modern primates (order Primates) are a diverse group of placental mammals (Archibald and Rose 2005) that includes two main groups, the strepsirrhines (Strepsirrhini) and the haplorhines (Haplorhini), inhabiting tropical and subtropical regions of South America, Africa, and Asia, and drier, more open country,

S. Nummela (✉)
Department of Biosciences, University of Helsinki,
POB 65 (Viikinkaari 1), Helsinki 00014, Finland
e-mail: sirpa.nummela@helsinki.fi

R.M. Quam et al. (eds.), *Primate Hearing and Communication*, Springer Handbook of Auditory Research 63, DOI 10.1007/978-3-319-59478-1_2

particularly in sub-Saharan Africa (Ramsier and Quam, Chap. 1). The living strepsirrhines include two main groups: lemuroids and lorisoids. The lemuroids (Lemuroidea) are endemic to Madagascar, and most species in this group are cathemeral or nocturnal. The lorisoids (Lorisoidea) are found in Africa and Asia, and all are nocturnal (Zimmermann, Chap. 5). The modern haplorhines include three main groups: tarsiers, platyrrhines, and catarrhines. The nocturnal tarsiers (Tarsiiformes) live in SE Asia. The platyrrhines (Platyrrhini, New World monkeys) live in Central and South America, and all but one species are diurnal. Finally, the catarrhines (Catarrhini, Old World monkeys and apes) are diurnal and found in Africa and Asia. The taxonomic diversity of primates is high among mammals, with great morphological and behavioral variety (Fleagle 2013; Henke and Tattersall 2015).

The primate auditory system is much like that in other mammals in general. The peripheral auditory system is situated in the temporal bone. Its three parts, the outer, middle, and inner ears (Fig. 2.1), form a well-functioning sensory system responsible for receiving sound energy from the surrounding environment, transforming this sound energy into mechanical vibrations, and converting this mechanical energy by transduction to nerve impulses (Møller 1974; Rosowski 2003). These nerve impulses are sent forward via the auditory nerve to the brain where they are perceived as sound (Geisler 1998; Ramsier and Rauschecker, Chap. 3). In this chapter, first the peripheral auditory system and its functional morphology in modern primates will be presented separately for the outer, middle, and inner ears (Sects. 2.2–2.4). This is followed by a consideration of the evolutionary history of primate hearing based on data from fossil specimens and modern primates with known audiograms, including the possible role of hearing in the origin of primates (Sect. 2.5). Finally, the sensory ecology of primates is discussed by relating the role of the sense of hearing to other sensory modalities, mainly vision and olfaction (Sect. 2.6).

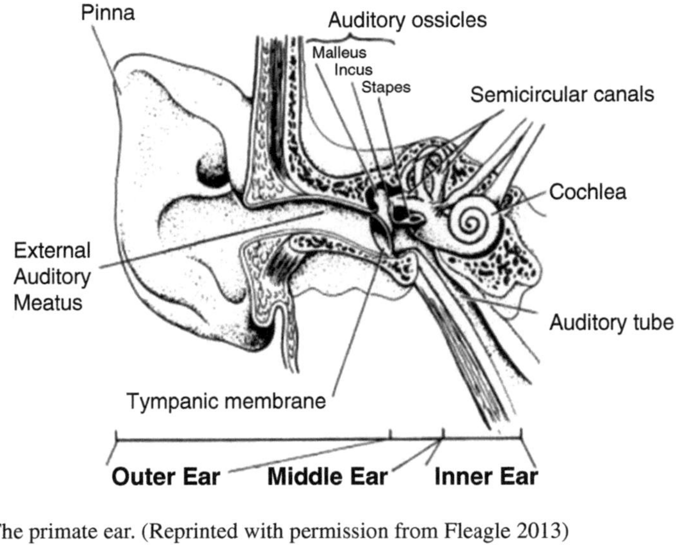

Fig. 2.1 The primate ear. (Reprinted with permission from Fleagle 2013)

Early comparative studies of mammalian/primate auditory anatomy were carried out in the nineteenth century (Hyrtl 1845; Doran 1879) and early twentieth century (van Kampen 1905; van der Klaauw 1931). During the decades since 1960, work by several researchers has moved from pure anatomical descriptions to augmenting the understanding of the functions of mammalian/primate auditory structures. The functional anatomy of the mammalian ear was promoted by Henson (1961, 1974), while Masterton et al. (1969) focused on the evolution of high-frequency hearing among mammals, including primates. Furthermore, Fleischer (1973, 1978) established a basis for future comparative studies in morphology and evolution of the middle and inner ear structures of a wide range of mammalian taxa. Cartmill (1975) and MacPhee (1979, 1981) were early pioneers in the study of the primate auditory region.

Previous reviews have described the diversity among mammalian ears and hearing. Differences in the contribution of each of the temporal bone elements to the construction of the ear region in different mammalian orders were presented in compilations by Novacek (1977, 1993). Research on mammalian hearing and the peripheral auditory system is firmly grounded in the work of Rosowski (1992, 2013). The relationship between the morphological variation of the mammalian ear and hearing characteristics was also studied by Hemilä et al. (1995) and by Nummela (1995). Vertebrate auditory physiology, evolution, and development are presented widely by Manley et al. (2004) and Manley (2014). A comparative view on hearing in the animal kingdom can be found in the review by Köppl et al. (2014). Experimental work on mammalian hearing and on the role of high-frequency hearing in sound localization combined morphological and experimental data (R. S. Heffner and H. E. Heffner 1992a; H. E. Heffner and R. S. Heffner 2014) that expanded the understanding of the significance of the auditory sense in an organism's behavioral ecology.

Primate auditory morphology and its evolution since the Paleocene epoch received detailed attention through the research efforts of Coleman and Ross (2004) and Coleman and Boyer (2012). The primate cochlear labyrinth and its relationship to hearing abilities were studied by Kirk and Gosselin-Ildari (2009) and Armstrong et al. (2011). Auditory capacities in fossil hominins were estimated from the anatomy of the outer and middle ears by Martínez et al. (2004) and Quam et al. (2015). For vertebrates in general, the evolution of the ear and hearing is treated by Clack et al. (2016), and vertebrate sound production and acoustic communication is discussed by Suthers et al. (2016).

The functional approach to sensory research often asks the question: What limits the organism's ability to receive sensory information? Within auditory science, the ultimate goal is to understand how animals hear and how they process information that they receive through their auditory organs (Friedland 2006; Yost et al. 2008). In addition to their auditory physiology, the morphology of auditory structures is useful for revealing ontogenetic changes (Basch et al. 2016; Maier and Ruf 2016), constructing phylogenetic and systematic scenarios (Gunz et al. 2012; Stoessel et al. 2016a, b), studying ecological interactions (Dominy et al. 2001), and discovering possible trade-offs between hearing and other sensory systems (Dominy et al. 2004; Nummela et al. 2013).

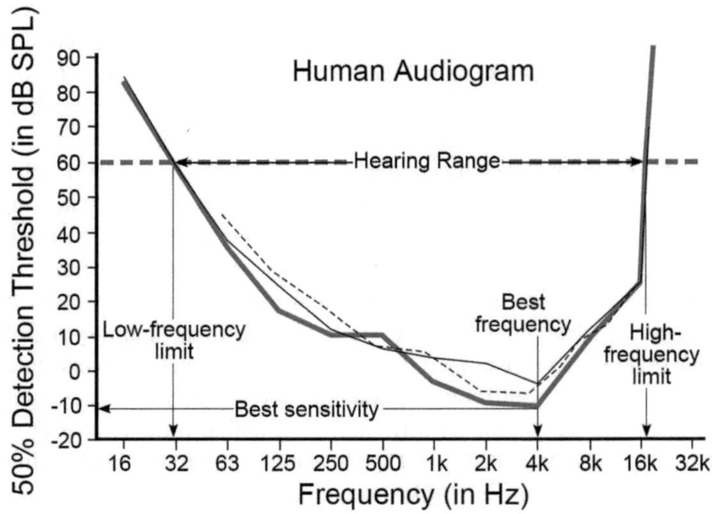

Fig. 2.2 Average human audiograms from three separate studies. The *thick gray line* indicates the audiograms measured with the conditions described by Jackson et al. (1999). Standard audiograms from Sivian and White (1933) (*dashed line*) and from Davis (1960) (*thin line*) are indicated for comparison. Common audiometric parameters used for describing hearing sensitivity are indicated, with the hearing range customarily set between high- and low-frequency limits at 60 dB SPL. (Reprinted with permission from R. S. Heffner 2004)

The two primary methods for measuring the hearing sensitivity of an animal are behavioral testing and auditory brainstem recordings to obtain audiograms, and comparisons of these methods are a matter of discussion (Ramsier and Dominy 2010; Ramsier and Rauschecker, Chap. 3). A comprehensive selection of vertebrate audiograms can be found in Fay (1988). Here, a selection of behavioral audiograms for primates will be presented. A human behavioral audiogram (Fig. 2.2) shows the general characteristics that are often of interest in an animal's hearing. These include the overall hearing range, the low- and high-frequency limits (the lowest and highest frequency heard at 60 dB SPL), the best sensitivity (in dB), and the best frequency (the frequency heard at the best sensitivity). Behavioral audiograms are shown separately for a few strepsirrhines, platyrrhines, and catarrhines (Figs. 2.3, 2.4 and 2.5). The behaviorally measured auditory parameters are often used for finding possible correlations between hearing sensitivity and auditory morphology in primates and determining how different structures affect the hearing characteristics of a species; the outer, middle, and inner ears all are known to contribute to this (Ruggero and Temchin 2002; Vater et al. 2004). This is discussed more in Sects. 2.2, 2.3.3, and 2.4.

Hearing is a particularly important sensory modality since it allows perception of phenomena that might be out of sight. To take advantage of any potential information carried by sound (e.g., vocalizations), an organism has to perform at least one of three basic auditory tasks: sound detection, sound localization, and sound identification (R. S. Heffner 2004; Yost et al. 2008). Although animals are

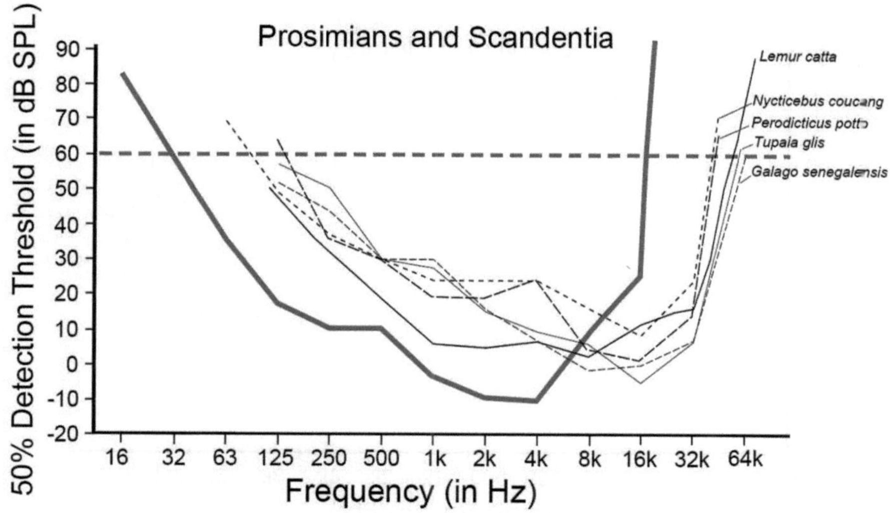

Fig. 2.3 Average audiograms for strepsirrhines (formerly known as prosimians): ring-tailed lemur (*Lemur catta*), lesser bushbaby (*Galago senegalensis*), potto (*Perodicticus potto*), and Sunda slow loris (*Nycticebus coucang*). The audiogram of the tree shrew (*Tupaia glis*; Scandentia), a species closely related to primates, is also shown. The *thick gray line* is the human audiogram. (Modified and reprinted with permission from R. S. Heffner 2004)

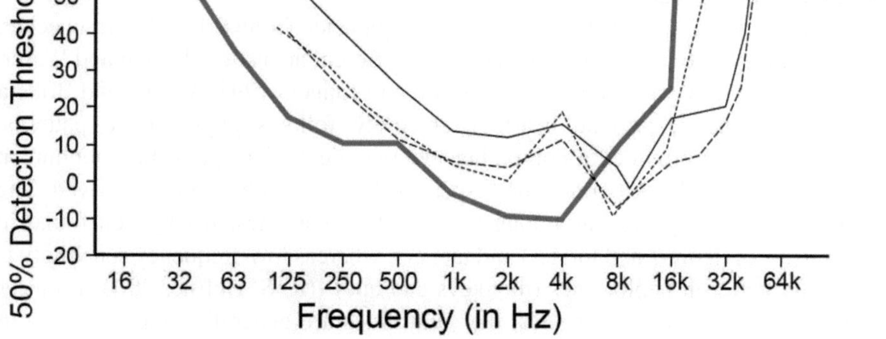

Fig. 2.4 Average audiograms for New World monkeys: squirrel monkey (*Saimiri sciureus*), owl monkey (*Aotus trivirgatus*), and marmoset (*Callithrix jacchus*) with the human audiogram (*thick gray line*) for comparison. (Reprinted with permission from R. S. Heffner 2004)

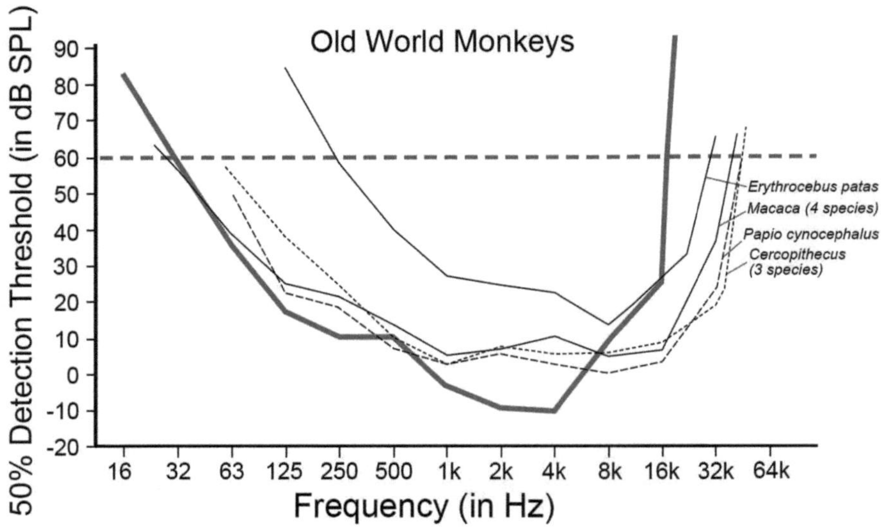

Fig. 2.5 Average audiograms for Old World monkeys: patas monkey (*Erythrocebus patas*), yellow baboon (*Papio cynocephalus*), four macaques (*Macaca fascicularis, M. fuscata, M. mulatta, and M. nemestrina*), and three cercopithecines (*Cercopithecus aethiops*, *C. mitis*, and *C. neglectus)* with the human audiogram (*thick gray line*) for comparison. (Reprinted with permission from R. S. Heffner 2004)

capable of performing these practically simultaneously, sound detection is still the most basic of these tasks since it underlies the other two. Of the various ways to measure and describe hearing sensitivity, a behavioral audiogram provides insights into what an animal actually hears (see Ramsier and Rauschecker, Chap. 3), and the behavior of an entire organism is generally considered the target for selective pressure (R. S. Heffner 2004).

Primates make use of different sound frequencies (Ramsier and Rauschecker, Chap. 3), they live in different habitats with different acoustics (Brown and Waser, Chap. 4), and they vary in body size (Smith and Jungers 1997; Mattila and Bokma 2008). In general, primate hearing sensitivity follows phylogenetic patterns. Coleman (2009) provided a comprehensive overview of all published nonhuman primate behavioral audiograms, both speaker derived and headphone derived. The best high-frequency hearing among the extant primates tested has been found in strepsirrhines (lemurs and lorises) and tarsiers, whereas low-frequency sensitivity is generally better in haplorhines (monkeys and apes) (R. S. Heffner 2004; Ramsier et al. 2012a). Within haplorhines, the catarrhines are generally more sensitive to lower frequencies than are the platyrrhines, and high-frequency sensitivity is reduced in apes compared with monkeys. Furthermore, monkeys and apes (except humans) often show two peaks of maximum sensitivity, whereas lemurs and lorises, as well as humans, generally have only a single peak in sensitivity in their audiograms (Coleman 2009; Ramsier and Rauschecker, Chap. 3).

2.2 Outer Ear and Interaural Distance

Mammals are the only vertebrates with a distinct outer ear that consists of a pinna and the ear canal (external auditory meatus). Both of these structures show large morphological variation among mammals. The pinna varies in its height and width, location on the head, and size in relation to the head size. The outer ear canal varies in its overall geometry, length, and cross-sectional area. A larger pinna can collect a larger amount of sound energy and is also more suitable for the long wavelengths of low frequencies (R. S. Heffner et al. 1982; Ahlborn 2004). The level of amplification of the incoming sound performed by the outer ear depends directly on the size of the pinna and, especially, on the length of the ear canal (Dallos 1973; Zwislocki 1975).

All primates possess cartilaginous pinnae, which show considerable diversity in prominence, morphology, and mobility (Fig. 2.6). There is also variation in the placement of the pinnae on the head, how much the pinnae protrude above the head, and whether they are covered by fur or not (Packer and Sarmiento 1984). Relative to head size, the pinna is larger in strepsirrhines than in haplorhines, with the largest pinnae found in aye-ayes (*Daubentonia madagascariensis*) (Coleman and Ross 2004) and the smallest ones in orangutans (*Pongo pygmaeus*) (Schultz 1973). In strepsirrhines, the auricular musculature is better developed, and the pinnae are generally more mobile and protrude more above the head than in haplorhines (Coleman and Ross 2004; Fleagle 2013). The large morphological variation of the primate pinna (Fig. 2.6) generally follows phylogeny but also has ecological patterns. Strepsirrhines and tarsiers, both of which include many nocturnal species, have relatively tall and narrow pinnae, whereas platyrrhines and catarrhines, which are mostly diurnal species, have shorter, wider, and more symmetrical pinnae. While no functional significance has been attributed to these shape differences, they may still be useful in primate systematics (Coleman and Ross 2004; Coleman and Colbert 2010).

The length of the bony portion of the outer ear canal was measured on dry skulls by Masali et al. (1992) for small samples of humans, chimpanzee (*Pan troglodytes*), gorilla (*Gorilla gorilla*), and orangutan. The cartilaginous portion of the ear canal was estimated to form one-third of the total length in humans, and the total ear canal length was calculated as one-and-a-half times the length of the bony portion for all the ape species. The fundamental resonance frequency of the outer ear canal (the frequency of maximum dB gain) was estimated from its length and was around 3.0 kHz in humans, while the apes ranged between approximately 1.5–2.0 kHz.

The correspondence between the fundamental resonance frequency of the external auditory meatus and the frequency of best hearing sensitivity can be examined. The best frequency for humans is around 3 kHz (Fig. 2.2) and the fundamental resonance frequency is 3.1 kHz. For the chimpanzee, the best frequency is around 2 kHz (Elder 1934; Masterton et al. 1969) and the fundamental resonance frequency is 2.15 kHz (Masali et al. 1992). A close association between the best-heard frequencies and the central portion of the frequency range of human language is well-known (Shaw 1974), and the fundamental resonance frequency of the ear canal also corresponds to the modal pitch of long-distance calls in the other ape species.

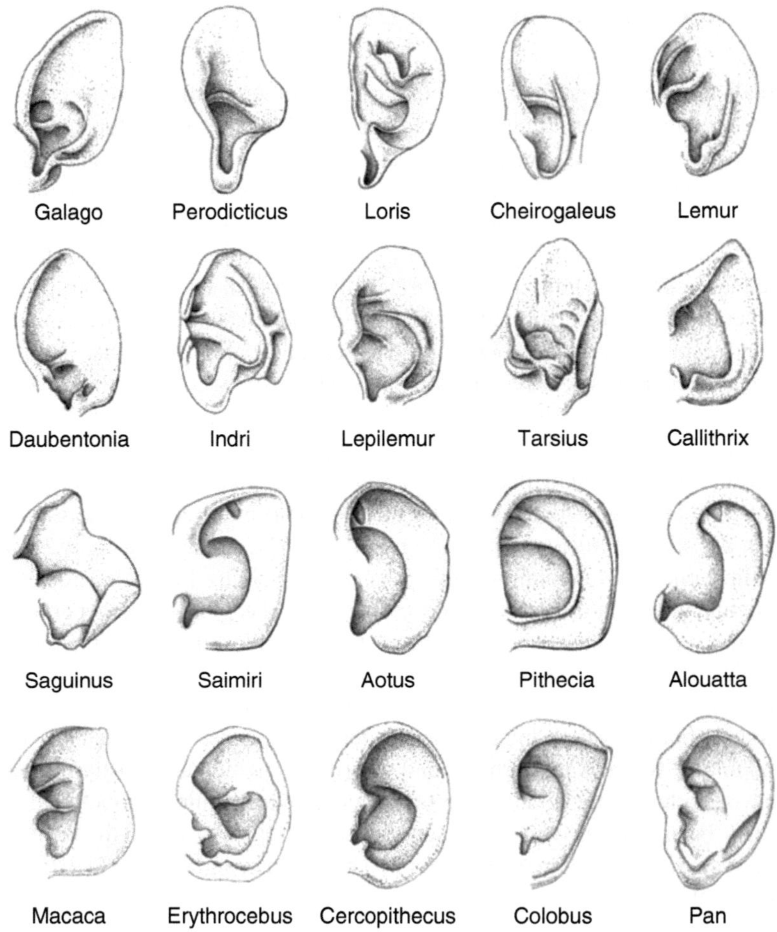

Fig. 2.6 Variation of the primate ear pinna morphology for diverse taxa. The pinnae here are scaled to the same size. Lorisoids: galago (*Galago* sp.), potto (*Perodicticus potto*), and a loris. Lemuroids: dwarf or mouse lemur (*Cheirogaleus* sp.), ring-tailed lemur (*Lemur catta*), aye-aye (*Daubentonia madagascariensis*), indri (*Indri indri*), sportive lemur (*Lepilemur* sp.). Tarsiiformes: tarsier (*Tarsius* sp.). New World monkeys: marmoset (*Callithrix* sp.), tamarin (*Saguinus* sp.), *squirrel monkey* (*Saimiri* sp.), owl monkey (*Aotus* sp.), saki monkey (*Pithecia* sp.), howling monkey (*Alouatta* sp.). Old World monkeys and apes: a macaque (*Macaca* sp.), patas monkey (*Erythrocebus patas*), guenon (*Cercopithecus* sp.), colobus monkey (*Colobus* sp.), and chimpanzee (*Pan troglodytes*). (Modified and reprinted with permission from Coleman and Ross 2004)

The lower frequencies of best sensitivity in apes (compared with humans) might be advantageous in a forested environment where low frequencies propagate better (Masali et al. 1992; Brown and Waser, Chap. 4).

Sound localization is of vital importance for the survival of an animal and has most likely been under selective pressure during the evolution of mammals (H. E. Heffner and R. S. Heffner 2016). In addition to head movements, mammals can use

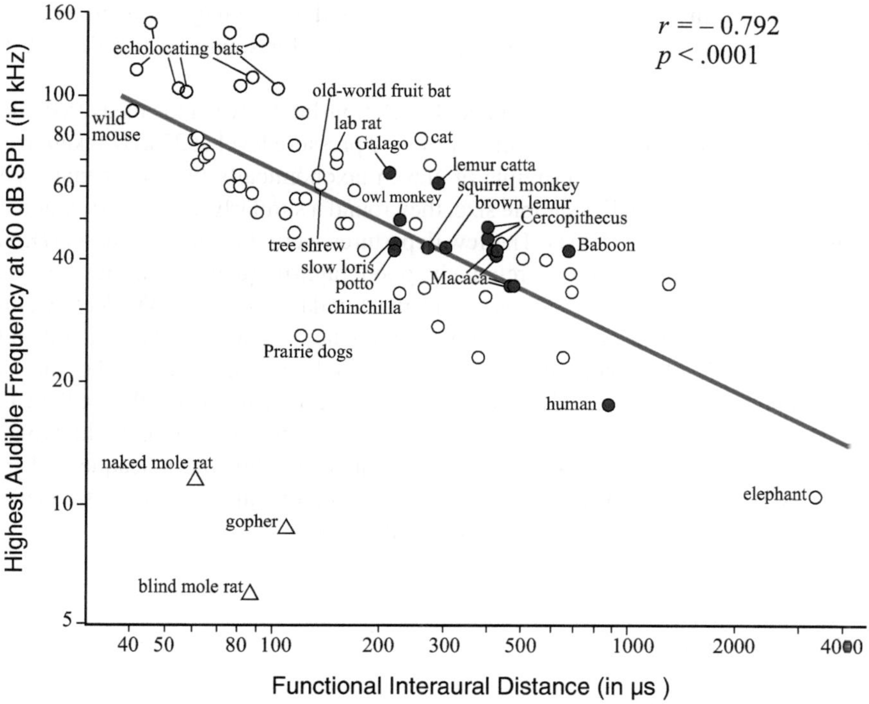

Fig. 2.7 Relationship between maximum functional interaural distance (the time for a sound in air or water to travel from one ear canal to the other) and the high-frequency hearing limit at 60 dB SPL for over sixty mammals. *Filled circles*: primates, with species referenced earlier and the brown lemur (*Eulemur fulvus*), three guenons (*Cercopithecus* sp.), and the yellow baboon; *open circles*: selection of other mammals; *open triangles*: subterranean mammals (not included in the statistical analysis). (Reprinted with permission from R. S. Heffner 2004)

their movable pinnae when locating a sound source and directing the eyes to it, thus enabling the animal to add another sense to the sound localization task (Masterton et al. 1969; R. S. Heffner and H. E Heffner 1992a; also see Sect. 2.6). Large animals with a large interaural distance can use the interaural time difference for sound localization, but for small animals with a small head, including many primate species, the interaural intensity difference is especially useful at higher frequencies.

Figure 2.7 shows how the high-frequency hearing limit is related to the maximum functional interaural distance (interaural distance divided by sound velocity, in microseconds) in over sixty different mammals. The sixteen primate species included do not deviate from the general mammalian pattern in this respect (Masterton et al. 1969; R. S. Heffner and H. E Heffner 1992a). Better high-frequency hearing is correlated with smaller interaural distance. With some exceptions among other vertebrates, only mammals hear frequencies over 10 kHz (Fay 1988).

Indeed, the ability to hear high frequencies gives the advantage of using spectral cues if an animal has a pinna to produce these cues for sound localization (R. S. Heffner 2004).

Coleman and Colbert (2010) studied the relationship between the measured interaural distance and frequency sensitivity (sound pressure level at 32 kHz taken from an animal's audiogram) in a sample of eleven taxonomically diverse primate species. Even with this smaller sample size, they found a similarly strong correlation as found by R. S. Heffner (2004). The few departures were the yellow baboon (*Papio cynocephalus*), which showed relatively good high-frequency hearing, and the Japanese macaque (*Macaca fuscata*) and the Sunda slow loris (*Nycticebus coucang*), which both had somewhat poor high-frequency sensitivity relative to interaural distance. In this particular study, these findings may simply be due to methodological differences. Coleman and Colbert (2010) suggested that in the case of baboons, some other ecological factors related to communication could explain the exceptionally good sensitivity to high frequencies. For the macaque and the loris, other anatomical or ecological factors may be impairing the high-frequency sensitivity that is predicted by the interaural distance.

2.3 Middle Ear

2.3.1 Temporal Bone

The anatomical divisions of the primate temporal bone (squamous, petrous, and tympanic) have large contacts with each other, and the temporal bone articulates with other cranial bones as well (Novacek 1977, 1993). As a result, the ear is not acoustically isolated from the skull, and sound waves can travel through the skull. Primates are the only mammals with an auditory bulla formed solely by the petrous part (petrosal) of the temporal bone (Cartmill et al. 1981; MacPhee 1981). The auditory bulla surrounds the middle ear cavity (Table 2.1; Fig. 2.8). In strepsirrhines, the petrosal is expanded into a balloon-like protrusion and forms a comparatively inflated auditory bulla. Lemurs have a single-chambered middle ear cavity, whereas lorises exhibit a two-chambered middle ear cavity with additional air-filled space in the form of an anterior accessory cavity, but there is no diverticulum (Cartmill 1972, 1975). In tarsiers, which are small nocturnal haplorhines, the auditory bulla is fairly large and located almost in the center of the cranial base, close to the foramen magnum. As a result, the left and right bullae are near one another. As for other haplorhines, in many New World monkeys the petrosal is similar to that in strepsirrhines, whereas in Old World monkeys and apes (including humans), the petrosal has a rough texture on its ventral surface and does not form a balloon-like structure. Haplorhines have a two-chambered middle ear cavity with a posterior accessory cavity with air-filled spaces off of the epitympanic recess and a diverticulum off of the Eustachian tube (Packer and Sarmiento 1984; MacPhee and Cartmill 1986).

Table 2.1 Ear characteristics among different primate groups. For further information, see Sect. 2.3.1

	Strepsirrhines		Haplorhines		
	Lemuroids	Lorisoids	Tarsiers	New World monkeys (platyrrhines)	Old World monkeys and apes (catarrhines)
Petrosal	Forms the auditory bulla in strepsirrhines and haplorhines				
Auditory bulla	Balloon-like and inflated		Enlarged and situated close to the foramen magnum	Balloon-like and inflated (in most species)	Not balloon-like, rough texture ventrally
Middle ear cavity	Single chambered	Two-chambered: anterior accessory cavity with air-filled spaces, but no diverticulum	Two chambered: posterior accessory cavity with air-filled spaces off the epitympanic recess and a diverticulum off the Eustachian tube		
Tympanic	Forms complete bony ring or bony tube in all primates				
Bony ring/ bony tube	Bony ring unfused to the bulla (free floating)	Bony ring fused to the bulla	Bony tube fused to the bulla	Bony ring fused to the bulla	Bony tube fused to the bulla
Outer ear canal	Almost entirely cartilaginous, no bony tube		Elongated tympanic forms the bony part	Almost entirely cartilaginous, no bony tube	Elongated tympanic forms the bony part

In primates, as in moles, many rodents, and elephants, the insertion site for the tympanic membrane takes the form of a complete bony ring (Table 2.1; Fig. 2.9), whereas in some other mammals (e.g., artiodactyls and perissodactyls) the insertion is U-shaped (van der Klaauw 1931; Fleischer 1973). Within strepsirrhines, the tympanic bone consists solely of a bony ring. In lemurs, this ring is unfused (free floating) within the middle ear cavity (Fig. 2.8), but it is still surrounded by the relatively large, single-chambered auditory bulla, whereas in lorises, the tympanic ring is fused to the wall of the bulla (Cartmill 1975). Within haplorhines, New World monkeys resemble lorises in having a laterally fused tympanic ring. In catarrhines, as well as in tarsiers, the tympanic bone is likewise fused to the bulla wall, but instead of forming a single ring, the tympanic forms a bony tube. This elongated bony tube extends laterally toward the external ear opening at the side of the skull and forms the outer ear canal. Strepsirrhines and platyrrhines do not possess this bony tube but have only a tympanic ring, and their external auditory meatus is almost entirely cartilaginous (Table 2.1).

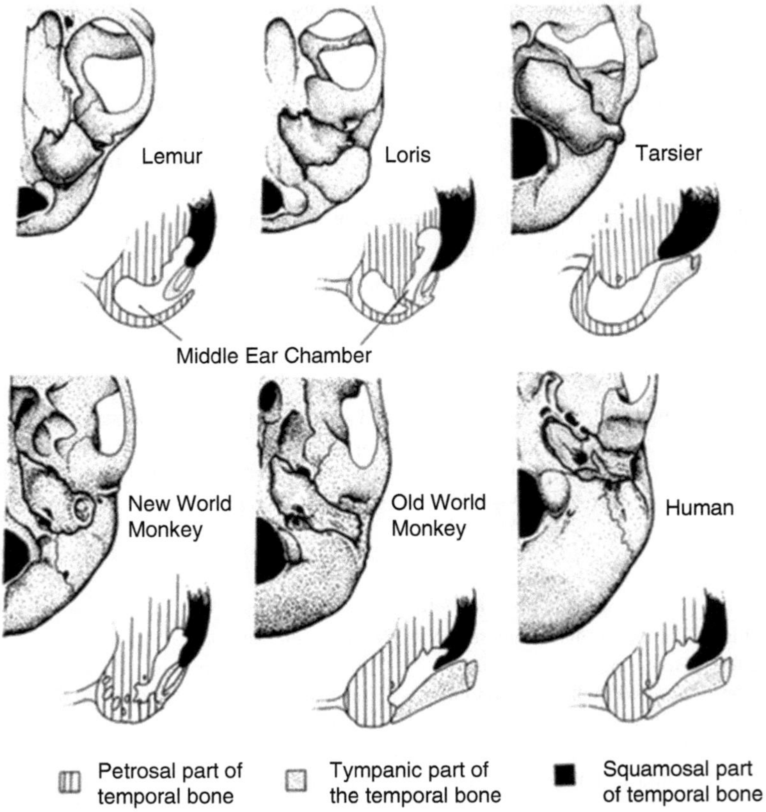

Fig. 2.8 Variations in the temporal bone anatomy of primates: ventral view (*above*) and cross-sectional view of the ear region (*below*). The petrous, the tympanic, and the squamosal portions vary considerably across different taxa (also see Table 2.1). In lemurs, the tympanic (ectotympanic) bone supporting the eardrum is ring-shaped, suspended within the tympanic cavity, and surrounded by the petrous portion. In lorises, the tympanic is connected to the wall of the bulla and the bulla cavity is divided. The tympanic bone is an elongated bony tube on the lateral side of the skull in tarsiers and in Old World monkeys and apes (including humans). In New World monkeys the tympanic forms a tympanic ring that is fused to the auditory bulla laterally. (Modified and reprinted with permission from Fleagle 2013)

2.3.2 Middle Ear Structure and Function

The primate middle ear has the general mammalian anatomy with three middle ear ossicles (malleus, incus, and stapes) forming an ossicular chain in the cavity between the tympanic membrane laterally and the oval window of the cochlea medially (Figs. 2.1 and 2.9). In primates, as in most mammals, the malleus is attached to the tympanic membrane at its lower, slender part called the manubrium. The anterior process of the malleus (also called the gonial or the processus gracilis) makes contact with the tympanic ring anteriorly. The looseness/stiffness of this contact varies

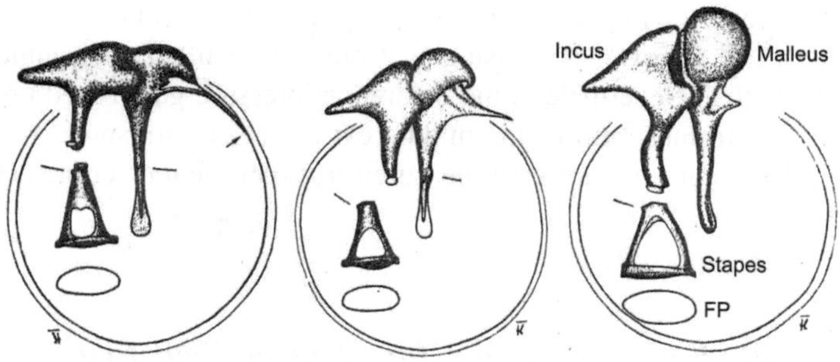

Fig. 2.9 Middle ear ossicles from the left ear of three primate species as viewed from inside the middle ear cavity. *Left*: Greater bushbaby (strepsirrhine, lorisoid); *middle*: macaque (haplorhine, Old World monkey); *right*: chimpanzee (haplorhine, ape). The contours of the tympanic membrane are shown for each as well as the area of the stapes footplate (FP). Tympanic membrane diameters measured perpendicular to the manubrium of the malleus are 5.0 mm, 7.5 mm, and 10.2 mm, respectively. (Reprinted with permission from Fleischer 1973)

among different mammals. In turn, the malleus head and the incus body articulate with one another (incudomalleolar joint). The incus has a small lenticular process at the tip of its long process, and this lenticular process articulates with the stapes head (incudostapedial joint). The medial end of the stapes (the footplate) is attached to the oval window of the cochlea.

In sound transmission, from the lower acoustic impedance of air in the ear canal to the much higher acoustic impedance of the fluid-filled inner ear (Rosowski 1994; Hemilä et al. 1995), the middle ear acts as an acoustic impedance-matching device that compensates for the loss in energy associated with this change in medium. The mass and stiffness of the ossicles are the main factors limiting transmission at higher frequencies; the middle ear cavity volume and the tightness of the connections between the malleus and tympanic membrane and at the joints between the ossicles are the main limiting factors at lower frequencies. Smaller middle ears with smaller tympanic membranes, lighter ossicles, and tighter connections between them are better at transmitting higher frequencies while larger middle ears with larger tympanic membranes, heavier ossicles, and looser connections between them are better suited for transmitting lower frequencies (Møller 1974; Rosowski and Relkin 2001).

This impedance matching in land mammals is accomplished by the *area ratio*, which is the ratio between the tympanic membrane area and the oval window area, and by the *lever ratio*, which is the ratio between the malleus lever arm length and the incus lever arm length (Rosowski 1994; Nummela et al. 2007). The *impedance transformer ratio* is the product of these two ratios: [(area ratio) × (lever ratio)^2]. Although the middle ear is often seen as a pressure-enhancing mechanism between the surrounding air and the inner ear fluid, it is evident that the earliest land vertebrates did not have such an impedance-matching mechanism in their ears (Manley and Sienknecht 2013). Additionally, the middle ear transmission mechanism is not solely responsible for the hearing sensitivity of mammals. Rather, the outer, middle, and inner ears together are important for determining the overall shape of an animal's audiogram (Ruggero and Temchin 2002; Hemilä et al. 2010).

In an ideal case, the impedance-matching system of the middle ear would offset the loss of sound energy when changing from the air-filled environment to the fluid-filled cochlea. Nevertheless, the theoretical pressure gain based on the ideal impedance transformer ratio of the middle ear does not correspond to empirical results of middle ear pressure gain measured in experimental settings (Rosowski and Relkin 2001).

2.3.3 Middle Ear Morphology and Hearing Sensitivity

Detailed anatomical descriptions on the comparative morphology of mammalian ossicles, including twenty-five primate species from all major taxonomic groups, can be found in Doran (1879), who separated primate ossicles into four groups (as did Hyrtl 1845): (1) humans, (2) apes, (3) Old World monkeys, and (4) New World monkeys and strepsirrhines. Likewise, Masali et al. (1992) identified two grouping for primate ossicles: (1) the Old World primate type, and (2) the New World primate and strepsirrhine type. This grouping mirrors the temporal bone similarities between these two groups (Sect. 2.3.1; Table 2.1).

Based on a wide, systematic survey of the morphology of the mammalian peripheral auditory region, including twenty-five primate species, Fleischer (1973, 1978) studied the evolutionary history of the mammalian middle ear and, particularly, of the malleus-incus complex. Detailed morphological descriptions and quantitative models for ossicular chain function were also presented. Fleischer identified four middle ear types across mammals: the *ancestral ear*, the *microtype ear*, the *transitional ear* (also called the intermediate ear), and the *freely mobile ear*. He suggested that an ancestral middle ear type had given rise to the other three types. A central role in grouping the middle ear types was given to the anterior process of the malleus. This structure varies in size and in strength of attachment to the tympanic ring. Hence, the different ear types deviate from each other by how loose or tight the connections are between different parts.

In the ancestral ear and the microtype ear, the malleus is tightly attached to the tympanic ring with its long anterior process, and the connection between the malleus and the incus is firm (although not ossified) with the joint surfaces being tightly interlocked. These kinds of ears are best suited for transmitting high-frequency sounds and are found in small mammals, such as bats and rodents. The two other types, the transitional ear and the freely mobile ear, are found among primates (Fig. 2.9). In the transitional middle ear (e.g., greater bushbaby, *Otolemur crassicaudatus*, for which the old name *Galago crassicaudatus* was used by Fleischer 1973), the anterior process of the malleus has only a loose contact with the tympanic ring, and the joint between the malleus and incus is not as tight as in the microtype ear. In the freely mobile middle ear (e.g., macaque, *Macaca* sp., and chimpanzee), the anterior process of the malleus has been reduced and is loose from the tympanic ring; the joint between the malleus and incus is also loose with more mobility between the ossicles than in the transitional ear or the microtype ear. The head of the malleus is larger in the freely mobile ear, and the incus is relatively larger as compared to the malleus than in the other two types (Fig. 2.9).

Based on the sizes of the tympanic membrane and stapes footplate in over fifty mammalian species, including three primates (bushbaby, macaque, and chimpanzee), Fleischer (1973) concluded that the middle ear size is not totally dependent on body size (mass). Nevertheless, strong correlations between both tympanic membrane and stapes footplate area with body mass were found across mammals (Hunt and Korth 1980; Rosowski and Graybeal 1991). The size variation in both the tympanic membrane and the oval window is clearly negatively allometric to body mass in mammals in general, including primates (Rosowski 1994). In addition, a strong correlation was found between ossicular mass and body mass across both placental and marsupial taxa (Nummela 1995; Nummela and Sánchez-Villagra 2006). In contrast, the relationships between different middle ear structures themselves (i.e., tympanic membrane and oval window areas; malleus and incus lever arms and masses) are highly isometric among mammals, including primates (Rosowski 1992; Hemilä et al. 1995). Within primates, Coleman et al. (2010) showed that the areas of the tympanic membrane and stapes footplate are strongly correlated with body mass across a large number of strepsirrhine and catarrhine taxa. They also found a strong correlation between the tympanic membrane and the stapes footplate areas for these same species.

Rosowski (1992) divided twenty mammals, including three primate species, into three groups according to their middle ear type (microtype, transitional, or freely mobile type) *sensu* Fleischer (1973, 1978) and established their low- and high-frequency hearing limits (at 60 dB SPL) and their best frequency on the basis of the behavioral audiogram for each species. These audiometric parameters were then compared with anatomical measurements to establish correlations between ear structure and function. Each mammalian group had a distinct hearing range with only limited overlap. The strongest correlations were found between the best frequency and the tympanic membrane area and for the low- and high-frequency limits with the footplate area. Thus, increases in body size and corresponding increases in the sizes of the tympanic membrane and the stapes footplate lead to a lowering of the best frequency and the low- and high-frequency limits.

It is evident that the middle ear structures and auditory capacities of extant mammals vary widely, and attempts to correlate these two can reveal significant relationships, which in turn can be used to predict the auditory capacities of fossil mammals for which only structural data are available (Nummela et al. 2004; Quam et al. 2013). Plassmann and Brändle (1992) developed a general model to explain auditory function from structural measurements that was applicable to a wide diversity of mammalian species with different ear dimensions, including tympanic membrane radius, the middle ear volume radius, the cross-sectional area and length of the external auditory meatus, and the frequency of best sensitivity. The tympanic membrane area and the middle ear volume have identical resonance frequencies in each species studied, and it is possible to predict the tympanic membrane size on the basis of the frequency range of best sensitivity and vice versa. Similarly, Hemilä et al. (1995) have shown that the behavioral high-frequency limit is inversely related to the cube root of the combined mass of the malleus and incus across twenty-eight mammalian species, including nine primates. These models can potentially be applied to fossil specimens to predict aspects of their unknown audiograms (see Quam, Martínez, Rosa, and Arsuaga, Chap. 8).

The impedance transformer ratio was studied for thirty-three primate taxa by Coleman and Ross (2004). The area ratio was found not to differ significantly between haplorhines and strepsirrhines, reflecting the largely isometric relationship between these two variables (Sect. 2.3.2). However, clear differences exist in the lever ratio, with strepsirrhines showing a higher lever ratio than haplorhines. These differences are independent of variation in body size but seem to reflect differences in malleus lever arm length, suggesting that differences in the impedance transformer ratio among primates are primarily driven by differences in length of the manubrium of the malleus. Morphological variation of extant hominid (gorilla, chimpanzee, and human) ossicles was studied in detail with a large sample by Quam et al. (2014). The lever ratio was found to be close (gorilla) or equal (chimpanzee) to the mean lever ratio found for haplorhines by Coleman and Ross (2004), whereas in humans the lever ratio turned out to be clearly lower; in fact, humans have the lowest lever ratio among primates (for functional inferences of the hominid lever ratios, see Quam, Martínez, Rosa, and Arsuaga, Chap. 8).

Size variation in middle ear structures in a large number of primate taxa were used to study the relationship between ear morphology and hearing sensitivity (Coleman and Colbert 2010). The sound pressure threshold level at 250 Hz was used as a proxy for low-frequency sensitivity, and the sound pressure threshold level at 32 kHz was used as a proxy for high-frequency sensitivity. Additionally, several other audiometric parameters (taken from behaviorally measured audiograms like those described in Sect. 2.1) were considered. Of the middle ear structures measured, the tympanic membrane area, the oval window area, the ossicular mass, and various middle ear cavity volumes seemed to exert a considerable influence over low-frequency sensitivity. Structures that seemed to play a role in determining high-frequency sensitivity included the ossicular mass and the interaural distance. These findings agree with the studies by Plassmann and Brändle (1992) and Hemilä et al. (1995) and show that primates follow the general pattern between interaural distance and high-frequency hearing (Masterton et al. 1969; R. S. Heffner and H. E. Heffner 1992a).

2.4 Inner Ear

The primate inner ear is a typical mammalian inner ear with the cochlear part and the vestibular part housed within the bony labyrinth of the petrosal portion of the temporal bone. Early quantitative inner ear data on the cochlea and the oval and round windows were presented by Hyrtl (1845) based on cranial and embryological material from several mammalian orders, including a few primate species, together with ideas of sound conduction through the tympanic cavity and hearing ranges of animals related to ear size.

The interspecific variation of the cochlea has raised the natural question of how this variation might correlate with different hearing capacities of mammalian species (West 1985; Echteler et al. 1994). Morphometric studies of the primate cochlea have been accelerated since high-resolution computed tomography was developed to be applied to the bony labyrinth (Spoor and Zonneveld 1995). While it is evident through quantitative data that hearing characteristics of mammals largely depend on the properties of the peripheral auditory system, considerable differences among the outer ears (see Sect. 2.2), the middle ears (Sect. 2.3), and the inner ears have been shown (Vater et al. 2004; Ekdale 2013). Echteler et al. (1994) pointed out that the mammalian cochleae can be divided into generalized and specialized categories, and this should be taken into account when predicting hearing abilities on the basis of cochlear morphology. However, in general, the cochlear labyrinth volume, the number of spiral turns, and the basilar membrane length were found to correlate with the octave range and the high- and low-frequency limits of hearing in several mammalian species. The number of spiral turns in the cochlea is also a good indicator for the hearing range provided that cochlear specializations of the species are known (Echteler et al. 1994; Vater and Kössl 2011).

Manoussaki et al. (2006) suggested that the curvature gradient of the cochlea might be related to low-frequency sensitivity. The radius of curvature of the cochlea decreases toward the apex and is smallest in the region where low-frequency sounds are analyzed, and the cochlea's mechanical response to low frequencies is enhanced. In an analysis with thirteen mammals (including humans), Manoussaki et al. (2008) demonstrated that the ratio between the radii of curvature of the basal turn and the apical turn correlates strongly with low-frequency hearing limits.

Focusing on primates, Kirk and Gosselin-Ildari (2009) measured the cochlear labyrinth volume as a proxy for the basilar membrane length in a taxonomically diverse sample of thirty-three primate species, and they found that the mean cochlear labyrinth volume was strongly negatively allometric when related to body mass. This suggests that cochlear size has increased during the course of primate evolution (Armstrong et al. 2011; also see Sect. 2.5). Three-dimensional virtual reconstructions of the cochlear labyrinth were also published (Fig. 2.10) for ten primate taxa for which audiogram data are available. The high- and low-frequency limits of hearing were strongly negatively correlated with cochlear labyrinth volume in these taxa, independent of body mass or phylogeny (Kirk and Gosselin-Ildari 2009). Thus, the highest and lowest audible frequencies in primates shift downward as cochlear labyrinth volume increases, especially the high-frequency limit of hearing, even when body mass is held constant.

Coleman and Colbert (2010) used cochlear length as a proxy for basilar membrane length, measuring from the edge of the oval window to the apex along the outside of the cochlea. They found a strong positive relationship with body mass for extant primates. Haplorhines had relatively longer cochleae than other primate groups, extant or extinct. The cochlear length also showed a clear correlation with low-frequency sensitivity (Coleman and Colbert 2010), but no relationship between cochlear length and high-frequency sensitivity was found. This seems to indicate that primates have increased their low-frequency hearing by extending the apical

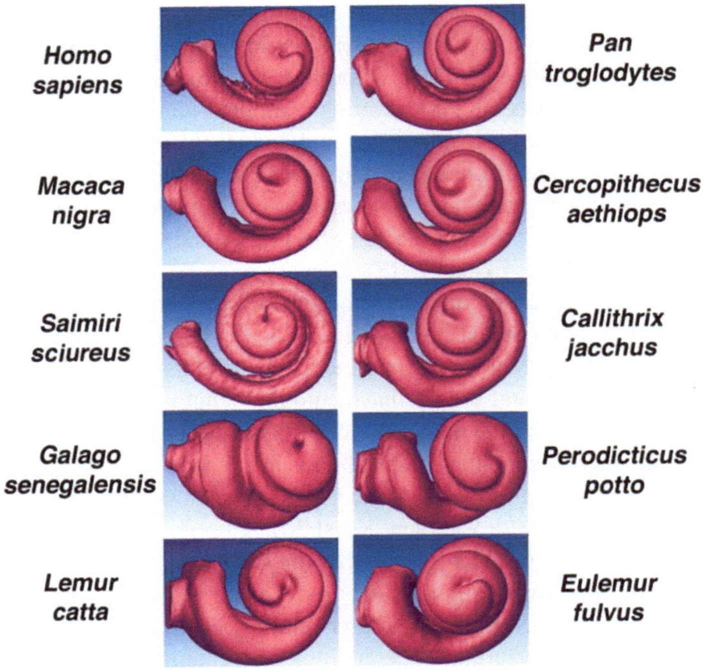

Fig. 2.10 Three-dimensional renderings of the cochlea of ten primate taxa for which audiograms are known. The cochleae are scaled to the same size. The round window for each cochlea faces left. (Modified and reprinted with permission from Kirk and Gosselin-Ildari 2009)

end of the basilar membrane (and presumably the stiffness gradient), which in turn provides space for more hair cells. It is evident that low-frequency sensitivity can be best predicted based on the cochlear length (Coleman and Boyer 2012).

Among living haplorhines, only tarsiers have developed relatively long cochleae while still retaining a well-developed secondary bony lamina that supports the outer edge of the basilar membrane (Fig. 2.11). These features apparently confer both heightened low- and high-frequency sensitivity to this unique genus (Coleman et al. 2010; Ramsier et al. 2012a). The development of secondary laminae does not appear to be directly related to either cochlear length or the number of cochlear spiral turns. Tarsiers show the greatest expression of secondary lamina development, yet they have similar cochlear lengths and a similarly high number of cochlear turns as most monkeys and apes that lack laminae. In some primate groups, the loss of secondary bony laminae may reflect an increase in body size (together with poorer high-frequency sensitivity). However, small-bodied New World monkeys, like tamarins (*Saguinus* sp.) and marmosets (*Callithrix* sp.), apparently lack secondary bony laminae, while similarly sized strepsirrhines, like bushbabies and slender lorises (*Loris* sp.), possess them (Coleman and Boyer 2012), suggesting a phylogenetic component to the expression of this feature as well.

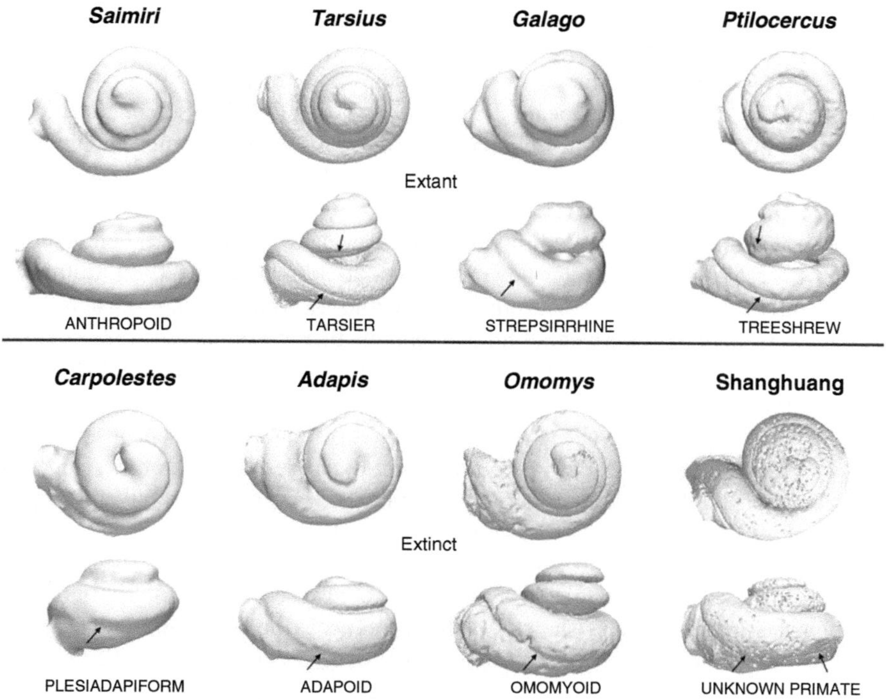

Fig. 2.11 Endocasts based on CT scans of cochleae of mammals, scaled to the same size. From *left to right*, *upper row*: squirrel monkey, tarsier, bushbaby, and treeshrew; *lower row*: plesiadapiform, adapoid, omomyoid, and the Shanghuang petrosal. The tarsius cochlea has a higher number of spiral turns than the other taxa shown here. The well-developed secondary bony lamina is shown with *black arrows* in all taxa except in the squirrel monkey, in which there is no evidence of it. (Reprinted with permission from Coleman and Boyer 2012)

2.5 Evolution of Hearing in Primates

The relationships described in Sects. 2.3 and 2.4 between morphological features of the ear, including bony structures, and aspects of the experimentally determined audiogram can be applied to fossil specimens to predict aspects of the hearing abilities in extinct species. This approach provides insights into their lives and makes it possible to understand the evolution of hearing.

2.5.1 *Early Mammalian Hearing*

An early seminal study into the evolution of mammalian hearing suggested that early mammals were capable of perceiving high frequencies and that ability was related directly to their small head size (Masterton et al. 1969). That idea has been subsequently corroborated by a large number of behavioral audiograms on diverse mammalian species with a large range in head sizes (H. E. Heffner and R. S. Heffner 2008, 2014). In addition, based on correlations between auditory structures and function, Rosowski and Graybeal (1991) and Rosowski (1992) also concluded that early mammals had good high-frequency hearing. This auditory pattern may represent an adaptation to the nocturnal niches presumably occupied by early mammals (Jerison 1973; Gerkema et al. 2013).

Several studies have revealed more about the hearing abilities and the nocturnal niche of early mammals (Kielan-Jaworowska et al. 2004; Ji et al. 2009). Eutherian (placental) mammals emerged during the Mesozoic era, 160 million years ago (Ma). The earliest mammals were small-bodied animals that fed on insects and occupied nocturnal niches in a low-light environment for most of their history (Allin and Hopson 1992; Luo et al. 2016). Early mammals went through a nocturnal bottleneck that had a tremendous influence on the evolution of mammalian sensory systems, and all extant mammals, including diurnal species, are descendants of these nocturnal precursors. Unlike other terrestrial vertebrates, however, these small and cryptic early mammals de-emphasized vision and, instead, specialized in the senses that could be used in low-light conditions, particularly high-frequency hearing and greatly expanded olfaction (Heesy and Hall 2010; also see Sect. 2.6). Similarly, the likely ancestors of primates show good high-frequency hearing but relatively poor low-frequency sensitivity (Coleman and Boyer 2012). Since high-frequency hearing was present in early mammals, the good high-frequency hearing of modern strepsirrhines would appear to represent a primitive mammalian feature (Coleman and Boyer 2012; H. E. Heffner and R. S. Heffner 2016; Zimmermann, Chap. 5).

2.5.2 *Early Primate Hearing*

The first primates evolved at the beginning of the Cenozoic era, approximately 60 Ma (Ramsier and Quam, Chap. 1). It is evident that hearing in primates evolved in several stages through the first half of the Cenozoic (60–50 Ma), and the earliest primates are also believed to have been nocturnal (Cartmill 1974; Heesy and Ross 2001). Microcomputerized tomography (μCT) studies of the cochlea in fossil and living primates have revealed variations in several cochlear structures, including the size of the oval window, the length of the cochlea, and the development of the bony lamina (Fig. 2.11). The low- and high-frequency sensitivities (SPL at 250 Hz and SPL at 32 kHz, respectively) were predicted for many early primate species based on cochlear dimensions and morphology (Coleman et al. 2010; Coleman and Boyer 2012; see Sect. 2.4).

The Plesiadapiforms were small- to medium-sized primates that lived in North America and Europe 60–54 Ma (Silcox et al. 2015). These taxa were characterized by a small oval window and a relatively short cochlea that housed a moderate- to well-developed secondary bony lamina. Those traits suggest good high-frequency sensitivity but relatively poor low-frequency sensitivity, which is somewhat intermediate between primitive extant mammals (e.g., tree shrews and hedgehogs) and extant strepsirrhines (Coleman and Boyer 2012). The size (volume) of the cochlear labyrinth in early primates and extant mammals closely related to primates was found to be smaller than the cochlear labyrinth size in modern primates, lending support to the idea that the cochlear labyrinth size started to increase first within the Euprimates (strepsirrhines and haplorhines) (Armstrong et al. 2011).

The emergence of the strepsirrhines and haplorhines, representing the major taxonomic division in extant primates today, occurred during the Eocene epoch in North America and Europe between 50 Ma and 35 Ma. These taxa show an increase in cochlear length and a reduction of the secondary bony lamina, indicating that low-frequency sensitivity increased and high-frequency sensitivity decreased modestly and that the strepsirrhine auditory pattern had largely emerged by this time (Coleman and Boyer 2012).

Further lengthening of the cochlea and reduction (or loss) of the secondary bony lamina occurred in haplorhines (except in tarsiers). Evidence from fossil Old World Monkeys from Africa (Oligocene epoch, about 30 Ma) and fossil New World monkeys from South America (Miocene epoch, about 20–16 Ma) suggests that cochlear elongation was completed by the early Miocene. Indeed, three early platyrrhine taxa (*Homunculus*, *Dolichocebus*, and *Tremacebus*) resemble modern platyrrhines in many of their auditory structures (e.g., middle ear area ratio) (Coleman et al. 2010).

Thus, it is evident that some primates had developed low-frequency hearing similar to that of modern species by at least the early Miocene. Low-frequency sensitivity of fossil New World Monkeys was likely very similar to that of their extant counterparts, and they may have had similar high-frequency sensitivity as modern taxa as well. These data are in accordance with the idea that a few characteristics of fossil platyrrhines are still retained in their modern descendants (Fleagle 2013). Given that the fossils studied are most likely not directly ancestral to any specific extant New World taxa (Kay et al. 2008), it is likely that good low-frequency hearing dates back at least 20 Ma.

The good high-frequency hearing in early mammals and the nocturnal bottleneck in the early Cenozoic era (Meng and Fox 1995; Kielan-Jaworowska et al. 2004) suggest that the earliest primates already had good high-frequency hearing (Coleman and Boyer 2012). This ability has presumably been enhanced by increases in social complexity in primates (Ramsier et al. 2012b). However, sensitivity to low frequencies among the earliest primates was likely not as developed as in species that evolved later in the Miocene. The advent of good low-frequency hearing made it possible to take advantage of the fact that low frequencies carry much further than high frequencies (Brown and Waser, Chap. 4). This may indicate something about the acoustic environment and the type of habitat where the Eocene primates lived.

2.5.3 Body Size, Ear Dimensions, and Hearing

Figure 2.12 shows the high- and low-frequency hearing limits for a sample of various mammals ranging in body size and taxonomic diversity. Body size plays an important role in auditory morphology (see Sects. 2.3 and 2.4), and changes in the

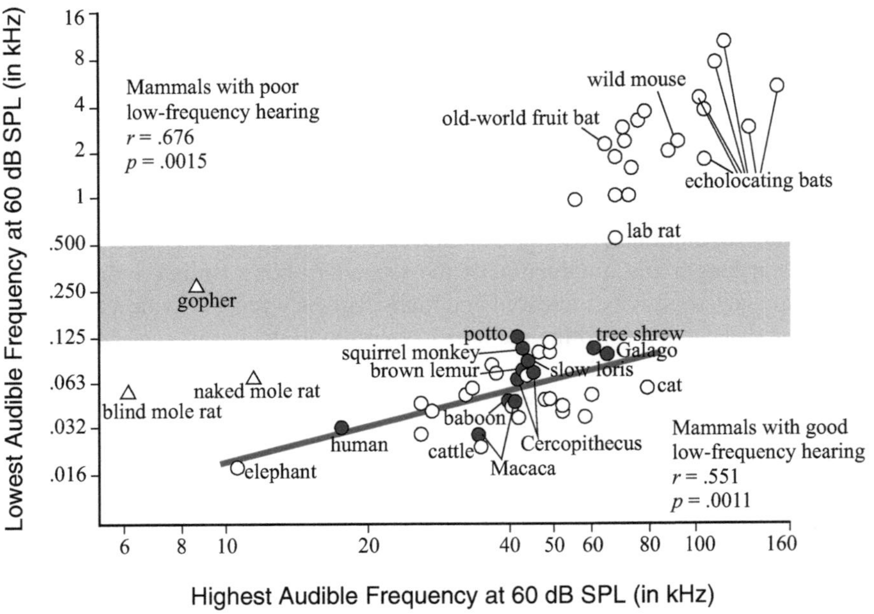

Fig. 2.12 The highest and lowest frequencies detected at 60 dB SPL by various mammalian taxa. Regarding the low-frequency hearing limits, these taxa show a distribution in two groups, with a gap between them (indicated by the *gray shading*). Primates are in the group with good low-frequency hearing, although clearly separated from subterranean species; primates have relatively good high-frequency hearing, though clearly separated from bats and rodents. *Filled circles*: primates and the tree shrew; *open circles*: selection of other mammals; *open triangles*: subterranean mammals (not included in the statistical analysis). (Reprinted with permission from R. S. Heffner 2004)

oval window area and in high-frequency sensitivity throughout primate evolution seem to be largely related to overall increases in body size. Thus, increases in body size (and head size) were likely paralleled by increases in the size of the stapes footplate. In some primate groups, the loss of the secondary bony lamina may also reflect this trend toward larger body size in mammals (together with poorer high-frequency sensitivity). However, here the relationship is less clear; the absence or presence of the secondary lamina is not strictly tied to body size (see Sect. 2.4). All primate fossils older than 30 Ma show some development of a secondary bony lamina, and this structure was already present in ancestral therian mammals (Ruf et al. 2009).

Increases in body size through time have clearly resulted in overall increases in cochlear length, but changes in body size alone cannot explain these patterns. *Plesiadapis cookei*, a fossil primate from the Paleocene, has a cochlear length of 19 mm and a body mass of more than 2 kg, whereas the extant platyrrhine owl monkey (*Aotus* sp.) has a cochlear length of 22.4 mm and a body mass of less than 1 kg. This lack of correlation might suggest that head size is a more important indicator of cochlear length than body mass, at least in some taxa.

Assessing the correlation of cochlear labyrinth volume and low- and high-frequency hearing, it was found that some differences in primate hearing between fossil strepsirrhines and haplorhines were already present by the Eocene, reflecting the differences that characterize living strepsirrhines and haplorhines today (Kirk et al. 2014; Godinot 2015). Changes in the cochlear length also reveal some interesting phylogenetic and temporal patterns in primate auditory morphology, and only the geologically youngest fossils show cochlear length values that approach those found in extant taxa.

2.6 Hearing, Sensory Ecology, and Primate Origins

Coevolution between vision and hearing is well-known among mammals (Seyfarth and Cheney 2009), and one good example of this coevolution involves the ability to localize a sound source (R. S. Heffner 2004; Ramsier and Rauschecker, Chap. 3). Auditory perception frequently initiates visual activity, and the task of localizing a sound source is to inform the visual system where to look (R. S. Heffner and H. E. Heffner 1992b; H. E. Heffner and R. S. Heffner 2016). Cooperation of eyes and ears is crucial for sound localization and shaping the map of auditory space in the brain so that it matches the neural representations of the other sensory modalities (King 1999). Vision and hearing also provide sensory input relevant for locomotion (Ankel-Simons 2007).

Nummela et al. (2013) studied hearing, vision, and olfaction in more than 100 extant mammalian species, including 6 strepsirrhine and 12 haplorhine primates, by measuring functionally relevant structures (tympanic membrane area, vertical eye orbit diameter, and cribriform plate area) and normalizing them to body mass. The primate taxa included show variations in ecological niche, diet, and body mass. The results indicate coevolution between vision and hearing in both arboreal and terrestrial species, corroborating the general mammalian pattern that animals with large eyes (in relation to body size) tend to have large ears, too. For both arboreal and terrestrial groups there seems to be a trade-off between vision and olfaction and, similarly, between hearing and olfaction. In other words, eyes and ears relate to noses in a similar way, although arboreal mammals do tend to have larger eyes and smaller noses than terrestrial mammals. Thus, there is cooperation between vision and hearing in primates, but a trade-off exists between vision and hearing on the one hand and olfaction on the other (Nummela et al. 2013).

This question of trade-offs between different sensory modalities is related to primate origins and phylogeny (Ross and Kay 2004; Silcox et al. 2015), since primate sensory biology forms part of most theories of primate origins (Sayers 2015). The *arboreal hypothe*sis of primate origins holds that the grasping abilities, visual adaptation of stereoscopic vision, and brain characteristics arose in response to life in the trees. However, there are animals that show that arboreality clearly predates the rise of primates, including colugos (Dermoptera) and tree shrews (Scandentia), the two closest relatives of primates (Sayers 2015; Silcox et al. 2015).

The *visual predation hypothesis* of primate origins states that certain anatomical features indicate that the initial evolutionary divergence of primates involved a specialization for visual predation on insects (Cartmill 1974). This emphasis was challenged by studies on nocturnal primates that did not support the visual predation theory since many living nocturnal primates are not specialized in feeding on insects (Zimmermann, Chap. 5). The earliest primates most likely did not solely consume insects either, but rather shifted the focus of their diet away from insect predation in favor of flowering plants. Insects are high in protein and fat but are small and thus profitable only when available in large quantities, which most likely had a role in early primate evolution (Rothman et al. 2014).

The evolution and dispersal of angiosperms (flowering plants) approximately coincided with the appearance of primates around the Paleocene-Eocene boundary (Sussman 1991; Sussman et al. 2013). The *angiosperm coevolution hypothesis* of primate origins holds that the grasping and visual adaptations of primates came to be useful when reaching for terminal branches and the flowers, fruits, and seeds found there (Sussman and Raven 1978). It has been suggested that the diversity of angiosperm seed size and fleshy fruits commenced around 80 Ma and peaked in the Eocene around 55–50 Ma (Eriksson 2016). However, terminal branches of the flowering plants were also exploited by plesiadapiforms (a group of fossil primates) and by colugos and tree shrews, the two closest living relatives of primates. It is more likely that the coevolution with angiosperms took place across this larger group, since the increased benefits of seed dispersal would have been significant, and angiosperms also show mutualistic relationships with frugivorous birds, rodents, and bats. Of these three hypotheses of primate origins, only the visual predation hypothesis relies on good hearing, given that sound localization is one component of the ability to locate prey visually (see Sect. 2.2). The characteristics and changes in hearing abilities of primates during their (early) evolution may have provided other advantages than those directly related to feeding, for example, to sound communication within various groups and various habitats.

The multitude of resources available on the newly evolved rain forest trees probably led to the morphological adaptations seen in modern primates, while the prevalence of visual predation in extinct primates has been questioned, since living primates are typically neither specialized visual predators (Silcox et al. 2015) nor insectivores (Sussman 1991). Among extant species, the nocturnal dwarf lemurs (*Cheirogaleus* sp.), whose diet mainly consists of fruits and flowers (e.g., Mittermeier et al. 2010), have been regarded as good living models for the early primates (Sussman et al. 2013). Furthermore, olfaction and hearing are emphasized over vision as methods of capturing prey among many primates, and the lorises, "...the most orbitally convergent primates..." (p. 1057 in Silcox et al. 2015), use scent to detect their slow-moving and often smelly prey (Sussman et al. 2013).

Based on its relatively small orbit size, an early primate from the Eocene in Asia, *Teilhardina asiatica*, was suggested to be diurnal, leading to the argument that the last common ancestor of strepsirrhines and haplorhines would have been a diurnal, visually oriented, insectivorous predator (Ni et al. 2004). While orbit size is a reasonable proxy for visual activity pattern, ancestral primates may have shown only moderate enlargement of their eye orbits due to the small eyes of their early mammalian ancestors. Some Eocene strepsirrhines did have much smaller orbits than modern diurnal lemurs, but the brain in early primates was also less than half the size of the brains of their modern relatives, suggesting a lower capacity to process sensory information, including visual input (Martin 2004). Furthermore, the markedly larger infraorbital foramen of *T. asiatica* compared to modern primates (large in primitive nocturnal mammals, typically reduced in diurnal mammals) suggests nonvisual orientation by well-developed tactile whiskers.

Based on evidence from other stem primates (Silcox et al. 2009), the increased brain size that is characteristic of living primates (relative to other mammalian orders) evolved within the Euprimates rather than in the basal primate ancestor. This increase in brain size is associated with an increase in the cochlear labyrinth volume (Kirk 2006) along with specializations in other sensory systems, including vision (Hall et al. 2012).

Like brain size, the size of a sensory organ is under conflicting evolutionary pressures. Increases in size of the sensory organs lead to increased sensitivity and improves the signal-to-noise ratio, but a larger sensory organ also necessitates increased metabolic investment from the organism. Body size correlates with many important aspects of biology, and evolutionary changes in body size can have important implications for life style and adaptations, as well as for the accomplishments of the sensory systems of an animal. Primates appear to have undergone a rapid increase in body size early in the evolution of the primate order (Cooper and Purvis 2010). There are also indications of independent evolution of brain size and cerebral organization between Old World monkeys and hominoids (Gonzales et al. 2015). Combining these data with more knowledge from auditory systems will help build an integrated view of the evolution of hearing in mammals, and in primates.

2.7 Conclusions

The primate peripheral auditory organ resembles that of other terrestrial mammals both anatomically and functionally, and hearing sensitivity in primates often follows phylogenetic patterns (Ramsier and Rauschecker, Chap. 3). Low-frequency sensitivity is better in monkeys and apes than in lemurs and lorises, and within the haplorhines, catarrhines (Old World monkeys and apes) are more sensitive to lower frequencies than are platyrrhines (New World monkeys). High-frequency sensitivity is reduced in apes compared with monkeys, and the best high-frequency hearing among the extant primates tested has been found in lemurs and lorises. The audiograms of monkeys and apes (except humans) often show two peaks of maximum sensitivity, whereas lemurs and lorises (and humans) generally have only a single peak sensitivity in their audiograms.

It is evident that hearing in primates evolved through several stages during the first half of the Cenozoic Era (Ramsier and Quam, Chap. 1). The hearing pattern in the ancestors of primates was characterized by good high-frequency hearing but relatively poor low-frequency sensitivity. The primitive condition for primates is thought to be nocturnality. The morphological variation of the oval window area, the development of the bony lamina of the cochlea, and the cochlear length can be determined from primate species that extend from 60 Ma to recent times using CT. Low-frequency and high-frequency hearing thresholds for fossil primates can be predicted based on comparisons between morphological auditory data from fossil and modern species and experimental data on modern primate hearing.

Plesiadapiforms are fossil stem primates that lived in North America and Europe 60–54 Ma. They were characterized by a small oval window and a relatively short cochlea that housed moderate- to well-developed secondary bony laminae. These traits suggest that plesiadapiforms had good high-frequency hearing but relatively poor low-frequency hearing, perhaps intermediate between extant strepsirrhines and primitive living mammals like tree shrews and hedgehogs (Ravizza et al. 1969).

Indeed, the origin of strepsirrhines and haplorhines, the two major taxonomic groupings in extant primates (appearing 50–35 Ma), is marked by an increase in cochlear length and reduction of the secondary bony lamina. This suggests that low-frequency hearing sensitivity increased and high-frequency sensitivity decreased modestly. Based on a few fossil primate taxa, the strepsirrhine hearing pattern first emerged during the Eocene (about 50 Ma), or perhaps even in the late Cretaceous, based on molecular evidence (Springer et al. 2003; Steiper and Seiffert 2012).

Cochlear elongation and the reduction (or loss) of secondary bony laminae continued in haplorhines. Evidence from early haplorhines in North America and Europe suggests that cochlear elongation might have taken place independently in haplorhines and strepsirrhines relative to the basal primate condition. Unlike other haplorhines, tarsiers retained the secondary bony spiral lamina, an ancestral feature in their inner ear. Evidence from fossil catarrhines (Old World Monkeys) in Africa and fossil platyrrhines (New World Monkeys) in South America suggests that the cochlear elongation process was completed by the early Miocene (about 25 Ma), and fossil platyrrhine taxa share many auditory characteristics with modern species.

It is evident that primates had developed low-frequency hearing, similar to that of modern forms, around this same time, and both the low-frequency and high-frequency sensitivity of fossil platyrrhines was likely very similar to living species. These data are in accordance with the idea that a few characteristics of fossil platyrrhines are retained in their modern descendants. Given that the fossil taxa studied most likely are not directly ancestral to any extant platyrrhine species, good low-frequency hearing, and perhaps good high-frequency hearing, dates back at least 20 Ma.

The potential for good high-frequency hearing in the early mammals and the nocturnal bottleneck in the early Cenozoic together suggest that the earliest primates already had good high-frequency sensitivity. If the earliest primates had good high-frequency hearing, their low-frequency hearing was most likely poorer than later Miocene species. Increased low-frequency hearing may have been advantageous for long-distance communication since low frequencies can travel much longer distances than high frequencies without degrading. This may also reveal something about the acoustic environment and the type of habitat that the earliest Cenozoic primates occupied.

Acknowledgments Many thanks to the editors Rolf Quam, Marissa Ramsier, Arthur Popper, and Richard Fay for inviting me to contribute to this book and for their valuable support and advice. I also thank Rickye Heffner, Simo Hemilä, Gimseong Koay, and Tom Reuter for help and discussions while preparing this manuscript. Financial support was provided by the Academy of Finland and by the Ella and Georg Ehrnrooth Foundation.

Compliance with Ethics Requirements
Sirpa Nummela declares that she has no conflict of interest.

References

Ahlborn, B. K. (2004). *Zoological physics: Quantitative models of body design, actions, and physical limitations of animals*. Berlin: Springer Science+Business Media.

Allin, E. F., & Hopson, J. A. (1992). Evolution of the auditory system in Synapsida ("Mammal-like reptiles" and primitive mammals) as seen in the fossil record. In D. B. Webster, R. R. Fay, & A. N. Popper (Eds.), *The evolutionary biology of hearing* (pp. 587–614). New York: Springer-Verlag.

Ankel-Simons, F. (2007). *Primate anatomy: An introduction* (3rd ed.). Burlington, MA: Academic Press.

Archibald, J. D., & Rose, K. D. (2005). Womb with a view: The rise of placentals. In K. D. Rose & J. D. Archibald (Eds.), *The rise of placental mammals: Origins and relationships of the major extant clades* (pp. 1–8). Baltimore, MD: Johns Hopkins University Press.

Armstrong, S. D., Block, J. I., Houde, P., & Silcox, M. T. (2011). Cochlear labyrinth volume in euarchontoglirans: Implications for the evolution of hearing in primates. *The Anatomical Record*, 294, 263–266.

Basch, M. L., Brown II, R. M., Jen, H.-I., & Groves, A. K. (2016). Where hearing starts: The development of the mammalian cochlea. *Journal of Anatomy*, 228, 233–254.

Cartmill, M. (1972). Arboreal adaptations and the origin of the order Primates. In R. Tuttle (Ed.), *The functional and evolutionary biology of primates* (pp. 97–122). Chicago, IL: Aldine-Atherton.

Cartmill, M. (1974). Rethinking primate origins. *Science*, 184, 436–443.

Cartmill, M. (1975). Strepsirhine basicranial structures and the affinities of the Cheirogaleidae. In W. P. Luckett & F. S. Szalay (Eds.), *Phylogeny of the primates—A multidisciplinary approach* (pp. 313–354). New York: Plenum Press.

Cartmill, M., MacPhee, R. D. E., & Simons, E. L. (1981). Anatomy of the temporal bone in early anthropoids, with remarks on the problem of anthropoid origins. *American Journal of Physical Anthropology*, 56, 3–21.

Clack, J. A., Fay, R. R., & Popper, A. N. (Eds.). (2016). *The evolution of the vertebrate ear—Evidence from the fossil record*. New York: Springer International Publishing.

Coleman, M. N. (2009). What do primates hear? A meta-analysis of all known non-human primate behavioral audiograms. *International Journal of Primatology*, 30, 55–91.

Coleman, M. N., & Boyer, D. M. (2012). Inner ear evolution in primates through the Cenozoic: Implications for the evolution of hearing. *The Anatomical Record*, 295, 615–631.

Coleman, M. N., & Colbert, M. W. (2010). Correlations between auditory structures and hearing sensitivity in non-human primates. *Journal of Morphology*, 271, 511–532.

Coleman, M. N., & Ross, C. F. (2004). Primate auditory diversity and its influence on hearing performance. *The Anatomical Record Part A: Discoveries in Molecular, Cellular, and Evolutionary Biology*, 281A, 1123–1137.

Coleman, M. N., Kay, R. F., & Colbert, M. W. (2010). Auditory morphology and hearing sensitivity in fossil New World monkeys. *The Anatomical Record*, 293, 1711–1721.

Cooper, N., & Purvis, A. (2010). Body size evolution in mammals: Complexity in tempo and mode. *The American Naturalist*, 175, 727–738.

Dallos, P. (1973). *The auditory periphery: Biophysics and physiology*. New York: Academic Press.
Davis, H. (1960). Physics and psychology of hearing. In H. Davis (Ed.), *Hearing and deafness* (pp. 29–60). New York: Holt, Rinehart, & Winston.
Dominy, N. J., Lucas, P. W., Osorio, D., & Yamashita, N. (2001). The sensory ecology of primate food perception. *Evolutionary Anthropology*, 10, 171–186.
Dominy, N. J., Ross, C. J., & Smith, T. D. (2004). Evolution of the special senses in primates: Past, present, and future. *The Anatomical Record Part A: Discoveries in Molecular, Cellular, and Evolutionary Biology*, 281A, 1078–1082.
Doran, A. H. G. (1879). Morphology of the mammalian ossicula auditus. *The Transactions of the Linnean Society of London. 2nd Series: Zoology*, 1, 371–497. + plates 58–64. doi:10.1111/j.1096-3642.1878.tb00663.x.
Echteler, S. M., Fay, R. R., & Popper, A. N. (1994). Structure of the mammalian cochlea. In R. R. Fay & A. N. Popper (Eds.), *Comparative hearing: Mammals* (pp. 134–171). New York: Springer-Verlag.
Ekdale, E. G. (2013). Comparative anatomy of the bony labyrinth (inner ear) of placental mammals. *PLoS One*, 8(6), e66624.
Elder, J. H. (1934). Auditory acuity of the chimpanzee. *Journal of Comparative Physiology and Psychology*, 17, 157–183.
Eriksson, O. (2016). Evolution of angiosperm seed disperser mutualisms: The timing of origins and their consequences for coevolutionary interactions between angiosperms and frugivores. *Biological Reviews*, 91, 168–186.
Fay, R. R. (1988). *Hearing in vertebrates: A psychophysics databook*. Winnetka, IL: Hill-Fay Associates.
Fleagle, J. G. (2013). *Primate adaptation and evolution* (3rd ed.). San Diego, CA: Academic Press.
Fleischer, G. (1973). Studien am Skelett des Gehörorgans der Säugetiere, einschliesslich des Menschen. *Säugetierkundliche Mitteilungen (München)*, 21, 131–239.
Fleischer, G. (1978). Evolutionary principles of the mammalian middle ear. *Advances in Anatomy, Embryology, and Cell Biology*, 55(5), 1–70.
Friedland, D. R. (2006). Structure and function in the auditory system: From cochlea to cortex. *The Anatomical Record Part A: Discoveries in Molecular, Cellular, and Evolutionary Biology*, 288A, 326–330.
Geisler, C. D. (1998). *From sound to synapse. Physiology of the mammalian ear*. Oxford: Oxford University Press.
Gerkema, M. P., Davies, W. I. L., Foster, R. G., Menaker, M., & Hut, R. A. (2013). The nocturnal bottleneck and the evolution of activity patterns in mammals. *Proceedings of the Royal Society of London B: Biological Sciences*, 280, 20130508.
Godinot, M. (2015). Fossil record of the primates from the Paleocene to the Oligocene. In W. Henke & I. Tattersall (Eds.), *Handbook of paleoanthropology* (Vol. 2. 2nd ed, pp. 1137–1259). Berlin: Springer Science+Business Media.
Gonzales, L. A., Benefit, B. R., McCrossin, M. L., & Spoor, F. (2015). Cerebral complexity preceded enlarged brain size and reduced olfactory bulbs in Old World monkeys. *Nature Communications*, 6, 7580. doi:10.1038/ncomms8580.
Gunz, P., Ramsier, M., Kuhrig, M., Hublin, J.-J., & Spoor, F. (2012). The mammalian bony labyrinth reconsidered, introducing a comprehensive geometric morphometric approach. *Journal of Anatomy*, 220, 529–543.
Hall, M. I., Kamilar, J. M., & Kirk, E. C. (2012). Eye shape and the nocturnal bottleneck of mammals. *Proceedings of the Royal Society of London B: Biological Sciences*, 279, 4962–4968.
Heesy, C. P., & Hall, M. I. (2010). The nocturnal bottleneck and the evolution of mammalian vision. *Brain, Behavior and Evolution*, 75, 195–203.
Heesy, C. P., & Ross, C. F. (2001). Evolution of activity patterns and chromatic vision in primates: Morphometrics, genetics, and cladistics. *Journal of Human Evolution*, 40, 111–149.
Heffner, H. E., & Heffner, R. S. (2008). High-frequency hearing. In P. Dallos & D. Oertel (Eds.), *The senses: A comprehensive reference, Audition* (Vol. 3, pp. 55–60). San Diego, CA: Academic Press.

Heffner, H. E., & Heffner, R. S. (2014). The behavioral study of mammalian hearing. In A. N. Popper & R. R. Fay (Eds.), *Perspectives on auditory research* (pp. 269–285). New York: Springer Science+Business Media.

Heffner, H. E., & Heffner, R. S. (2016). The evolution of mammalian sound localization. *Acoustics Today*, 12, 20–27, 35.

Heffner, R. S. (2004). Primate hearing from a mammalian perspective. *The Anatomical Record Part A: Discoveries in Molecular, Cellular, and Evolutionary Biology*, 281A, 1111–1122.

Heffner, R. S., & Heffner, H. E. (1992a). Evolution of sound localization in mammals. In D. B. Webster, R. R. Fay, & A. N. Popper (Eds.), *The evolutionary biology of hearing* (pp. 691–715). New York: Springer-Verlag.

Heffner, R. S., & Heffner, H. E. (1992b). Visual factors in sound localization in mammals. *Journal of Comparative Neurology*, 317, 219–232.

Heffner, R., Heffner, H., & Stichman, N. (1982). Role of the elephant pinna in sound localization. *Animal Behaviour*, 30, 628–630.

Hemilä, S., Nummela, S., & Reuter, T. (1995). What middle ear parameters tell about impedance matching and high frequency hearing. *Hearing Research*, 85, 31–44.

Hemilä, S., Nummela, S., & Reuter, T. (2010). Anatomy and physics of the exceptional sensitivity of dolphin hearing (Odontoceti: Cetacea). *Journal of Comparative Physiology A: Sensory, Neural, and Behavioral Physiology*, 196, 165–179.

Henke, W., & Tattersall, I. (Eds.). (2015). *Handbook of paleoanthropology* (Vol. 2, 2nd ed.). Berlin: Springer Science+Business Media.

Henson, O. W. (1961). Some morphological and functional aspects of certain structures of the middle ear in bats and insectivores. *University of Kansas Science Bulletin*, 42, 151–255.

Henson, O. W. (1974). Comparative anatomy of the middle ear. In W. D. Keidel & W. D. Neff (Eds.), *Handbook of sensory physiology, The auditory system* (Vol. V/1, pp. 39–110). New York: Springer-Verlag.

Hunt Jr., R. M., & Korth, W. W. (1980). The auditory region of Dermoptera: Morphology and function relative to other living mammals. *Journal of Morphology*, 164, 167–211.

Hyrtl, J. (1845). *Vergleichend-anatomische Untersuchungen über das innere Gehörorgan des Menschen und der Säugethiere*. Prague: Verlag von Friedrich Ehrlich.

Jackson, L. S., Heffner, R. S., & Heffner, H. E. (1999). Free-field audiogram of the Japanese macaque (*Macaca fuscata*). *The Journal of the Acoustical Society of America*, 106, 3017–3023.

Jerison, H. J. (1973). *Evolution of the brain and intelligence*. New York: Academic Press.

Ji, Q., Luo, Z.-X., Zhang, X., Yuan, C.-X., & Xu, L. (2009). Evolutionary development of the middle ear in Mesozoic therian mammals. *Science*, 326, 278–281.

Kay, R. F., Fleagle, J. G., Mitchell, T. R. T., Colbert, M., et al. (2008). The anatomy of *Dolichocebus gaimanensis*, a stem platyrrhine monkey from Argentina. *Journal of Human Evolution*, 54, 323–382.

Kielan-Jaworowska, Z., Cifelli, R. L., & Luo, Z.-X. (2004). *Mammals from the age of dinosaurs: Origins, evolution, and structure*. New York: Columbia University Press.

King, A. J. (1999). Sensory experience and the formation of a computational map of auditory space in the brain. *BioEssays*, 21, 900–911.

Kirk, E. C. (2006). Effects of activity pattern on eye size and orbital aperture size in primates. *Journal of Human Evolution*, 51, 159–170.

Kirk, E. C., & Gosselin-Ildari, A. D. (2009). Cochlear labyrinth volume and hearing abilities in primates. *The Anatomical Record*, 292, 765–776.

Kirk, E. C., Daghighi, P., Macrini, T. E., Bhullar, B.-A. S., & Rowe, T. B. (2014). Cranial anatomy of the Duchesnean primate *Rooneyia viejaensis*: New insights from high resolution computed tomography. *Journal of Human Evolution*, 74, 82–95.

Köppl, C., Manley, G. A., Popper, A. N., & Fay, R. R. (Eds.). (2014). *Insights from comparative hearing research*. New York: Springer Nature.

Luo, Z.-X., Schultz, J. A., & Ekdale, E. G. (2016). Evolution of the middle and inner ears of Mammaliaforms: The approach to mammals. In J. A. Clack, R. R. Fay, & A. N. Popper (Eds.),

The evolution of the vertebrate ear—Evidence from the fossil record (pp. 139–174). New York: Springer Science+Business Media.
MacPhee, R. D. E. (1979). Entotympanics, ontogeny, and primates. *Folia Primatologica*, 31, 23–47.
MacPhee, R. D. E. (1981). Auditory regions of primates and Eutherian insectivores. *Contributions to Primatology*, 18, 1–282.
MacPhee, R. D. E., & Cartmill, M. (1986). Basicranial structures and primate systematics. In D. R. Swindler & J. Erwin (Eds.), *Comparative primate biology: Systematics, evolution, and anatomy* (Vol. 1, pp. 210–275). New York: Alan R. Liss.
Maier, W., & Ruf, I. (2016). Evolution of the mammalian middle ear. *Journal of Anatomy*, 228, 270–283.
Manley, G. A. (2014). Fundamentals of hearing in amniote vertebrates. In A. N. Popper & R. R. Fay (Eds.), *Perspectives on auditory research* (pp. 321–341). New York: Springer Nature.
Manley, G. A., & Sienknecht, U. J. (2013). The evolution and development of middle ears in vertebrates. In S. Puria, R. R. Fay, & A. N. Popper (Eds.), *The middle ear: Science, otosurgery, and technology* (pp. 7–30). New York: Springer Science+Business Media.
Manley, G. A., Popper, A. N., & Fay, R. R. (Eds.) (2004). *Evolution of the vertebrate auditory system*. New York: Springer Science+Business Media.
Manoussaki, D., Dimitriadis, E. K., & Chadwick, R. S. (2006). Cochlea's graded curvature effect on low frequency waves. *Physical Review Letters*, 96, 088701.
Manoussaki, D., Chadwick, R. S., Ketten, D. R., Arruda, J., et al. (2008). The influence of cochlear shape on low-frequency hearing. *Proceedings of the National Academy of Sciences of the United States of America*, 105, 6162–6166.
Martin, R. D. (2004). Chinese lantern for early primates. *Nature*, 427, 22–23.
Martínez, I., Rosa, M., Arsuaga, J., Jarabo, P., et al. (2004). Auditory capacities in Middle Pleistocene humans from the Sierra de Atapuerca in Spain. *Proceedings of the National Academy of Sciences of the United States of America*, 101(27), 9976–9981.
Masali, M., Borgognini Tarli, S., & Maffei, M. (1992). Auditory ossicles and the evolution of the primate ear: biomechanical approach. In J. Wind, B. Chiarelli, B. Bichakjian, A. Nocentini, & A. Jonker (Eds.), *Language origin: A multidisciplinary approach* (pp. 67–86). Dordrecht: Kluwer Academic.
Masterton, B., Heffner, H., & Ravizza, R. (1969). The evolution of human hearing. *The Journal of the Acoustical Society of America*, 45, 966–985.
Mattila, T. M., & Bokma, F. (2008). Extant mammal body masses suggest punctuated equilibrium. *Proceedings of the Royal Society of London B: Biological Sciences*, 275, 2195–2199.
Meng, J., & Fox, R. C. (1995). Osseous inner ear structures and hearing in early marsupials and placentals. *Zoological Journal of the Linnean Society*, 115, 47–71.
Mittermeier, R. A., Louis Jr., E. E., Richardson, M., Schwitzer, C., et al. (2010). *Lemurs of Madagascar* (3rd ed.). Arlington, VA: Conservation International.
Møller, A. R. (1974). Function of the middle ear. In W. D. Keidel & W. D. Neff (Eds.), *Handbook of sensory physiology: The auditory system* (Vol. V/1, pp. 491–517). Berlin: Springer-Verlag.
Ni, X., Wang, Y., Hu, Y., & Li, C. (2004). A euprimate skull from the early Eocene of China. *Nature*, 427, 65–68.
Novacek, M. J. (1977). Aspects of the problem of variation, origin, and evolution of the eutherian auditory bulla. *Mammal Review*, 7, 131–150.
Novacek, M. J. (1993). Patterns of diversity in the mammalian skull. In J. Hanken & B. K. Hall (Eds.), *The skull: Patterns of structural and systematic diversity* (Vol. 2, pp. 438–545). Chicago: The University of Chicago Press.
Nummela, S. (1995). Scaling of the mammalian middle ear. *Hearing Research, 85*, 18–30.
Nummela, S., & Sánchez-Villagra, M. R. (2006). Scaling of the marsupial middle ear and its functional significance. *Journal of Zoology*, 270, 256–267.
Nummela, S., Thewissen, J. G. M., Bajpai, S., Hussain, S. T., & Kumar, K. (2004). Eocene evolution of whale hearing. *Nature*, 430, 776–778.

Nummela, S., Thewissen, J. G. M., Bajpai, S., Hussain, T., & Kumar, K. (2007). Sound transmission in archaic and modern whales: Anatomical adaptations for underwater hearing. *The Anatomical Record*, 290, 716–733.

Nummela, S., Pihlström, H., Puolamäki, K., Fortelius, M., et al. (2013). Exploring the mammalian sensory space: Co-operations and trade-offs among senses. *Journal of Comparative Physiology A: Sensory, Neural, and Behavioral Physiology*, 199, 1077–1092.

Packer, D. J., & Sarmiento, E. E. (1984). External and middle ear characteristics of primates, with reference to tarsier-anthropoid affinities. *American Museum Novitates*, 2787, 1–23.

Plassmann, W., & Brändle, K. (1992). A functional model of the auditory system in mammals and its evolutionary implications. In D. B. Webster, R. R. Fay, & A. N. Popper (Eds.), *The evolutionary biology of hearing* (pp. 637–653). New York: Springer-Verlag.

Quam, R. M., de Ruiter, D. J., Masali, M., Arsuaga, J.-L., & Martínez, I. (2013). Early hominin auditory ossicles from South Africa. *Proceedings of the National Academy of Sciences of the United States of America*, 110(22), 8847–8851.

Quam, R. M., Coleman, M. N., & Martínez, I. (2014). Evolution of the auditory ossicles in extant hominids: Metric variation in African apes and humans. *Journal of Anatomy*, 225, 167–196.

Quam, R. M., Martínez, I., Rosa, M., Bonmatí, A., et al. (2015). Early hominin auditory capacities. *Science Advances*, 1, e1500355.

Ramsier, M. A., & Dominy, N. J. (2010). A comparison of auditory brainstem responses and behavioral estimates of hearing sensitivity in *Lemur catta* and *Nycticebus coucang*. *American Journal of Primatology*, 72, 217–233.

Ramsier, M. A., Cunningham, A. J., Moritz, G. L., Finneran, J. J., et al. (2012a). Primate communication in the pure ultrasound. *Biology Letters*, 8, 508–511.

Ramsier, M. A., Cunningham, A. J., Finneran, J. J., & Dominy, N. J. (2012b). Social drive and the evolution of primate hearing. *Philosophical Transactions of the Royal Society of London B: Biological Sciences*, 367, 1860–1868.

Ravizza, R. J., Heffner, H. E., & Masterton, B. (1969). Hearing in primitive mammals. 2. Hedgehog (*Hemiechinus auritus*). *Journal of Auditory Research*, 9, 8–11.

Rosowski, J. J. (1992). Hearing in transitional mammals: Predictions from the middle-ear anatomy and hearing capabilities of extant mammals. In D. B. Webster, R. R. Fay, & A. N. Popper (Eds.), *The evolutionary biology of hearing* (pp. 615–631). New York: Springer-Verlag.

Rosowski, J. J. (1994). Outer and middle ear. In A. N. Popper & R. R. Fay (Eds.), *Comparative hearing: Mammals* (pp. 172–247). New York: Springer-Verlag.

Rosowski, J. J. (2003). The middle and external ears of terrestrial vertebrates as mechanical and acoustic transducers. In F. G. Barth, J. A. C. Humphrey, & T. W. Secomb (Eds.), *Sensors and sensing in biology and engineering* (pp. 59–69). New York: Springer Science+Business Media.

Rosowski, J. J. (2013). Comparative middle ear structure and function in vertebrates. In S. Puria, R. R. Fay, & A. N. Popper (Eds.), *The middle ear: Science, otosurgery, and technology* (pp. 31–64). New York: Springer Science+Business Media.

Rosowski, J. J., & Graybeal, A. (1991). What did *Morganucodon* hear? *Zoological Journal of the Linnean Society*, 101, 131–168.

Rosowski, J. J., & Relkin, E. M. (2001). Introduction to the analysis of middle-ear function. In A. Jahn & J. Santos-Sacchi (Eds.), *Physiology of the ear* (2nd ed., pp. 161–190). San Diego, CA: Singular.

Ross, C. F., & Kay, R. F. (2004). Anthropoid origins: Retrospective and prospective. In C. F. Ross & R. F. Kay (Eds.), *Anthropoid origins: New visions* (pp. 710–725). New York: Kluwer Academic.

Rothman, J. M., Raubenheimer, D., Bryer, M. A. H., Takahashi, M., & Gilbert, C. C. (2014). Nutritional contributions of insects to primate diets: Implications for primate evolution. *Journal of Human Evolution*, 71, 59–69.

Ruf, I., Luo, Z.-X., Wible, J. R., & Martin, T. (2009). Petrosal anatomy and inner ear structures of the Late Jurassic *Henkelotherium* (Mammalia, Cladotheria, Dryolestoidea): Insight into the early evolution of the ear region in cladotherian mammals. *Journal of Anatomy*, 214, 679–693.

Ruggero, M. A., & Temchin, A. N. (2002). The roles of the external, middle, and inner ears in determining the bandwidth of hearing. *Proceedings of the National Academy of Sciences of the United States of America*, 99(20), 13206–13210.

Sayers, K. (2015). Models of primate evolution. *eLS*, 1–10. doi:10.1002/9780470015902.a0026406.

Schultz, A. H. (1973). The skeleton of the Hylobatidae and other observations on their morphology. In D. M. Rumbaugh (Ed.), *Gibbon and siamang: Anatomy, dentition, taxonomy, molecular evolution and behavior* (Vol. 2, pp. 1–54). Basel: Karger.

Seyfarth, R. M., & Cheney, D. L. (2009). Seeing who we hear and hearing who we see. *Proceedings of the National Academy of Sciences of the United States of America*, 106, 669–670.

Shaw, A. G. (1974). The external ear. In W. D. Keidel & W. D. Neff (Eds.), *Handbook of sensory physiology: The auditory system* (Vol. V/1, pp. 455–490). New York: Springer-Verlag.

Silcox, M. T., Dalmyn, C. K., & Bloch, J. I. (2009). Virtual endocast of *Ignacius graybullianus* (Paromomyidae, Primates) and brain evolution in early Primates. *Proceedings of the National Academy of Sciences of the United States of America*, 106, 10987–10992.

Silcox, M. T., Sargis, E. J., Bloch, J. I., & Boyer, D. M. (2015). Primate origins and supraordinal relationships: Morphological evidence. In W. Henke & I. Tattersall (Eds.), *Handbook of paleoanthropology* (Vol. 2, 2nd ed., pp. 1053–1081). Heidelberg, Berlin: Springer Science+Business Media.

Sivian, L. J., & White, S. D. (1933). On minimum audible sound fields. *The Journal of the Acoustical Society of America*, 4, 234–288.

Smith, R. J., & Jungers, W. L. (1997). Body mass in comparative primatology. *Journal of Human Evolution*, 32, 523–559.

Spoor, F., & Zonneveld, F. (1995). Morphometry of the bony labyrinth: A new method based on high-resolution computed tomography. *Journal of Anatomy*, 186, 271–286.

Springer, M. S., Murphy, W. J., Eizirik, E., & O'Brien, S. J. (2003). Placental mammal diversification and the Cretaceous-Tertiary boundary. *Proceedings of the National Academy of Sciences of the United States of America*, 100, 1056e1061.

Steiper, M. E., & Seiffert, E. R. (2012). Evidence for a convergent slowdown in primate molecular rates and its implications for the timing of early primate evolution. *Proceedings of the National Academy of Sciences of the United States of America*, 109, 6006–6011.

Stoessel, A., David, R., Gunz, P., Schmidt, T., et al. (2016a). Morphology and function of Neandertal and modern human ear ossicles. *Proceedings of the National Academy of Sciences of the United States of America*, 113, 11489–11494.

Stoessel, A., Gunz, P., David, R., & Spoor, F. (2016b). Comparative anatomy of the middle ear ossicles of extant hominids—Introducing a geometric morphometric protocol. *Journal of Human Evolution*, 91, 1–25.

Sussman, R. W. (1991). Primate origins and the evolution of angiosperms. *American Journal of Primatology*, 23, 209–223.

Sussman, R. W., & Raven, P. H. (1978). Pollination of lemurs and marsupials: An archaic coevolutionary system. *Science*, 200, 731–736.

Sussman, R. W., Rasmussen, D. T., & Raven, P. H. (2013). Rethinking primate origins again. *American Journal of Primatology*, 75, 95–106.

Suthers, R. A., Fitch, W. T., Fay, R. R., & Popper, A. N. (Eds.). (2016). *Vertebrate sound production and acoustic communication*. New York: Springer Science+Business Media.

van der Klaauw, C. J. (1931). The auditory bulla in some fossil mammals, with a general introduction to this region of the skull. *Bulletin of the American Museum of Natural History*, 62, 1–352.

van Kampen, P. N. (1905). Die Tympanalgegend des Säugetierschädels. *Morphologisches Jahrbuch*, 24, 321–722.

Vater, M., & Kössl, M. (2011). Comparative aspects of cochlear functional organization in mammals. *Hearing Research*, 273, 89–99.

Vater, M., Meng, J., & Fox, R. C. (2004). Hearing organ evolution and specialization: Early and later mammals. In G. A. Manley, A. N. Popper, & R. R. Fay (Eds.), *Evolution of the vertebrate auditory system* (pp. 256–288). New York: Springer Science+Business Media.

West, C. D. (1985). The relationship of the spiral turns of the cochlea and the length of the basilar membrane to the range of audible frequencies in ground dwelling mammals. *The Journal of the Acoustical Society of America*, 77, 1091–1101.

Yost, W. A., Popper, A. N., & Fay, R. R. (Eds.). (2008). *Auditory perception of sound sources*. New York: Springer Science+Business Media.

Zwislocki, J. J. (1975). The role of the external and middle ear in sound transmission. In D. B. Tower (Ed.), *The nervous system: Human communication and its disorders* (Vol. 3, pp. 45–55). New York: Raven Press.

Chapter 3
Primate Audition: Reception, Perception, and Ecology

Marissa A. Ramsier and Josef P. Rauschecker

Abstract The auditory system of nonhuman primates shows evidence of many similarities to humans, such as specializations for the processing of vocalizations overall, processing species-specific vocalizations in particular, and in some cases, the recognition of specific individuals based on call structure. Additionally, nonhuman primates are similar to humans in their excellent localization acuity. Nonhuman primates show differences from humans, though, and not only in the subtleties of the aforementioned abilities. With respect to overall auditory sensitivity, primates have traditionally been portrayed as unspecialized, although there is variation between species. Species in the semiorder Strepsirrhini are, on average, more adept at detecting higher frequencies, whereas the Haplorhini are, on average, more adept at detecting lower frequencies. In addition, a well-supported allometric model explains that smaller headed species with smaller interaural distances need to utilize high-frequency cues for sound localization. Overall auditory sensitivity, particularly to high frequencies, also has been related to increased sociality in some primates. The lack of identification of additional broad trends and relationships between audition and ecology may be partially attributed to the limited dataset, which lacks representation from several major taxonomic subgroups. Additionally, order-wide trends may be minimal given the many possible reasons why enhanced or reduced sensitivity to certain frequency regions may be beneficial for different species. These are just a few of the many facets of primate audition that need to be explored in more depth through additional data gathered via continually evaluated and refined methodologies.

M.A. Ramsier (✉)
Department of Anthropology, Humboldt State University,
1 Harpst Street, Arcata, CA 95521, USA
e-mail: marissa.ramsier@humboldt.edu

J.P. Rauschecker
Department of Neuroscience, Georgetown University,
3900 Reservoir Road NW, Washington, DC 20057, USA

Institute for Advanced Study, TUM, Munich-Garching, Munich 80333, Germany
e-mail: rauschej@georgetown.edu

R.M. Quam et al. (eds.), *Primate Hearing and Communication*, Springer
Handbook of Auditory Research 63, DOI 10.1007/978-3-319-59478-1_3

Keywords Acoustic communication • Audiogram • Bioacoustics • Hearing • Localization • Monkey • Neural processing • Psychoacoustics • Sensory ecology

3.1 Introduction

The special senses are central to the behavior, ecology, and ultimately the survival and reproductive success of primates (Dominy et al. 2001). Through the auditory sense, primates are able to locate sound sources and derive information about the surrounding environment both at close distances and, in general, farther away than the other senses permit. For example, primates can use gustatory and tactile senses to evaluate food sources only up close (Laska et al. 2007) and the tactile sense to communicate only when in direct contact (Weber 1973). Olfaction is useful at close and intermediate ranges and for extended periods of time, including when food resources are obscured by vegetation and leaf litter (Irwin et al. 2007); however, forest substrates are discontinuous, and scent is not useful for immediately conveying time-sensitive information about resources and threats from afar. Enhanced vision is one of the hallmarks of primate evolution (Crompton 1995), and it can be utilized at both close and far distances. However, using vision to communicate across long distances can be challenging when vegetation is dense or at night (Bearder et al. 2006).

Sound can be used to communicate under varying circumstances. Audition allows primates to detect predators and alarm calls of nearby animals and even identify specific predator types and locations (e.g., Blumstein 2002; Zuberbühler 2007). Audition also allows primates to detect vocal signals from conspecifics that indicate divisible food resources; for example, when toque macaques (*Macaca sinica*) locate abundant food sources, they give specific calls that evoke rapid direct approach from dispersed group members (Dittus 1984). Primates utilize a variety of acoustic cues, such as the sound of rustling leaves, to locate prey (Goerlitz and Siemers 2007), and vocalizations also facilitate social behavior and mating practices (Semple and McComb 2000).

Of the acoustic signals and cues present in primate habitats, vocalizations have been a topic of particularly intensive research, owing in part to their usefulness for identifying species and behaviors even from a distance (e.g., Gautier 1988; Snowdon 1993; Snowdon, Chap. 6; Zuberbühler, Chap. 7) and for studying the evolution of communication in humans (e.g., Owren 2003; Nishimura 2008; Quam, Martínez, Rosa, and Arsuaga, Chap. 8). Variations in primate vocal acoustics have been associated with behavior (e.g., Sekulic and Chivers 1986; Zimmermann, Chap. 5) as well as ecological and habitat conditions (e.g., Masters 1991; Brown et al. 1995; Brown and Waser, Chap. 4). Thus, it is reasonable to suspect that, as the receiving end of vocalizations, the relative auditory sensitivity of primates to varying types of signals, and the frequencies included therein, vary in relation to vocal acoustics, behavior, and ecology. Such relationships are documented in other organisms and are found to be complex and variable. For example, Vélez et al. (2015) found that among nine species of sparrows (Passeriformes), those that had more complex song structure had greater auditory sensitivity to high frequencies than sparrows with pure-trilled or tonal call structure. Some species of freshwater fish may also have evolved enhanced auditory sensitivity as an adaptation to take advantage of quiet ambient noise levels in still waters (Amoser and Ladich 2005).

A few relationships between overall auditory sensitivity and behavioral ecology have been reported for primates. A longstanding model explains variations in auditory sensitivity, specifically to high frequencies, as a function primarily of sound localization acuity (R. S. Heffner 2004) (Sect. 3.4.3.1). Brown and Waser (1984) report that blue monkeys (*Cercopithecus mitis*) are particularly adept at detecting low frequencies associated with their low-frequency *long calls* and forested environments (Sect. 3.4.5). Ramsier et al. (2012a) reported a correlation between enhanced auditory sensitivity and sociality among strepsirrhine primates (Sect. 3.4.3.2).

Multiple studies have focused on the neural processing and perception of vocalizations by primates, providing a comparative context for understanding the evolution of speech, language, and social communication in humans (e.g., Ghazanfar and Santos 2004; Rauschecker and Scott 2009) (Sect. 3.2). However, studies of primate vocal communication and ecology do not discuss audition to any significant degree—vocalizations and audition are generally treated separately in the literature—in large part due to the lack of auditory data on many primates of interest and the tendency of auditory studies to take a clinical or biomedical approach. Similarly, the two key areas of primate audition—overall auditory sensitivity (range of audible frequencies reported as an audiogram) and neural processing and perception—are largely treated separately in the literature.

Measuring the overall auditory sensitivity of nonhuman primates is a complicated process that traditionally has involved months of training in laboratory settings (H. E. Heffner and R. S. Heffner 2014). Since the 1930s, audiograms derived using behaviorally based testing methods have been reported for more than twenty primate species; however, major primate taxonomic groups and hundreds of species are still unstudied (Fay 1988; Coleman 2009) (Sect. 3.4.1). Accordingly, few widespread trends in primate auditory sensitivity have been identified in the literature, leading to the supposition that primate auditory sensitivity is unspecialized in terms of range and relative sensitivity to various frequencies (R. S. Heffner 2004) (Sect. 3.4.3). At the same time, the neurobiological literature describes primates as auditory specialists in terms of auditory processing and perception, such as having species-specific vocalizations (e.g., Ghazanfar and Santos 2002; Rauschecker and Scott 2009). Taken together, these findings point to the similarity of nonhuman primates to humans in their auditory capabilities. It has become convention (or necessity), therefore, to largely disregard the potential influence of interspecies variations in auditory sensitivity when studying bioacoustic communication among nonhuman primates, a practice that is reinforced by the close evolutionary relationship between humans and nonhuman primates and the tendency to anthropomorphize nonhuman primate behaviors (Asquith 2011). Field workers may be left with little choice but to assume that what is loud or quiet to the human observer is also loud or quiet to the animals being observed when, in reality, this may not be the case. Sounds that humans may not be able to hear well or at all may affect or be utilized by nonhuman primates in ways that are not fully understood (Barber et al. 2010; Kight and Swaddle 2011).

An increasing number of studies seek to build on the solid foundation of decades of behaviorally based auditory testing to better understand the ecological implications of variations in primate auditory sensitivity. This has involved an exploration of physiologically derived auditory testing techniques for constructing audiograms (e.g., Ramsier and Dominy 2010) (Sect. 3.3.3), detailed neurophysiological and anatomical studies (e.g., Micheyl et al. 2005; Coleman and Colbert 2010; Nummela, Chap. 2), and computer modeling (e.g., Quam et al. 2015). These studies have demonstrated that at least some species do indeed have specialized neural structures and processing abilities that share similarities with humans. In addition, the sensitivity of primates to different frequencies may be more variable than previously thought; for example, species that have been described as relatively quiet may in fact be communicating in a realm outside of the range of human hearing (e.g., Ramsier et al. 2012b; Gursky 2015).

This chapter begins with an overview of auditory neurobiological processing and perception in primates (Sect. 3.2). The chapter then discusses ways in which overall auditory sensitivity is conceptualized and measured among primates (Sect. 3.3) and then reviews the current data for primates along with potential explanations for variations (Sect. 3.4). The chapter concludes with implications for future research (Sect. 3.5).

3.2 Auditory Processing and Perception in Primates

3.2.1 The Path of Sound: From Cochlea to Auditory Cortex

After sound is captured by the outer ear, transformed into mechanical energy in the middle ear, and translated into electrical impulses within the cochlea of the inner ear (Nummela, Chap. 2), the central auditory system is responsible for transmitting those signals to various brain centers for processing to determine sound source location, to identify features of the source (e.g., species or sex of an individual that produced a communication call) and, ultimately, to determine the sound's meaning. Neuroanatomical structures and their physiological workings affect the complexity of information that can be acoustically communicated and the efficiency and specificity of sound localization. A common feature of the primate auditory system is its map-like "tonotopic" organization, wherein specific neurons or groups of neurons fire most strongly in response to particular temporal and spectral characteristics of stimuli along a tonotopic or cochleotopic frequency axis. The following section focuses on pathways for, and processing of, locus cues and vocalizations. It is in these abilities that the specialized nature of the primate auditory system may be indicated.

Within the fluid-filled spiral cochlea of the inner ear, the organ of Corti winds up the basilar membrane of the cochlear duct—this organ is the sensory structure responsible for converting fluidborne vibrations into electrical impulses that can be interpreted by the brain (Webster et al. 1992; Geisler 1998). Sound-induced movement of the basilar membrane causes movement of mechanoreceptor hair cells on the organ of Corti. Like mammals in general, the primate cochlea is tonotopically organized in that the hair cells at the basal cochlea are more sensitive to high frequencies, and the hair cells at the apex are more sensitive to low frequencies. This occurs largely by virtue of cochlear mechanics, whereby traveling waves peak at certain locations along the basilar membrane in a frequency-dependent manner (von Békésy 1960). Bipolar neurons have cell bodies that lie in the spiral ganglion, which is a string of tens of thousands of neurons along the central axis (modiolus) of the cochlea, and they are the first neurons in the auditory system to fire an action potential. They supply all of the brain's auditory input (Nayagam et al. 2011). The dendrites of bipolar neurons make synaptic contact with the base of hair cells, and their axons form the auditory portion of the vestibulocochlear nerve. The first major center of auditory neural processing is the cochlear nucleus (with a ventral and a dorsal subdivision). Figure 3.1a depicts the pathway of sound (and its neural representations) from the cochlea to the primary auditory cortices of the temporal lobe of the cerebrum, including the major (generally tonotopically organized) relay stations along this path. The pathway is similar in humans and nonhuman primates, such as the common marmoset (*Callithrix jacchus*) (Aitkin and Park 1993) or the rhesus macaque (*Macaca mulatta*) (e.g., Hackett 2011), as well as generally similar within the mammals (Webster et al. 1992; Geisler 1998).

In primates, conscious awareness of sound takes place within the various divisions of the auditory cortex (Fig. 3.1b). Within the auditory cortex, acoustic signals first travel to one or more of the primary cortical areas, which are most responsive to pure tones (Ghazanfar and Santos 2004). There are at least two widely agreed on primary cortical areas (A1 and R), but possibly there are as many as three or four (e.g., Kaas and Hackett 2000). Signals then travel to one or more of the surrounding seven (or so) auditory cortical belt areas and subsequently enter the prefrontal cortex of the frontal lobe, either directly from the belt or through functionally specific auditory parabelt areas in auditory and/or auditory-related fields in the superior temporal gyrus (Romanski et al. 1999; Kaas and Hackett 2000; Rauschecker and Tian 2000; Poremba et al. 2003; Hackett 2011; Rauschecker and Romanski 2011).

Like other major partitions of the primate auditory pathway, portions of the human and nonhuman primate auditory cortices work in a map-like fashion to represent frequency. For example, rhesus macaques and common marmosets have a tonotopic map on auditory area A1 (Aitkin et al. 1986; Micheyl et al. 2005). Individual fibers carry information from (and neurons are most responsive to) particular tones, with response strength decreasing sharply as frequencies depart from the preferred frequency. This organization is also present in most other mammals (e.g., cats: Imig and Adrian 1977).

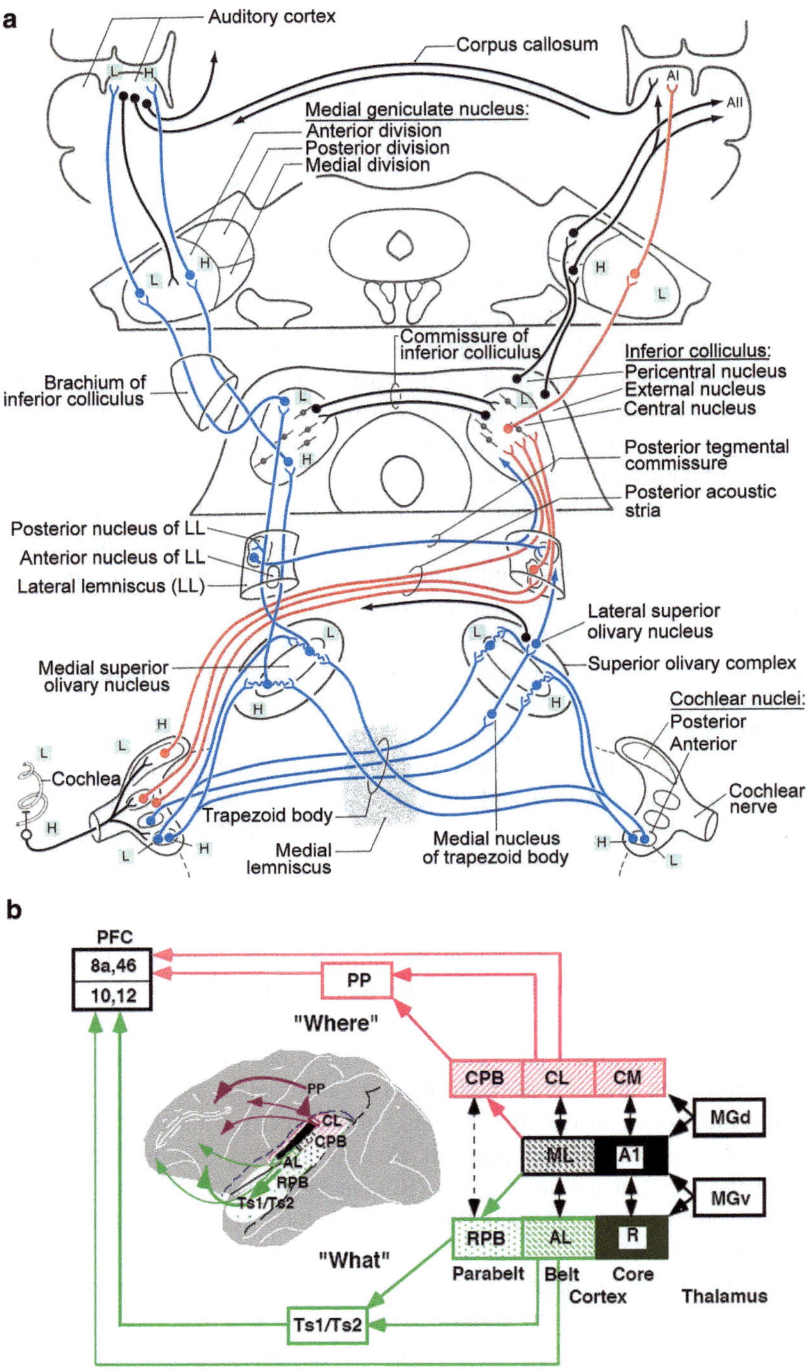

Fig. 3.1 Neuroanatomy of the auditory system in primates. (**a**) Ascending auditory pathway from the cochlea to the auditory cortices. Fibers in *blue* originate from neurons in the ventral cochlear nucleus, form the lemniscal pathway (*LL*), and eventually pass through the ventral division of the medial geniculate nucleus on their way to primary auditory cortex. Fibers in *red* originate from the dorsal cochlear nucleus and form the extralemniscal pathway. Low-frequency (*L*) and high-frequency (*H*) pathways are present throughout. (**b**) Cortical pathways for auditory processing in

3.2.2 Alternate Pathways for Spectral and Spatial Information

Neural processing of localization cues begins at the superior olivary nuclei of the medulla-pons junction and the inferior colliculus of the auditory midbrain. Later, at the cortical level in human and nonhuman primates, functional divergence of object-related (what) and spatial (where) information takes place after the primary auditory cortex in the superior temporal plane (Rauschecker and Tian 2000). More specifically, in humans, divergence takes place at the planum temporale, after which object-related spectral information is processed in the anterolateral planum temporale, planum polare, lateral Heschl's gyrus, and the superior temporal gyrus anterior to Heschl's gyrus (Warren and Griffiths 2003). Spatial information is processed in the posteromedial planum temporale and in the parietal and frontal lobes (Bushara et al. 1999). In macaques (*Macaca* sp.), divergence occurs in the belt areas (along the superior temporal gyrus): object-related spectral information proceeds from the anterior lateral belt through fields in the anteroventral superior temporal region into ventrolateral prefrontal cortex, whereas spatial information proceeds from the caudolateral belt and through fields in the posterodorsal superior temporal lobe and the posterior parietal cortex into dorsolateral prefrontal cortex (e.g., Romanski et al. 1999; Tian et al. 2001). Mostly based on clinical stroke studies, the posterior part of superior temporal gyrus (STG) in humans has classically been considered as specialized for speech processing ("Wernicke's area"). Given reports from human imaging that anterior regions of STG are at least as selective for the perception of words as posterior regions (DeWitt and Rauschecker 2012), a redefinition of posterior STG as an area specializing in sensorimotor integration and control seems appropriate (Rauschecker 2011). This would include a role in spatial processing as well as in speech production and perception.

An important aspect of the primate central auditory system is its redundancy. For example, in the macaque lateral belt, signals are largely segregated into spatial (caudolateral belt) and nonspatial (anterior lateral belt) information; however, the streams obviously interact (Kaas and Hackett 1999; Romanski et al. 1999). Some neurons in the primate caudolateral belt respond to both location and specific calls, and the middle lateral belt is approximately equally selective for both call type and

(Continued) the macaque. Corticocortical projections of the central auditory system run along two segregated pathways: a ventral pathway (*green*) runs from the anterolateral belt (*area AL*) along the anterior superior temporal cortex to the ventrolateral prefrontal cortex, while a dorsal pathway (*red*) extends from the caudolateral belt (*area CL*) to superior temporal cortex and inferior parietal cortex and ends in dorsolateral prefrontal cortex. Discrete thalamic input to the two pathways is provided from different medial geniculate (*MG*) nuclei: The ventral part (*MGv*) projects only to the core fields *A1* and *R*, whereas the dorsal part (*MGd*) projects to primary auditory cortex (*A1*) and the caudomedial field (*CM*) (Rauschecker et al. 1997). Likewise, feedforward projections from *AL* and *CL* are largely separated and target the rostral parabelt (*RPB*) and caudal parabelt (*CPB*) regions, respectively (Hackett et al. 1998). Additional pathways involve the middle lateral area (*ML*), posterior parietal cortex (*PP*), and *RPB* areas on the surface of the rostral superior temporal gyrus (*Ts1/Ts2*) (Pandya and Sanides 1973). Prefrontal cortex projections (*PFC*) are segregated in Brodmann areas 10 and 12 versus 8a and 46, respectively (Romanski et al. 1999). (**a** was modified and reprinted with permission from Henkel 2006; **b** was modified from Rauschecker and Romanski 2011; reproduced with permission from the original source, Rauschecker and Tian 2000) sound source location (Tian et al. 2001). Furthermore, each side of the brain receives and processes impulses from both ears, although in primates (human and nonhuman) the left cerebral hemisphere may have greater selectivity for processing temporal information, and the right cerebral hemisphere may have greater selectivity for processing spectral information (Joly et al. 2012; Ortiz-Rios et al. 2015).

3.2.3 Encoding Signals

In humans, the cortical region around Heschl's gyrus, which also contains primary auditory cortex, is responsible for pitch perception (Schneider et al. 2005). A cortical area analogous to this region has been described for nonhuman primates (Bendor and Wang 2005). In their study on common marmosets, Bendor and Wang demonstrate that an area (restricted to low frequencies) on the border between two of the primary cortical areas (A1 and R) and adjacent to the anterior auditory cortical belt (AL and ML) contains pitch-selective neurons (also see Tomlinson and Schwarz 1988). Each neuron or group of neurons responds best to a specific pitch, whether it is generated by an actual pure tone or by a "missing fundamental" frequency represented by its spectral envelope.

Temporal relationships of signals and signal elements are important for identifying target proximity and location and distinguishing between calls (e.g., Ghazanfar and Santos 2004). In many cases, temporal alteration may affect representation more than spectral manipulation (Nagarajan et al. 2002; Ghazanfar and Santos 2004). Some neurons in the auditory midbrain respond selectively to order and spacing combinations (Wollberg and Newman 1972). This is demonstrated by the differential processing of temporally expanded and compressed vocalizations by the common marmoset (Wang et al. 1995) (Sect. 3.2.5). Other neurons in the auditory midbrain respond selectively to duration of frequency modulation or rates of amplitude modulation (e.g., Casseday et al. 1994). In another example, researchers presented a series of alternating high- and low-frequency tones to awake long-tailed macaques (*Macaca fascicularis)* and found that increasing the frequency separation, presentation rate, and tone duration improved the spatial differentiation of tonal responses on A1's tonotopic map (p. 1656 in Fishman et al. 2004).

Studies on auditory cortex in anesthetized primates (e.g., common marmosets: Wang et al. 1995; squirrel monkeys, *Saimiri sciureus*: Bieser 1998) have reported that neurons mainly detect signal changes (onsets or transients). By contrast, when recording from primary auditory cortical and lateral belt neurons in awake common marmosets, Wang et al. (2005) found that responses are not only phasic but also tonic, indicating that some neurons respond continuously to spectrally and temporally *optimal* parts of the signal. Thus, cortical responses may be phasic (onset or offset), persistent tonic, inhibitory, and/or excitatory depending on stimulus frequency, intensity, location, and duration, similar to simple and complex cells in visual cortex (Tian et al. 2013). Since responses in anesthetized animals to pure tones are generally only phasic, they may not represent the full range of cortical responses/firing patterns. Considering this, studies of awake rather than anesthetized animals (e.g., Recanzone et al. 2000; Malone et al. 2002) may be preferable, depending on research questions and methods.

3.2.4 *Are Primate Brains Specialized for Processing Vocalizations?*

The human brain has long been claimed to have specialized neural structures, such as Wernicke's area, for processing speech and, perhaps, others for interpreting meaning and auditory imagery (Fisher and Marcus 2006), but the notion of areas specialized for *speech perception* is undergoing some revision. Although primates show evidence of homologous neuroanatomical pathways and structures, a topic of debate is whether the nonhuman primate central auditory system contains regions that are (or, even as a whole, is) specialized for processing vocalizations. First, it is important to distinguish between auditory brain areas being *sensitive* versus *selective*. That an area is *vocalization sensitive* means that its neurons respond especially well to all vocalizations. That an area is *vocalization selective* means that single or groups of neurons within that area each respond to different vocalizations: some neurons may respond preferentially to contact calls, whereas others may respond to predator warning calls. Based on neurophysiological experiments, authors such as Rauschecker et al. (1995) and Tian et al. (2001) argue convincingly that certain regions of the primate lateral belt may be vocalization selective. However, during the above experiments, responses to vocalizations were not consistently compared with responses to relevant nonvocal complex sounds in the same neurons. Thus, it is possible that neurons in the primate lateral belt are vocalization sensitive but not selective, and such selectivity is not generated until later in higher processing regions.

Many authors have reviewed vocal communication and parallels with human language in primates. In their study on speech segmentation in cotton-top tamarins (*Saguinus oedipus*), Hauser et al. (2001) demonstrate that nonhuman primates are able to recognize different sequences of syllables in a speech stream. Humans use this ability to calculate statistical probabilities of sequence occurrence (transitional probabilities) for the segmentation and identification of words in an unknown language (e.g., Chomsky 1975). Interestingly, many authors have pointed out that some facets of speech that are central to speech perception in humans, such as syllable onsets, formant frequencies, glottal-pulse periods, and the spectral profiles of consonants and vowels, are already encoded in peripheral hearing not only of primates but of mammals as a whole (e.g., Delgutte 1997; Lieberman 2006).

However, although the mammalian ear may be well-equipped to encode aspects of speech important to human perception, this does not mean that primates are specialized to process the meaning of these features. Many attempts have been made to understand the differences and similarities between human and nonhuman primates with regard to auditory-vocal processing. Because of the complex nature of identified (or as yet unidentified) relationships, Owren and Rendall (2001) rightly warn that, at present, comparisons between (and models of) human language and nonhuman primate vocalizations need to be approached cautiously (also see Ghazanfar and Santos 2004).

3.2.5 Potential Specializations for Processing Species-Specific Vocalizations

Although the communication systems of nonhuman primates do not match humans in either their combinatorial power or the recursive structure of human speech and language, the primate auditory cortex displays similarities with humans, particularly in having a hierarchical structure with tonotopic mapping and specialized streams for processing specific types of information (Rauschecker and Scott 2009). The primate central auditory system shows evidence of specialization for processing location as well as complex bioacoustic communication signals such as conspecific vocalizations. In fact, acoustic sensitivity may decrease when frequencies are not heard in sequences corresponding to biologically meaningful stimuli such as species-specific calls.

The acoustically distinct vocalizations of primate species are well documented and those vocalizations can even be utilized, in some cases, to assess phylogenetic relationships (e.g., Zimmermann 1990). Behavioral studies in the wild provide substantial evidence that primates are able to recognize conspecifics, kin groups, and individuals based on variations in vocal acoustics (e.g., Chapman and Weary 1990). Neurobiological experiments measuring the responses of auditory cortical areas to natural vocalizations versus artificially manipulated or synthesized vocalizations provide a basis for understanding how at least some nonhuman primate species are able to distinguish conspecifics based on their calls. In both human and nonhuman primates, conspecific vocalizations are received in both the left and right cerebral hemispheres, but processing is focused in specific areas of the left hemisphere where some single neurons or groups of neurons may respond particularly well to distinct vocalizations (Ghazanfar and Santos 2004; Poremba et al. 2004). Studies on rhesus macaques demonstrate that the lateral belt systematically represents tones and frequencies and is especially responsive to complex signals such as species-specific vocalizations (e.g., Rauschecker et al. 1995; Romanski et al. 1999). Studies on squirrel monkeys found that neurons in the auditory cortex responded to frequency modulations in both natural and synthesized vocalizations, but responses were greater for natural, strongly amplitude-modulated vocalizations, possibly owing to their syllable-like divisions (Bieser 1998; Ghazanfar and Santos 2004).

In a study on common marmosets, neurons in the primary auditory cortex responded preferentially to normal versus time-reversed, compressed, or expanded conspecific vocalizations. When the same marmoset vocalizations were presented to cats, the evoked responses were relatively small and roughly equal for normal and time-reversed examples (Wang et al. 1995; Wang and Kadia 2001). A behavioral experiment where *long calls* were played back to cotton-top tamarins found that individuals were more likely to respond to whole rather than parts of conspecific calls (Ghazanfar et al. 2001; Snowdon, Chap. 6). Additional studies indicate that among some animals, neuronal responses to temporally correct combinations of tones are stronger than the summed response to the individual signals presented separately (Viemeister and Wakefield 1991; Alder and Rose 1998). Other studies have shown that, whereas squirrel monkey and cotton-top tamarin auditory cortical areas respond more strongly to conspecific vocalizations than to those of other species, time reversing and pitch shifting did not significantly alter the results, indicating order/spectral insensitivity (e.g., Glass and Wollberg 1983). The preferential processing of and response to species-specific calls may be preprogrammed or dependent on experience and may be related to recognizing signals that are similar to those that are self-produced (Brainard and Doupe 2002). *Correctness* likely varies at the species level (Alder and Rose 1998).

3.2.6 Interindividual Recognition

Unarguably, humans are able to distinguish between individual voices based on spectral and temporal cues. Two humans saying the same word or phrase (call) can be distinguished from one another. Conversely, tamarin and squirrel monkey studies suggest that the primate auditory system does not respond differently to variants (different examples from different individuals) of the same call (Ghazanfar and Hauser 1999). This suggests that primates may not be universally adept at recognizing individuals based on call structure (Ghazanfar and Santos 2004). However, Wang and colleagues (1995) report that in marmosets, auditory cortical representations from spectrotemporal variants of calls from different individuals were different but overlapping, suggesting some individual recognition might be possible.

Behavioral evidence also supports that primates can recognize individuals from their calls. For example, vervet monkeys (*Chlorocebus aethiops*) can organize individuals hierarchically and into kin groups based on individual calls (Cheney and Seyfarth 1990), and Waser (1977) provides evidence from playback studies that monkeys can recognize individuals based on their vocalizations. The results of these studies are perhaps not surprising, considering that individual recognition based on call structure has long been reported in birds (e.g., Thorpe 1968). It is completely unknown at present how the primate brain processes and stores these subtle differences.

3.3 Defining, Representing, and Measuring Overall Auditory Sensitivity in Primates

Comparative audiograms for primates have been gathered primarily via traditional behaviorally based testing and physiological techniques such as the auditory brainstem response (ABR) method (Sect. 3.3.3). Currently, data are available for only a small percentage of the hundreds of nonhuman primate species (Sect. 3.4), and much of the existing data is likely to be incomparable due to issues or inconsistencies with experimental design or data reporting, greater than average interindividual variation, unexpected results that do not fit preconceptions about variation in the order, or philosophical debates with regard to potential incompatibilities between behaviorally and physiologically derived data (Coleman 2009; H. E. Heffner and R. S. Heffner 2014). This section introduces the ways in which auditory sensitivity is defined and represented, the conceptual issues surrounding methods of data collection, and the comparability of the resulting data.

3.3.1 Defining and Representing Auditory Sensitivity in Primates

The term auditory sensitivity is utilized throughout this chapter as the broadest definition of the function of the sense—it can be conceived of herein as a representation of all sounds that are collected via the ear, are received (produce a neural response) in the brain, and have the potential of being utilized by the individual. Although the terms *auditory sensitivity* (audition) and *hearing* are often used interchangeably, the term hearing carries additional complex meanings related to perception and psychoacoustics.

The auditory sensitivity of primates can be represented as the range of audible frequencies, measured in hertz (Hz), that are detectable at varying amplitudes, measured in decibels (dB re 20 μPa). Frequencies below 20 Hz are defined as infrasound because they are below the range of human hearing, and frequencies above 20 kHz are defined as ultrasound, or above the range of human hearing. Auditory sensitivity can be represented graphically as an audiogram—a curve showing the lowest audible level (*threshold*, in dB) at each tested frequency. In this chapter, variation in auditory sensitivity within and between species is considered through the most common audiometric parameters: *frequency of best sensitivity*, defined as the frequency that can be detected at the lowest level (in dB); and the *low-frequency* and *high-frequency limits*, defined as the lowest and highest frequencies, respectively, detectable at reasonable amplitudes (conventionally 60 dB). The *audible range*, defined as the number of octaves between the low- and high-frequency limits, is also a common audiometric parameter, but it is not considered here since it is highly reliant on both the low- and high-frequency limits, and the former is not available for most subjects. Studies have also sought to formulate additional audiometric parameters to facilitate interspecific comparisons, such as the *absolute threshold level* at particular frequencies, or measures of overall sensitivity across the audiogram, or sensitivity within low-, mid-, and high-frequency areas (e.g., Coleman and Colbert 2010; Ramsier et al. 2012a); these parameters are yet to be widely adopted and thus are not considered further in this chapter.

3.3.2 Determining Threshold

When constructing an audiogram, the precision of the threshold measurement is highly dependant upon the frequency steps used and the accurate calibration of stimuli (Coleman 2009). A free-field speaker is generally considered the ideal transducer for delivering stimuli to primates. The use of headphones, from inserts to circumaural, is also relatively common when testing auditory sensitivity in humans and other animals, as headphones may help minimize interference from subject position, room noise, and electrical artifacts (Martin and Clark 2006). However, earphones that depress or bypass the pinnae may influence or negate the amplification effects of the pinnae (Sinyor and Laszlo 1973; Rosowski 1991). Thus, some workers express concern over the use of headphones, particularly insert varieties, over pinna amplification issues or concerns that delivering low-frequency signals through these devices can be problematic (R. S. Heffner 2004; Coleman 2009). Tables 3.1 and 3.2 show data gathered free-field and with headphones for several species. There seems to be good agreement in the high-frequency limit but more variation with the frequency of best sensitivity, which may be more strongly subject to methodological variations. More data are needed to fully evaluate pinna effects and the influence of transducer type on auditory thresholds. Another potential issue is that pure tone stimuli may only broadly represent auditory sensitivity, given that in at least some primates, neural responses to conspecific vocalizations are enhanced compared to nonspecific noise (Sect. 3.2.5).

3.3.3 Testing Methods

After decades of refinement, well-designed behavioral testing regimens produce what are generally considered to be ideal estimates of auditory sensitivity, as the behavior of whole animals is measured (H. E. Heffner and R. S. Heffner 2014). Beginning with Elder's (1934) audiogram for chimpanzees (*Pan troglodytes*), most existing data on primate audition have been gathered via behaviorally based methodologies, although very few have been collected in recent decades (Sect. 3.4) (Coleman 2009).

An alternative to behaviorally based testing is minimally invasive, physiologically based testing, such as the ABR method (Jacobson 1985), during which the responses of the auditory system are measured directly. The ABR method has been widely adopted within the biomedical and clinical realms (Burkard and Don 2007) and recently within primatology (Ramsier and Dominy 2010). The ABR method reliably estimates overall audiogram shape (dips and peaks in sensitivity) and the behaviorally derived high-frequency limit and frequency of best sensitivity. However, threshold levels for low-frequency stimuli may be underestimated by the ABR method, and additional data are needed to fully evaluate to what degree it is possible to compare absolute thresholds derived through each method.

3.4 Auditory Sensitivity Among Primates

3.4.1 Primate Audiograms

Reasonably complete and comparable audiograms for twenty-nine nonhuman primate species have been published using either traditional behavioral testing or the ABR method (Sect. 3.3.3). These data are considered together in this section, despite some debate over the degree to which data gathered with different methodologies (e.g., behavioral versus ABR, speaker versus headphones) can be compared (Sect. 3.3).

The sample of published audiograms represents both primate semiorders. The semiorder Strepsirrhini (Table 3.1) is the evolutionarily oldest primate clade and more closely reflects the ancestral primate condition (Masters et al. 2013; Zimmermann, Chap. 5). The Strepsirrhini includes two infraorders, the Lorisiformes and Lemuriformes. The Lorisiformes include relatively small-bodied, nocturnal, highly arboreal species from Africa and Asia; audiograms have been published for six species. The strepsirrhine infraorder Lemuriformes is more variable than Lorisiformes in body size, behavior, and ecology—it consists of small- to medium-bodied, arboreal to semiterrestrial, nocturnal, cathemeral, and diurnal species from the island of Madagascar. Audiograms have been published for nine taxa of Lemuriformes.

The semiorder Haplorhini (Table 3.2) includes primates that are more closely related to humans than are the strepsirrhines. The haplorhine suborder Tarsiiformes includes one infraorder (also Tarsiiformes) and multiple species of small-bodied, nocturnal, arboreal tarsiers (*Carlito* sp., *Cephalopachus* sp., *Tarsius* sp.) from Asia; an audiogram exists for one species (Ramsier et al. 2012b). Due to behavioral and morphological similarities with the semiorder Strepsirrhini, tarsiers were traditionally grouped with them (Masters et al. 2013). The haplorhine suborder Anthropoidea has two infraorders. Infraorder Platyrrhini consists of New World monkeys from Central and South America, which are medium-bodied arboreal species that generally are diurnal, with the exception of the nocturnal owl monkey (*Aotus* sp.). Comparable audiograms exist for three smaller bodied species, but none exist for the many larger bodied New World monkeys, such as howling monkeys (*Alouatta* sp.), spider monkeys (*Ateles* sp.), and capuchins (*Cebus* sp. and *Sapajus* sp.), nor the many species of tamarin (*Saguinus* sp.).

The Anthropoid infraorder Catarrhini consists of Old World monkeys, apes, and humans from Africa, Asia, and Europe. This is a highly diverse group that consists of medium- to large-bodied species that are all diurnal and range from terrestrial to highly arboreal. Much research in this group has focused on common laboratory species such as macaques. No comparable audiograms are published for the speciose Colobinae subfamily of monkeys nor for the apes other than the chimpanzee.

There is notable variation in the auditory sensitivity of the primate species tested to date. This can be conceptualized visually by comparing median behavioral audiograms for each infraorder (Fig. 3.2) and by examining audiometric parameters for both behavioral and ABR audiograms (Tables 3.1 and 3.2; Fig. 3.3).

Table 3.1 Auditory sensitivity in primate semiorder Strepsirrhini

Species	Method, transducer[a]	Best freq.[b] (kHz)	High freq.[c] (kHz)	Low freq.[d] (Hz)	References
Infraorder Lorisiformes					
Bushbaby (*Galago senegalensis*)	Beh, Spk	8	65.0	70	H. E. Heffner et al. (1969)
Slow loris (*Nycticebus coucang*)	ABR, Spk	16	42.6	-	Ramsier et al. (2012a)
	Beh, Spk	16	43	83	H. E. Heffner and Masterton (1970)
Pygmy slow loris (*Nycticebus pygmaeus*)	ABR, Spk	11.3	51.5	-	Ramsier et al. (2012a)
Potto (*Perodicticus potto*)	Beh, Spk	16	42.0	135	H. E. Heffner and Masterton (1970)
Infraorder Lemuriformes					
Aye-aye (*Daubentonia madagascariensis*)	ABR, Spk	11.3 (4)	65.6	-	Ramsier et al. (2012a)
Crowned lemur (*Eulemur coronatus*)	ABR, Spk	8	59.6	-	Ramsier et al. (2012a)
Collared lemur (*Eulemur fulvus collaris*)	ABR, Spk	8	57.4	-	Ramsier et al. (2012a)
Red-fronted lemur (*Eulemur fulvus rufus*)	ABR, Spk	11.3 (5.7)	63.7	-	Ramsier et al. (2012a)
Mongoose lemur (*Eulemur mongoz*)	ABR, Spk	8	54.2	-	Ramsier et al. (2012a)
Red-bellied lemur (*Eulemur rubriventer*)	ABR, Spk	5.7	45.1	-	Ramsier et al. (2012a)
Ring-tailed lemur (*Lemur catta*)	ABR, Spk	11.3 (5.7)	62.2	-	Ramsier et al. (2012a)
	Beh, Spk	8 (2)	58	57	Gillette et al. (1973)
Gray mouse lemur (*Microcebus murinus*)	ABR, Spk	7.9	44.6	-	Schopf et al. (2014)
Fork-marked lemur (*Phaner furcifer*)	Beh, Spk	16	60.0	150	Niaussat and Molin (1978)
Coquerel's sifaka (*Propithecus coquereli*)	ABR, Spk	11.3	49.7	-	Ramsier et al. (2012a)
Red-ruffed lemur (*Varecia rubra*)	ABR, Spk	11.3 (5.7)	59.0	-	Ramsier et al. (2012a)

[a]ABR, auditory brainstem response testing; Beh, behavioral testing; Spk, speakers
[b]Best freq., frequency of best sensitivity (numbers in parentheses = secondary peaks within 10 dB)
[c]High freq., highest frequency detectable at 60 dB
[d]Low freq., lowest frequency detectable at 60 dB

Table 3.2 Auditory sensitivity in primate semiorder Haplorhini

Species	Method, transducer[a]	Best freq.[b] (kHz)	High freq.[c] (kHz)	Low freq.[d] (Hz)	References
Suborder Tarsiiformes, Infraorder Tarsiiformes					
Philippine Tarsier (*Carlito syrichta*)	ABR, Spk	16 (1.4)	76–91	-	Ramsier et al. (2012a)
Suborder Anthropoidea, Infraorder Platyrrhini					
Owl monkey (*Aotus trivirgatus*)	Beh, Spk	10	44.5	-	Beecher (1974b)
Common marmoset (*Callithrix jacchus*)	Beh, Spk	7 (2)	28	-	Seiden (1957)
	Beh, Spk	7	44.9	-	Osmanski and Wang (2011)
Squirrel monkey (*Saimiri* sp.)	Beh, Spk	12 (2)	42.5	-	Fujita and Elliott (1965), Beecher (1974a)
	Beh, Phn	8	41	140	Green (1971, 1975)
Suborder Anthropoidea, Infraorder Catarrhini					
Blue monkey (*Cercopithecus mitis*)	Beh, Spk	4 (1, 2)	50.3	-	Brown and Waser (1984)
De Brazza's monkey (*Cercopithecus neglectus*)	Beh, Phn	5.7 (1.4)	43	61	Owren et al. (1988)
Vervet monkey (*Chlorocebus aethiops*)	Beh, Phn	1.4 (5.7)	45	69	Owren et al. (1988)
Grey-cheeked mangabey (*Lophocebus abligena*)	Beh, Spk	0.8 (8.0)	-	-	Brown (1986)
Long-tailed macaque (*Macaca fascicularis*)	Beh, Spk	16 (1)	-	-	Fugita and Elliott (1965)
	Beh, Phn	1 (8)	42	-	Stebbins et al. (1966)
Japanese macaque (*Macaca fuscata*)	Beh, Spk	4 (1)	36.5	28	Jackson et al. (1999)
	Beh, Phn	5.7 (1–1.4)	41	82	Owren et al. (1988)
Rhesus macaque (*Macaca mulatta*)	Beh, Spk	4 (16)	-	-	Behar et al. (1965), Fugita and Elliott (1965), Bennett et al. (1983)
	Beh, Phn	8 (1.4)	41	-	Pfingst et al. (1975, 1978), Lonsbury-Martin and Martin (1981)
	ABR, Spk	16 (4)	38.1	-	Lasky et al. (1999)
Pig-tailed macaque (*Macaca nemestrina*)	Beh, Phn	8 (1)	35	-	Stebbins et al. (1966), Gourevich (1970)
Chimpanzee (*Pan troglodytes*)	Beh, Phn	8 (1)	27	-	Elder (1934, 1935), Kojima (1990)
Yellow baboon (*Papio cynocephalus*)	Beh, Spk	8 (1)	41.0	-	Hienz et al. (1982)

[a]ABR, auditory brainstem response testing; Beh, behavioral testing; Phn, headphones; Spk, speakers
[b]Best freq., frequency of best sensitivity (numbers in parentheses = secondary peaks within 10 dB)
[c]High freq., highest frequency detectable at 60 dB
[d]Low freq., lowest frequency detectable at 60 dB

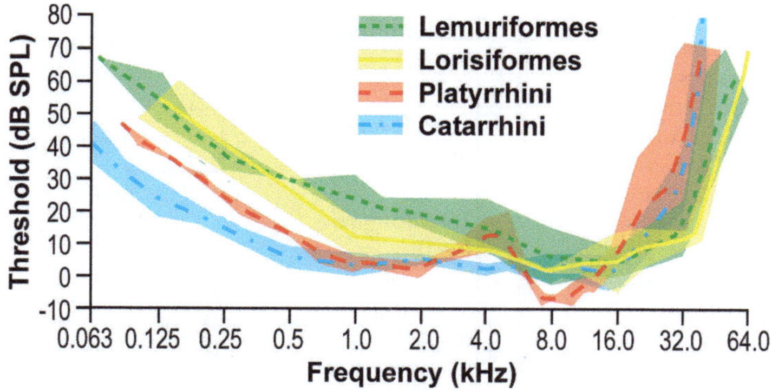

Fig. 3.2 Median (*lines*) and range (*shading*) of behavioral audiograms for the four major primate infraorders (based on Coleman 2009)

3.4.2 *Intraspecies Variation*

Coleman (2009) reviewed behaviorally based primate auditory studies and calculated the average within-study intraspecific variation in the threshold for each tested frequency to be ±4.2 dB around the mean (range ±0.95–9.25 dB), with slightly increased variation at frequencies greater than 8 kHz. There was a relationship between the number of individuals included in a study and the reported intraspecies variation—the average intraspecific variation for studies with four or more subjects was higher (mean ± 5.7 dB) than the overall average of ±4.2 dB. Given that studies tend to choose similar subjects (e.g., young adult males), it seems likely that intersubject variability is underestimated in tests of auditory sensitivity.

3.4.3 *Variation in High-Frequency Limit*

There is much variation in high-frequency limit (Tables 3.1 and 3.2; Fig. 3.3). On average, primates of the semiorder Strepsirrhini are more sensitive to high frequencies; within the Strepsirrhini, there is much overlap between the infraorders Lorisiformes and Lemuriformes, with the latter averaging the highest high-frequency limit. Monkeys and apes of the semiorder Haplorhini tend to be relatively less sensitive to high frequencies with the notable exception of the small, nocturnal tarsier, for which the high-frequency limit is the highest reported within the primate order (Table 3.2; Fig. 3.3).

3.4.3.1 High-Frequency Limit and Sound Source Localization

A long prevailing model explains variation in high-frequency auditory sensitivity among mammals as a product of head size and the need for localizing sound sources (Masterton et al. 1969; R. S. Heffner 2004). Auditory localization is the act of

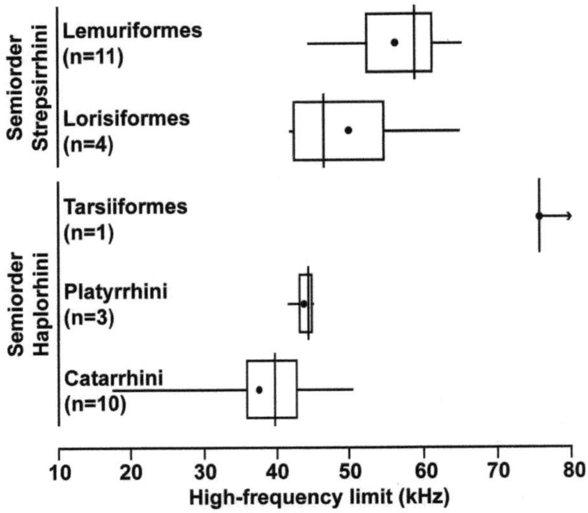

Fig. 3.3 The 60-dB high frequency limit among the primate infraorders. *Horizontal lines* show range, *box limits* show first and third quartiles, *vertical lines* show median, and *dots* show mean values. For the Tarsiiformes, the one data point is at least 76 kHz but could be higher, as represented by the *arrow*

determining the directional location of a sound source horizontally (azimuth) and in elevation (Blauert 1997; Popper and Fay 2005). How accurately an animal can localize sources is referred to as localization acuity. Most mammals, other than subterranean species, can localize within a window of 40° or less (R. S. Heffner and H. E. Heffner 1992; R. S. Heffner 2004). Data on Japanese macaques (*Macaca fuscata)* (Houben and Gourevitch 1979) and squirrel monkeys (Don and Starr 1972) suggest that nonhuman primates are very good localizers with acuity similar to that of cats, pigs, and opossums at around 4–6° azimuth (R. S. Heffner and H. E. Heffner 1988; R. S. Heffner 2004). Humans, like dolphins (Renaud and Popper 1975) and elephants (R. S. Heffner and H. E. Heffner 1982), are especially good localizers, with acuity of around 1° azimuth—in other words, humans can orient directly toward a sound source with almost perfect accuracy (Middlebrooks and Green 1991; R. S. Heffner 2004).

R. S. Heffner (2004) reported that auditory localization acuity is well-matched to the width of the field of best vision among mammals. The narrower the field of best vision is, the better the auditory localization acuity is so that the head can be oriented precisely to put the sound source in the subject's field of best vision. The especially good auditory localization ability of haplorhine primates, such as humans, corresponds with the presence of a very narrow field of best vision. This relationship is underlain by similarities in auditory and visual neural structures and mechanisms (Rauschecker 2015); for example, responses to stimuli coming from the area that is *attended to* are amplified, and responses to *peripheral stimuli* are attenuated (e.g., Bushara et al. 1999; Winkowski and Knudsen 2006). The first localization response allows the head to be turned for subsequent maximum auditory and visual localization acuity. Currently, there are insufficient comparative data on auditory and visual localization acuity among primates to fully investigate trends within the primate order, but further investigation would be interesting given that enhanced vision is one of the hallmarks of primate evolution (Martin and Ross 2005).

Many terrestrial vertebrates can detect the horizontal location of a sound's source with the aid of binaural cues—differences in the sound received at each ear (Geisler 1998; Heffner 2004). In general, a sound is perceived as more intense by the ear that is facing more directly toward the sound, at which it also arrives first. Interaural distance, the distance between the tympanic membranes, influences the effectiveness of binaural localization cues at different frequencies. Increasingly lower frequencies have increasingly longer wavelengths such that low-frequency sound waves may pass by the head (especially a small head) with little or no deflection, making low frequencies increasingly difficult or impossible to use for localization. Furthermore, interaural timing cues rely on low frequencies and decrease in usefulness as head size decreases (e.g., Klump and Eady 1956; Heffner 2004). Thus, the allometric model of auditory sensitivity explains high-frequency sensitivity as a negative function of interaural distance—smaller headed mammals are increasingly reliant on higher frequencies to enable localization through binaural and pinna cues (R. S. Heffner 2004). This is a well-supported model that explains general patterns observed among mammals.

R. S. Heffner (2004) concluded that primate hearing is not specialized in terms of audible frequency range but, rather, follows the typical mammalian pattern with smaller species capable of hearing higher frequencies than larger species. While this relationship holds across mammals and across the primate order in general (R. S. Heffner 2004; Ramsier et al. 2012a), interaural distance does not explain all variation among primates (Coleman 2009; Ramsier et al. 2012a). For example, a relationship between high-frequency sensitivity and interaural distance was not significant within the semiorder Strepsirrhini (Ramsier et al. 2012a). When all data (multimethod) from Tables 3.1 and 3.2 were considered, the relationship was significant among the Catarrhini and Lorisiformes but not among the Lemuriformes nor the Platyrrhini. When all primates were averaged, the relationship was not significant unless the Lemuriformes were averaged prior to order-wide analysis. Some individual primate species depart from the expected pattern as well. The yellow baboon (*Papio cynocephalus)*, for example, which is one of the largest primates for which data on auditory sensitivity exist, has a relatively elevated high-frequency limit—the opposite of what is predicted by the allometric model (Table 3.2).

R. S. Heffner (2004) noted that animals may take advantage of sensitivity to high frequencies that evolved in relation to localization acuity to communicate via high-frequency signals. However, high-frequency vocal communication is potentially a selective force in itself as well. Both small-headed and large-headed species may experience selective pressure to detect high-frequency sounds, such as those emitted by infants, insect prey, or smaller sympatric species (Sect. 3.4.3.2). Given that individual primates vary in their auditory sensitivity, such selective pressure could certainly operate in addition to, or in the absence of, selective pressure related to sound localization.

Examining limited data available at the time, R. S. Heffner (2004) concluded that intraspecies differences in interaural distance, even the twofold differences present between some dog breeds, did not seem to correlate with differences of equal magnitude in the high-frequency limit and suggested that the high-frequency limit is a species trait, not an individual trait. Along these lines, the lack of a significant relationship between interaural distance and high-frequency limit among the strepsirrhines (especially the lemurs) might be attributed to the close evolutionary relationship among some of the species (and subspecies) in the sample. Thus, the relationship between high-frequency limit and localization acuity may still hold at taxonomic levels above species, and other factors (perhaps after controlling for interaural distance) may further explain the evolution of variation in primate auditory sensitivity. These could be interesting areas for future research.

3.4.3.2 High-Frequency Limit, Behavior, and Ecology

Ramsier et al. (2012a) tested the auditory sensitivity of eleven strepsirrhine primate species and found a relationship between enhanced auditory sensitivity and group size, particularly to high frequencies, indicating that the more social species may benefit from enhanced acoustic communication with conspecifics if higher frequencies are used for communication. For some primate species, it may be particularly beneficial to emit and detect higher frequency alarm calls that are perhaps less audible to common aerial and terrestrial predators (Ramsier et al. 2012b). This model could partially explain why the yellow baboon, a highly social haplorhine species (Semple 2001; Barton et al. 1996), is sensitive to high frequencies despite its large head size and interaural distance. However, the haplorhines are, overall, characterized by relatively poor high-frequency and enhanced low-frequency auditory sensitivity, suggesting that haplorhines as a group may benefit from antipredator strategies other than emitting high-frequency alarm calls (Hill and Dunbar 1998). Lack of use of high-frequency localization cues (R. S. Heffner 2004) and reduced ability to produce high-frequency vocalizations (Fitch 1997) may have contributed to the particularly enhanced low-frequency auditory sensitivity of humans (see Quam, Martínez, Rosa, and Arsuaga, Chap. 8). Perhaps human ancestors relied more heavily on detecting low-frequency sounds produced by avian and felid predators, or perhaps they communicated directly with predators to deter them, similar to the African putty-nosed monkey (*Cercopithecus nictitans martini*) (Arnold et al. 2008).

3.4.4 Frequency of Best Sensitivity

The frequency of best sensitivity is an indication of the frequency at which a species hears best, and thus this audiometric parameter could provide clues to important selective pressures in a primate's environment. In the current dataset, the frequency of best sensitivity is higher on average in the strepsirrhines compared to the haplorhines (Fig. 3.4), following the overall pattern for high-frequency limit. Among the haplorhines, the catarrhines have the broadest overall range in frequency of best sensitivity (0.8–16 kHz), but nine of the eleven tested species have a frequency of best sensitivity (or a second peak in sensitivity) in the lower range of 0.8–4 kHz.

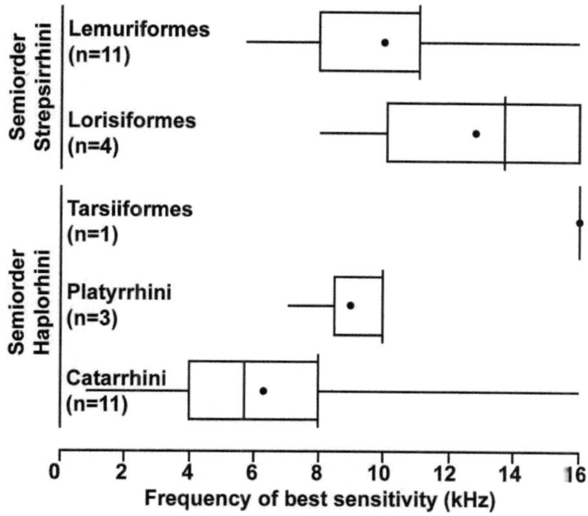

Fig. 3.4 Frequency of best sensitivity among the primate infraorders. *Horizontal lines* show ranges, *box limits* show first and third quartiles, *vertical lines* show medians, and *dots* show mean values

Some primates, particularly the platyrrhine monkeys, have a prominent dual peak of best sensitivity (a w-shaped audiogram, Fig. 3.2) (Coleman 2009). This pattern is not uncommon and is also found in other mammalian groups (e.g., Rice et al. 1992; Bohn et al. 2001). Among the platyrrhines, the higher peak (7–12 kHz) is the most sensitive and thus forms the actual frequency of best sensitivity; the lower (less sensitive) peak lies around 2 kHz, close to the lower frequency cluster found in catarrhines. Some authors speculate that a dip in sensitivity between the peaks in animal audiograms is an adaptation to enhance sound localization ability (e.g., Rice et al. 1992; R. S. Heffner 2004). Others hypothesize that the upper peak may be an adaptation for mother-infant communication (Bohn et al. 2001; Sterbing 2002). Given that acoustic communication can be affected by habitat (Brown and Waser, Chap. 4), perhaps the dual peak is also related to broad niche occupation (i.e., high and low strata, densely vegetated and open areas) or shifting niche occupation from the ancestral platyrrhine monkey to the extant species. This may be related to a larger evolutionary explanation, whereby the upper peak represents the ancestral primate condition (still conserved in the strepsirrhines), and the lower peak is a derived condition related to changing behavior, anatomy, and habitat acoustics. Such a pattern might have evolved partially as an adaptation to tune out loud ambient acoustical noise (biological and nonbiological in origin) or take advantage of "sound windows" in forest habitats (see Brown and Waser, Chap. 4). In any case, a larger sample and additional research, including re-evaluating how to report and compare the frequency of best sensitivity, could lead to interesting insights.

Importantly, identification of the frequency of best sensitivity is highly dependent on the frequencies tested—many studies have tested in octave steps, whereas others have been more specific with half-octave steps, intervals of 10 kHz, or other frequencies of interest. Also, the frequency of best sensitivity is sometimes determined within a narrow margin, with the best frequency differing by only 1–2 dB from a secondary peak (Tables 3.1 and 3.2), and this small difference may be within the margin of calibration or testing error (Coleman 2009). Thus, the frequency or frequencies that a species is most sensitive to are important to consider, but the values taken out of context of the whole audiogram should be compared cautiously.

3.4.5 Low-Frequency Limit

Interspecific variation in the low-frequency limit ranges from 28 Hz in the Japanese macaque to 150 Hz in the fork-marked lemur (*Phaner furcifer*) (Tables 3.1 and 3.2). The haplorhines have, on average, a lower limit than the strepsirrhines. One of the first studies considering both audition and ecology among primates was that of Brown and Waser (1984), which found that in blue monkeys low-frequency vocalizations were associated with enhanced low-frequency auditory sensitivity.

Currently, it is difficult to draw any broad conclusions about the low-frequency limit in primates given that the existing data overlap, data are unavailable for most species, and data for some species are based on a sample size of one. It is not currently clear whether the observed variation is significantly beyond what is normal for interindividual variation, due to physical limitations of the primate ear, or what is a product of selection. Patterns of existing variation and lack of more data may also reflect methodological issues—it can be particularly difficult to calibrate low-frequency acoustic stimuli in variable testing conditions that can include relatively loud low-frequency background noise.

3.5 Summary and Implications for Future Research

Anthropological, biological, and biomedical studies often use nonhuman primates as models for humans. However, human and nonhuman primates differ in their auditory capabilities (Sect. 3.4). Although researchers have identified 21 hearing-linked genes that differ between chimpanzees and humans (Clark et al. 2003), the intricacies and auditory consequences of these genetic differences are not yet fully understood, in part due to a relatively small sample of nonhuman primate auditory data (Sect. 3.4). Identifying the subtleties that separate human and nonhuman primate audition, and the biological relevance of such differences, will require continued effort to fully explore, integrate, and expand the current dataset. A major aspect of this exploration will be further evaluating and refining methods of data collection and analysis (Sect. 3.3).

With respect to auditory processing and perception, nonhuman primates show evidence of specialization for the processing of vocalizations overall, of species-specific vocalizations in particular, and, in some cases, of the ability to recognize specific individuals (Sect. 3.2). Whether or not the nonhuman primate auditory system is specialized for processing vocalizations in general is still a matter of debate, but studies do indicate that specialized cortical structures for processing vocalizations that were thought to be unique to humans actually have homologous counterparts in nonhuman primates. Numerous studies indicate that, like humans, at least some nonhuman primates are able to distinguish conspecifics, and possibly individuals, based on call structure. Studies on the differential processing of normal versus synthesized and spectrally or temporally modified calls both support and dispute that nonhuman primates possess these abilities. Compared to the majority of mammals, nonhuman primates are excellent sound source localizers, closely approaching humans in their high acuity (Sect. 3.4.3.1). The relationship between both excellent sound localization acuity and visual acuity and the similarities in underlying neural structures provide evidence for the coevolution of these two senses in primates. However, these data are based on relatively few species that are common to laboratory settings. The general trend seems to be that as more and more data are accumulated, the auditory abilities of nonhuman primates are increasingly indicated as being very close to those of humans. Identifying the subtleties that separate human and nonhuman primate auditory abilities, and the biological relevance of those differences, requires consideration of all available data. Additional research into the auditory processing and perception of other primate taxa is needed to fully evaluate patterns and evolutionary relationships within the order.

With respect to overall auditory sensitivity, primates have traditionally been portrayed as unspecialized, which may be a consequence of the overall generalized nature of the auditory sense among mammals. There is variation among the species, though, with respect to both order-wide trends and species that display auditory specializations. A review of the literature shows that strepsirrhines are, on average, more adept at detecting high frequencies, and the haplorhines are, on average, more adept at detecting low frequencies (Sect. 3.4). A few trends relating overall auditory sensitivity and behavior or ecology have been identified in the literature. The well-supported allometric model explains that smaller headed species with smaller interaural (between-ear) distances (such as strepsirrhines) particularly need to utilize high frequencies for sound localization (Sect. 3.4.3.1) (R. S. Heffner 2004). Overall auditory sensitivity, particularly to high frequencies, has also been related to increased sociality in lemurs (Sect. 3.4.3.2) (Ramsier et al. 2012a).

The lack of identification of additional broad trends and relationships may be partially attributed to the limited current dataset, which is lacking representation from major taxonomic subgroups. Additionally, or perhaps alternatively, order-wide trends may be minimal given the many possible reasons why enhanced or reduced sensitivity to low, mid, or high frequencies may be beneficial for different species. For example, small, nocturnal or insectivorous species, such as tarsiers and some strepsirrhines, may benefit from detecting the high-frequency signals of insect prey or by communicating in a high-frequency band that is less audible to potential avian or felid predators (Ramsier et al. 2012a, b) (Zimmermann, Chap. 5). For species subject to intensive predation pressure, the reception of alarm calls may be particularly vital to survival (Arnold et al. 2008; Ramsier et al. 2012a). Alternately, it may be more or less advantageous to detect the calls of infants, which tend to be particularly high in frequency, depending on a species' body size, behavior, and ecology (Snowdon and Hauserger 1997; Pistorio et al. 2006; Ey et al. 2007). In some species, it may be particularly advantageous to be attuned to low frequencies, and this may be attributed to factors such as the detection of low-frequency acoustic cues or the occupation of forested environments (Brown and Waser, Chap. 4). Some species may also benefit from reduced sensitivity to certain frequencies, such as those produced by forest insects, to enhance the detection of other important sounds.

The above are just a few of the many facets of primate audition that need to be explored in more depth, not only through additional data on auditory sensitivity gathered via continually evaluated and refined methodologies but also by more data documenting habitat acoustics (especially in disturbed habitats), anthropogenic noise, and acoustic signals and cues present in the wild and in captive facilities. These data could be a critical component to the survival of the endangered primates for which little or no data on auditory sensitivity currently exist.

Acknowledgments The authors wish to thank Rolf Quam, Arthur Popper, and Richard Fay for providing helpful feedback on this manuscript. MAR would also like to thank Nathaniel Dominy for support. Partial support to JPR was provided by the Technische Universität München-Institute for Advanced Study, funded by the German Excellence Initiative and the European Union Seventh Framework Programme under grant agreement no 291763.

Compliance with Ethics Requirements
Marissa A. Ramsier declares that she has no conflicts of interest.
Josef P. Rauschecker declares that he has no conflicts of interest.

References

Aitkin, L. M., Merzenich, M. M., Irvine, D. R., Clarey, J. C., & Nelson, J. E. (1986). Frequency representation in auditory cortex of the common marmoset (*Callithrix jacchus jacchus*). *Journal of Comparative Neurology*, 252, 175–185.

Aitkin, L. M., & Park, V. (1993). Audition and the auditory pathway of a vocal New World primate, the common marmoset. *Progress in Neurobiology*, 41, 345–367.

Alder, T. B., & Rose, G. J. (1998). Long-term temporal integration in the anuran auditory system. *Nature Neuroscience*, 1(6), 519–523.

Amoser, S., & Ladich, F. (2005). Are hearing sensitivities of freshwater fish adapted to the ambient noise of their habitats? *Journal of Experimental Biology*, 208, 3533–3542.

Arnold, K., Pohlner, Y., & Zuberbühler, K. (2008). A forest monkey's alarm call series to predator models. *Behavioral Ecology and Sociobiology*, 62, 549–559.

Asquith, P. J. (2011). Of bonds and boundaries: What is the modern role of anthropomorphism in primatological studies? *American Journal of Primatology*, 73, 238–244.

Barber, J. R., Crooks, K. R., & Fristrup, K. M. (2010). The costs of chronic noise exposure for terrestrial organisms. *Trends in Ecology and Evolution*, 25, 180–189.

Barton, R. A., Byrne, R. W., & Whiten, A. (1996). Ecology, feeding competition, and social structure in baboons. *Behavioral Ecology and Sociobiology*, 38, 321–329.

Bearder, S. K., Nekaris, K. A. I., & Curtis, D. J. (2006). A re-evaluation of the role of vision in the activity and communication of nocturnal primates. *Folia Primatologica*, 77, 50–71.

Beecher, M. D. (1974a). Pure-tone thresholds of the squirrel monkey (*Saimiri sciureus*). *The Journal of the Acoustical Society of America*, 55, 196–198.

Beecher, M. D. (1974b). Hearing in the owl monkey (*Aotus trivirgatus*): I. Auditory sensitivity. *Journal of Comparative and Physiological Psychology*, 86, 898–901.
Behar, I., Cronholm, J. N., & Loeb, M. (1965). Auditory sensitivity in the rhesus monkey. *Journal of Comparative and Physiological Psychology*, ***59***, 426–428.
Bendor, D., & Wang, X. (2005). The neuronal representation of pitch in primate auditory cortex. *Nature*, 436, 1161–1165.
Bennett, C. L., Davis, R. T., & Miller, J. M. (1983). Demonstration of presbycusis across related measures in a nonhuman primate species. *Behavioral Neuroscience*, 97, 602–607.
Bieser, A. (1998). Processing of twitter-call fundamental frequencies in insula and auditory cortex of squirrel monkeys. *Experimental Brain Research*, 122, 139–148.
Blauert, J. (1997). *Spatial hearing: The psychophysics of human sound localization*. Cambridge, MA: MIT Press.
Blumstein, D. T. (2002). The evolution of functionally referential alarm communication: Multiple adaptations; multiple constraints. *Evolution of Communication*, 3, 135–147.
Bohn, K. M., Moss, C. F., & Wilkinson, G. S. (2001). Audio-vocal matching in greater spear-nosed bats. *Bat Research News*, 42, 144–145.
Brainard, M. S., & Doupe, A. J. (2002). What songbirds teach us about learning. *Nature*, 417, 351–358.
Brown, C. H. (1986). The perception of vocal signals by blue monkeys and grey-cheeked mangabeys. *Experimental Biology*, 45, 145–165.
Brown, C. H., & Waser, P. M. (1984). Hearing and communication in blue monkeys (*Cercopithecus mitis*). *Animal Behaviour*, 32, 66–75.
Brown, C., Gomez, R., & Waser, P. (1995). Old world monkey vocalizations: Adaptation to the local environment? *Animal Behaviour*, 50, 945–961.
Burkard, R. F., & Don, M. (2007). The auditory brainstem response. In R. F. Burkard, M. Don, & J. Eggermont (Eds.), *Auditory evoked potentials: Basic principles and clinical applications* (pp. 229–253). Baltimore, MD: Lippincott Williams & Wilkins.
Bushara, K. O., Weeks, R. A., Ishii, K., Catalan, M. J., et al. (1999). Modality-specific frontal and parietal areas for auditory and visual spatial localization in humans. *Nature Neuroscience*, 2(8), 759–766.
Casseday, J. H., Ehrlich, D., & Covey, E. (1994). Neural tuning for sound duration: Role of inhibitory mechanisms in the inferior colliculus. *Science*, 264, 847–850.
Chapman, C. A., & Weary, D. M. (1990). Variability in spider monkeys' vocalizations may provide basis for individual recognition. *American Journal of Primatology*, 4, 279–284.
Cheney, D. L., & Seyfarth, R. M. (1990). *How monkeys see the world*. Chicago, IL: University of Chicago Press.
Chomsky, N. (1975). *The logical structure of linguistic theory*. New York: Plenum Press.
Clark, A. G., Glanowski, S., Nielsen, R., Thomas, P. D., et al. (2003). Inferring nonneutral evolution from human-chimp-mouse orthologous gene trios. *Science*, 302, 1960–1963.
Coleman, M. N. (2009). What do primates hear? A meta-analysis of all known nonhuman primate behavioral audiograms. *International Journal of Primatology*, 30, 55–91.
Coleman, M. N., & Colbert, M. W. (2010). Correlations between auditory structures and hearing sensitivity in non-human primates. *Journal of Morphology*, 271, 511–532.
Crompton, R. H. (1995). "Visual predation," habitat structure, and the ancestral primate niche. In L. Alterman & A. Doyle (Eds.), *Creatures of the dark: The nocturnal prosimians* (pp. 11–30). Boston, MA: Springer-Verlag.
Delgutte, B. (1997). Auditory processing of speech. In W. J. Hardcastle & J. Laver (Eds.), *The handbook of phonetic sciences* (pp. 507–538). Oxford: Blackwell.
DeWitt, I., & Rauschecker, J. P. (2012). Phoneme and word recognition in the auditory ventral stream. *Proceedings of the National Academy of Sciences of the United States of America*, 109(8), E505–E514.
Dittus, W. P. J. (1984). Toque macaque food calls: Semantic communication concerning food distribution in the environment. *Animal Behaviour*, 32, 470–477.

Dominy, N. J., Lucas, P. W., Osorio, D., & Yamashita, N. (2001). The sensory ecology of primate food perception. *Evolutionary Anthropology*, 10, 171–186.
Don, M., & Starr, A. (1972). Lateralization performance of squirrel monkey (*Samiri sciureus*) to binaural click signals. *Journal of Neurophysiology*, 35, 493–500.
Elder, J. H. (1934). Auditory acuity of the chimpanzee. *Journal of Comparative and Physiological Psychology*, 17, 157–183.
Elder, J. H. (1935). The upper limit of hearing in chimpanzee. *American Journal of Physiology—Legacy Content*, 112(1), 109–115.
Ey, E., Pfefferle, D., & Fisher, J. (2007). Do age- and sex-related variations reliably reflect body size in non-human primate vocalizations? A review. *Primates*, 48, 253–267.
Fay, R. R. (1988). *Hearing in vertebrates: A psychophysics databook*. Winnetka, IL: Hill-Fay Associates.
Fishman, Y. I., Arezzo, J. C., & Steinschneider, M. (2004). Auditory stream segregation in monkey auditory cortex: Effects of frequency separation, presentation rate, and tone duration. *The Journal of the Acoustical Society of America*, 116, 1656–1670.
Fisher, S. E., & Marcus, G. F. (2006). The eloquent ape: Genes, brains, and the evolution of language. *Nature Reviews Genetics*, 7, 9–20.
Fitch, W. T. (1997). Vocal tract length and formant frequency dispersion correlate with body size in rhesus macaques. *The Journal of the Acoustical Society of America*, 102, 1213–1222.
Fujita, S., & Elliott, D. (1965). Thresholds of audition in three species of monkey. *The Journal of the Acoustical Society of America*, 37, 139–144.
Gautier, J. P. (1988). Interspecific affinities among guenons as deduced from vocalizations. In A. Gautier-Hion, F. Bourliere, J. P. Gautier, & J. Kingdon (Eds.), *A primate radiation: Evolutionary biology of the African guenons* (pp. 194–226). Cambridge: Cambridge University Press.
Geisler, C. D. (1998). *From sound to synapse: Physiology of the mammalian ear*. Oxford: Oxford University Press.
Ghazanfar, A. A., & Hauser, M. D. (1999). The neuroethology of primate vocal communication: Substrates for the evolution of speech. *Trends in Cognitive Science*, 3, 377–384.
Ghazanfar, A. A., & Santos, L. R. (2002). Primates as auditory specialists. In A. A. Ghazanfar (Ed.), *Primate audition: Ethology and neurobiology* (pp. 1–12). Boca Raton, FL: CRC Press LLC.
Ghazanfar, A. A., & Santos, L. R. (2004). Primate brains in the wild: The sensory bases for social interactions. *Nature Reviews Neurocience*, 5, 603–616.
Ghazanfar, A. A., Flombaum, J. I., Miller, C. T., & Hauser, M. D. (2001). The units of perception in the antiphonal calling behavior of cotton-top tamarins (*Saguinus oedipus*): Playback experiments with long calls. *Journal of Comparative Physiology A: Sensory, Neural, and Behavioral Physiology*, 187, 27–35.
Gillette, R. G., Brown, R., Herman, P., Vernan, S., & Vernon, J. (1973). The auditory sensitivity of the lemur. *American Journal of Physical Anthropology*, 38, 365–370.
Glass, I., & Wollberg, Z. (1983). Responses of cells in the auditory cortex of awake squirrel monkeys to normal and reversed species-specific vocalizations. *Hearing Research*, 9, 27–33.
Goerlitz, H. R., & Siemers, B. M. (2007). Sensory ecology of prey rustling sounds: Acoustical features and their classification by wild grey mouse lemurs. *Functional Ecology*, 21, 143–153.
Gourevich, G. (1970). Detectability of tones in quiet and in noise by rats and monkeys. In W. C. Stebbins (Ed.), *Animal psychophysics: The design and conduct of sensory experiments* (pp. 67–97). New York: Appleton-Century-Crofts.
Green, S. (1971). *Two behavioral methods for measuring auditory sensitivity and loudness in the squirrel monkey (Saimiri sciureus)*. Ph.D. dissertation, University of Michigan, Ann Arbor, MI.
Green, S. (1975). Auditory sensitivity and equal loudness in the squirrel monkey. *Journal of the Experimental Analysis of Behavior*, 23, 255–264.
Gursky, S. (2015). Ultrasonic vocalizations by the spectral tarsier. *Tarsius spectrum Folia Primatologica*, 86, 153–163.

Hackett, T. A. (2011). Information flow in the auditory cortical network. *Hearing Research*, 271, 133–146.
Hackett, T. A., Stepniewska, I., & Kaas, J. H. (1998). Subdivisions of auditory cortex and ipsilateral cortical connections of the parabelt auditory cortex in macaque monkeys. *The Journal of Comparative Anatomy*, 394, 475–495.
Hauser, M. D., Newport, E. L., & Aslin, R. N. (2001). Segmentation of the speech stream in a nonhuman primate: Statistical learning in cotton-top tamarins. *Cognition*, 78, B53–B64.
Heffner, H. E., & Heffner, R. S. (2014). The behavioral study of mammalian hearing. In A. N. Popper & R. R. Fay (Eds.), *Perspectives on auditory research* (pp. 269–285). New York: Springer Science+Business Media.
Heffner, H. E., & Masterton, B. (1970). Hearing in primitive mammals: Slow loris (*Nycticebus coucang*) and potto (*Perodicticus potto*). *Journal of Comparative and Physiological Psychology*, 71, 175–182.
Heffner, H. E., Ravizza, R. J., & Masterton, B. (1969). Hearing in primitive mammals. IV. Bushbaby (*Galago senegalensis*). *The Journal of Auditory Research*, 9, 19–23.
Heffner, R. S. (2004). Primate hearing from a mammalian perspective. *The Anatomical Record Part A: Discoveries in Molecular, Cellular, and Evolutionary Biology*, 281A, 1111–1122.
Heffner, R. S., & Heffner, H. E. (1982). Hearing in the elephant (*Elephas maximus*): Absolute sensitivity, frequency discrimination, and sound localization. *Journal of Comparative and Physiological Psychology*, 96, 926–944.
Heffner, R. S., & Heffner, H. E. (1988). Sound localization acuity in the cat: Effect of azimuth, signal duration, and test procedure. *Hearing Research*, 36, 221–232.
Heffner, R. S., & Heffner, H. E. (1992). Evolution of sound localization in mammals. In D. B. Webster, R. R. Fay, & A. N. Popper (Eds.), *Evolutionary biology of hearing* (pp. 691–715). New York: Springer-Verlag.
Henkel, C. K. (2006). The auditory system. In D. E. Haines (Ed.), *Fundamental neuroscience for basic and clinical applications* (3rd ed.). Philadelphia, PA: Churchill Livingstone/Elsevier.
Hienz, R. D., Turkkan, J. S., & Harris, A. (1982). Pure tone thresholds in the yellow baboon (*Papio cynocephalus*). *Hearing Research*, 8, 71–75.
Hill, R. A., & Dunbar, R. I. M. (1998). An evaluation of the roles of predation rate and predation risk as selective pressures on primate grouping behaviour. *Behaviour*, 135(4), 411–430.
Houben, D., & Gourevitch, G. (1979). Auditory lateralization in monkeys: An examination of two cues serving directional hearing. *The Journal of the Acoustical Society of America*, 66, 1057–1063.
Imig, T. J., & Adrian, H. O. (1977). Binaural columns in the primary field (A1) of cat auditory cortex. *Brain Research*, 138, 241–257.
Irwin, M. T., Raharison, F. J. L., Rakotoarimanana, H., Razanadrakoto, E., et al. (2007). Diademed sifakas (*Propithecus diadema*) use olfaction to forage for the inflorescences of subterranean parasitic plants (Balanophoraceae: *Langsdorffia* sp. and Cytinaceae: *Cytinus* sp.) *American Journal of Primatology*, 69, 471–476.
Jackson, L. L., Heffner, R. S., & Heffner, H. E. (1999). Free-field audiogram of the Japanese macaque (*Macaca fuscata*). *The Journal of the Acoustical Society of America*, 106, 3017–3023.
Jacobson, J. T. (1985). An overview of the auditory brainstem response. In J. T. Jacobson (Ed.), *The auditory brainstem response* (pp. 3–12). San Diego, CA: College-Hill Press.
Joly, O., Ramus, F., Pressnitzer, D., Vanduffel, W., & Orban, G. A. (2012). Interhemispheric differences in auditory processing revealed by fMRI in awake rhesus monkeys. *Cerebral Cortex*, 22, 838–853. doi:10.1093/cercor/bhr150.
Kaas, J. H., & Hackett, T. A. (1999). 'What' and 'where' processing in auditory cortex. *Nature Neuroscience*, 2, 1045–1047.
Kaas, J. H., & Hackett, T. A. (2000). Subdivisions of auditory cortex and processing streams in primates. *Proceedings of the National Academy of Sciences of the United States of America*, 97, 11793–11799.
Kight, C. R., & Swaddle, J. P. (2011). How and why environmental noise impacts animals: An integrative, mechanistic review. *Ecology Letters*, 14, 1052–1061.

Klump, R. G., & Eady, H. R. (1956). Some measurements of interaural time difference thresholds. *The Journal of the Acoustical Society of America*, ***28***, 859–860.

Kojima, S. (1990). Comparison of auditory functions in the chimpanzee and human. *Folia Primatologica*, 55, 62–72.

Laska, M., Freist, P., & Krause, S. (2007). Which senses play a role in nonhuman primate food selection? A comparison between squirrel monkeys and spider monkeys. *American Journal of Primatology*, 69, 282–294.

Lasky, R. E., Soto, A. A., Luck, M. L., & Laughlin, N. K. (1999). Otoacoustic emission, evoked potential, and behavioral auditory thresholds in the rhesus monkey (*Macaca mulatta*). *Hearing Research*, 136, 35–43.

Lieberman, P. (2006). *Toward an evolutionary biology of language*. Cambridge, MA: Harvard University Press.

Lonsbury-Martin, B. L., & Martin, G. K. (1981). Effects of moderately intense sound on auditory sensitivity in rhesus monkeys: Behavioral and neural observations. *Journal of Neurophysiology*, 46, 563–587.

Malone, B. J., Scott, B. H., & Semple, M. N. (2002). Context-dependent adaptive coding of interaural phase disparity in the auditory cortex of awake macaques. *The Journal of Neuroscience*, 22, 4625–4638.

Martin, F. N., & Clark, J. G. (2006). *Introduction to audiology* (9th ed.). Boston, MA: Pearson, Allyn and Bacon.

Martin, R. D., & Ross, C. F. (2005). The evolutionary and ecological context of primate vision. In J. Kremers (Ed.), *The primate visual system: A comparative approach* (pp. 1–36). West Sussex, UK: John Wiley & Sons.

Masters, J. C. (1991). Loud calls of *Galago crassicaudatus* and *G. garnetti* and their relation to habitat structure. *Primates*, 32(2), 153–167.

Masters, J., Gamba, M., & Génin, F. (Eds.). (2013). *Leaping ahead: Advances in prosimian biology*. New York: Springer Science+Business Media.

Masterton, B., Heffner, H., & Ravizza, R. (1969). The evolution of human hearing. *The Journal of the Acoustical Society of America*, 45, 966–985.

Micheyl, C., Tian, B., Carlyon, R. P., & Rauschecker, J. P. (2005). Perceptual organization of tone sequences in the auditory cortex of awake macaques. *Neuron*, 48, 139–148.

Middlebrooks, J. C., & Green, D. M. (1991). Sound localization by human listeners. *Annual Review Psychology*, 42, 135–159.

Nagarajan, S. S., Cheung, S. W., Bedenbaugh, P., Beitel, R. E., et al. (2002). Representation of spectral and temporal envelope of twitter vocalizations in common marmoset primary auditory cortex. *Journal of Neurophysiology*, 87, 1723–1737.

Nayagam, B. A., Muniak, M. A., & Ryugo, D. K. (2011). The spiral ganglion: Connecting the peripheral and central auditory systems. *Hearing Research*, 278, 2–20.

Niaussat, M.-M., & Molin, D. (1978). Hearing and vocalization in a Malagasy lemur (*Phaner furcifer*). In D. J. Chivers & J. Herbert (Eds.), *Recent advances in primatology, Behavior* (Vol. 1, pp. 821–825). London: Academic Press.

Nishimura, T. (2008). Understanding the dynamics of primate vocalization and its implications for the evolution of human speech. In N. Masataka (Ed.), *The origins of language: Unraveling evolutionary forces* (pp. 111–131). Tokyo: Springer Science+Business Media.

Ortiz-Rios, M., Kusmierek, P., DeWitt, I., Archakov, D. A., et al. (2015). Functional MRI of the vocalization-processing network in the macaque brain. *Frontiers in Neuroscience*, 9, 113. doi:10.3389/fnins.2015.00113.

Osmanski, M. S., & Wang, X. (2011). Measurement of absolute auditory thresholds in the common marmoset (*Callithrix jacchus*). *Hearing Research*, 227, 127–133.

Owren, M. J. (2003). Vocal production and perception in nonhuman primates provide clues about early hominids and speech evolution. *ATR Symposium HIS Series*, 1, 1–19.

Owren, M. J., & Rendall, D. (2001). Sound on the rebound: Bringing form and function back to the forefront in understanding nonhuman primate vocal signaling. *Evolutionary Anthropology*, 10, 58–71.

Owren, M. J., Hopp, S. L., Sinnott, J. M., & Peterson, M. R. (1988). Absolute auditory thresholds in three Old World monkey species (*Cercopithecus aethiops, C. neglectus, Macaca fuscata*) and humans (*Homo sapiens*). *Journal of Comparative Psychology*, 102, 99–107.
Pandya, D. N., & Sanides, F. (1973). Architectonic parcellation of the temporal operculum in rhesus monkey and its projection pattern. *Zeitschrift Fur Anatomie Und Entwicklungsgeschichte*, 139, 127–161.
Pfingst, B. E., Hienz, R., & Miller, J. (1975). Reaction-time procedure for measurement of hearing. II. Threshold functions. *The Journal of the Acoustical Society of America*, **57**, 431–436.
Pfingst, B. E., Laycock, J., Flammino, F., Lonsbury-Martin, B., & Martin, G. (1978). Pure tone thresholds for the Rhesus monkey. *Hearing Research*, 1, 43–47.
Pistorio, A. L., Vintch, B., & Wang, X. (2006). Acoustic analysis of vocal development in a New World primate, the common marmoset (*Callithrix jacchus*). *The Journal of the Acoustical Society of America*, 120, 1655–1670.
Popper, A. N., & Fay, R. R. (2005). *Sound source localization*. New York: Springer Science+Business Media.
Poremba, A., Saunders, R. C., Crane, A. M., Cook, M., et al. (2003). Functional mapping of the primate auditory system. *Science*, 299, 568–572. doi:10.1126/science.1078900.
Poremba, A., Malloy, M., Saunders, R. C., Carson, R. E., et al. (2004). Species-specific calls evoke asymmetric activity in the monkey's temporal poles. *Nature*, 427, 448–451.
Quam, R., Martinez, I., Rosa, M., Bonmati, A., et al. (2015). Early hominin auditory capacities. *Science Advances*, 2(8), e1500355.
Ramsier, M. A., & Dominy, N. J. (2010). A comparison of auditory brainstem responses and behavioral estimates of hearing sensitivity in *Lemur catta* and *Nycticebus coucang*. *American Journal of Primatology*, 72, 217–233.
Ramsier, M. A., Cunningham, A. J., Finneran, J. J., & Dominy, N. J. (2012a). Social drive and the evolution of primate hearing. *Philosophical Transactions of the Royal Society of London B: Biological Sciences*, 367, 1860–1868.
Ramsier, M. A., Cunningham, A. J., Moritz, G. L., Finneran, J. J., et al. (2012b). Primate communication in the pure ultrasound. *Biology Letters*, 8, 508–511.
Rauschecker, J. P. (2011). An expanded role for the dorsal auditory pathway in sensorimotor integration and control. *Hearing Research*, 271, 16–25.
Rauschecker, J. P. (2015). Auditory and visual cortex of primates: A comparison of two sensory systems. *European Journal of Neuroscience Special Issue: Auditory Cortex*, 41, 579–585.
Rauschecker, J. P., & Romanski, L. (2011). Auditory cortical organization: Evidence for functional streams. In J. Winer & C. Schreiner (Eds.), *The auditory cortex* (pp. 99–116). New York: Springer Science+Business Media.
Rauschecker, J. P., & Scott, S. K. (2009). Maps and streams in the auditory cortex: Nonhuman primates illustrate human speech processing. *Nature Neuroscience*, 12, 718–724.
Rauschecker, J. P., & Tian, B. (2000). Mechanisms and streams for processing of "what" and "where" in auditory cortex. *Proceedings of the National Academy of Sciences of the United States of America*, **97**, 11800–11806.
Rauschecker, J. P., Tian, B., & Hauser, M. (1995). Processing of complex sounds in the macaque nonprimary auditory cortex. *Science*, 268, 111–114.
Rauschecker, J. P., Tian, B., Pons, T., & Mishkin, M. (1997). Serial and parallel processing in rhesus monkey auditory cortex. *Journal of Comparative Neurology*, 382, 89–103.
Recanzone, G. H., Guard, D. C., & Phan, M. L. (2000). Frequency and intensity response properties of single neurons in the auditory cortex of the behaving macaque monkey. *Journal of Neurophysiology*, 83(4), 2315–2331.
Renaud, D. L., & Popper, A. N. (1975). Sound localization by the bottlenose porpoise *Tursiops truncatus*. *Journal of Experimental Biology*, 63, 569–585.
Rice, J. J., May, B. J., Spirou, G. A., & Young, E. D. (1992). Pinna-based spectral cues for sound localization in the cat. *Hearing Research*, 58, 132–152.
Romanski, L. M., Tian, B., Fritz, J., Mishkin, M., et al. (1999). Dual streams of auditory afferents target multiple domains in the primate prefrontal cortex. *Nature Neuroscience*, 2(12), 1131–1136.

Rosowski, J. J. (1991). The effects of external and middle ear filtering on auditory threshold and noise induced hearing loss. *The Journal of the Acoustical Society of America*, 90, 124–135.
Schneider, P., Sluming, V., Roberts, N., Scherg, M., et al. (2005). Structural and functional asymmetry of lateral Heschel's gyrus reflects pitch perception preference. *Nature Neuroscience*, 8, 1241–1247.
Schopf, C., Zimmermann, E., Tünsmeyer, J., Kästner, S. B. R., Hubka, P., & Kral, A. (2014). Hearing and age-related changes in the gray mouse lemur. *Journal of the Association for Research in Otolaryngology*, 15(6), 993–1005.
Seiden, H. R. (1957). *Auditory acuity of the marmoset monkey (Hapale jacchus)*. Ph.D. dissertation, Princeton University. Available from the Princeton University library.
Sekulic, R., & Chivers, D. J. (1986). The significance of call duration in howler monkeys. *International Journal of Primatology*, 7, 183–190.
Semple, S. (2001). Individuality and male discrimination of female copulation calls in the yellow baboon. *Animal Behaviour*, 61, 1023–1028.
Semple, S., & McComb, K. (2000). Perception of female reproductive state from vocal cues in a mammal species. *Philosophical Transactions of the Royal Society of London B: Biological Sciences*, 267, 707–712.
Sinyor, A., & Laszlo, C. A. (1973). Acoustic behavior of the outer ear of the guinea pig and the influence of the middle ear. *The Journal of the Acoustical Society of America*, 54, 916–921.
Snowdon, C. T. (1993). A vocal taxonomy of callitrichids. In A. B. Rylands (Ed.), *Marmosets and tamarins: Systematics, behaviour, ecology* (pp. 78–94). Oxford: Oxford University Press.
Snowdon, C. T., & Hausberger, M. (1997). *Social influences on vocal development*. New York: Cambridge University Press.
Stebbins, W. C., Green, S., & Miller, F. L. (1966). Auditory sensitivity of the monkey. *Science*, 153, 1646–1647.
Sterbing, S. J. (2002). Postnatal development of vocalizations and hearing in the phyllostomid bat, *Carollia perspicillata*. *Journal of Mammalogy*, 83, 516–526.
Thorpe, W. H. (1968). Perceptual basis for group organization in social vertebrates, especially birds. *Nature*, 220, 124–128.
Tomlinson, R. W., & Schwarz, D. W. (1988). Perception of the missing fundamental in nonhuman primates. *The Journal of the Acoustical Society of America*, 84, 560–565.
Tian, B., Kusmierek, P., & Rauschecker, J. P. (2013). Analogues of simple and complex cells in rhesus monkey auditory cortex. *Proceedings of the National Academy of Sciences of the United States of America*, 110, 7892–7897.
Tian, B., Reser, D., Durham, A., Kustov, A., & Rauschecker, J. P. (2001). Functional specialization in rhesus monkey auditory cortex. *Science*, 292, 290–293.
Vélez, A., Gall, M. D., Fu, J., & Lucas, J. R. (2015). Song structure, not high-frequency song content, determines high-frequency auditory sensitivity in nine species of New World sparrows (Passeriformes: Emberizidae). *Functional Ecology*, 4, 487–497.
Viemeister, N. F., & Wakefield, G. H. (1991). Temporal integration and multiple looks. *The Journal of the Acoustical Society of America*, 90, 858–865.
von Békésy, G. (1960). *Experiments in hearing*. New York: McGraw-Hill.
Wang, X., & Kadia, S. C. (2001). Differential representation of species-specific primate vocalizations in the auditory cortices of marmoset and cat. *Journal of Neurophysiology*, 86, 2616–2620.
Wang, X., Lu, T., Snider, R., & Liang, L. (2005). Sustained firing in auditory cortex evoked by preferred stimuli. *Nature*, 435(7040), 341–346.
Wang, X., Merzenich, M. M., Beitel, R., & Schreiner, C. E. (1995). Representation of a species-specific vocalization in the primary auditory cortex of the common marmoset: Temporal and spectral characteristics. *Journal of Neurophysiology*, 74, 2685–2706.
Warren, J. D., & Griffiths, T. D. (2003). Distinct mechanisms for processing spatial sequences and pitch sequences in the human auditory brain. *The Journal of Neuroscience*, 23, 5799–5804.
Waser, P. M. (1977). Individual recognition, intragroup cohesion and intergroup spacing: Evidence from sound playback to forest monkeys. *Behaviour*, 60, 28–74.

Weber, I. (1973). Tactile communication among free-ranging langurs. *American Journal of Physical Anthropology*, 38, 481–486.

Webster, D. B., Popper, A. N., & Fay, R. R. (1992). *The mammalian auditory pathway: Neuroanatomy*. New York: Springer-Verlag.

Winkowski, D. E., & Knudsen, E. I. (2006). Top-down gain control of the auditory space map by gaze control circuitry in the barn owl. *Nature*, 439, 336–339.

Wollberg, Z., & Newman, J. D. (1972). Auditory cortex of squirrel monkey: Response patterns of single cells to species-specific vocalizations. *Science*, 175, 212–214.

Zimmermann, E. (1990). Differentiation of vocalizations in bushbabies (Galaginae, Prosimiae, Primates) and the significance for assessing phylogenetic relationships. *Journal of Zoological Systematics and Evolutionary Research*, 28, 217–239.

Zuberbühler, K. (2007). Monkey alarm calls. In S. McGraw, K. Zuberbühler, & R. Noë (Eds.), *Monkeys of the Taï Forest: An African primate community* (pp. 194–220). Cambridge, UK: Cambridge University Press.

Chapter 4
Primate Habitat Acoustics

Charles H. Brown and Peter M. Waser

Abstract Natural habitats are not recording studios. Calls emitted in nature encounter an irregular assortment of hard surfaces that reflect and scatter the wave front, producing complicated patterns of constructive and destructive interference. The propagated wave front is subsequently disturbed by wind, thermal gradients, and atmospheric absorption. Collectively, these phenomena result in an unpredictable and untidy acoustic environment. Furthermore, thunder, rain, crashing waves, or the relentless chatter of biotic sources can result in high ambient-noise levels that may mask the signal, overwhelm the recipient, and obliterate significant nuances and embellishments. Thus, vocal communication is hampered by attenuation, reverberation, distortion, and acoustic disturbances. Accordingly, the twin components of vocal communication, sound production and acoustic perception, may have undergone persistent selection to counter the most prominent impediments to both hearing and being heard. Primates have radiated from rain forest to grassland and other habitats, and each habitat differs acoustically. Hence, there is reason to believe that the duration, amplitude, pitch, and composition of primate vocal repertoires, the timing of emissions, and the placement and orientation of vocalizers is not haphazard, but each has become tuned to the acoustic parameters of the natal habitat to heighten the clarity of vocal exchanges. This chapter begins with an overview of the acoustic properties of rain forest, riverine forest, and savanna habitats occupied by East African primates, which is followed by reviews of how primate calls become distorted when propagated in natural habitats and how distortion scores have been used to explore the acoustic adaptation hypothesis. Finally, significant opportunities for additional research are highlighted.

Keywords Acoustic adaptation hypothesis • Ambient noise • Amplitude fluctuation • Animal vocalizations • Comparative bioacoustics • Distortion • Excess attenuation • Reverberation

C.H. Brown (✉)
Department of Psychology, University of South Alabama, Mobile, AL 36688, USA
e-mail: cbrown@southalabama.edu

P.M. Waser
Department of Biological Sciences, Purdue University, West Lafayette, IN 47907, USA
e-mail: waserpm@purdue.edu

R.M. Quam et al. (eds.), *Primate Hearing and Communication*, Springer Handbook of Auditory Research 63, DOI 10.1007/978-3-319-59478-1_4

4.1 Introduction

An evolutionary premium has been placed on the ability of organisms to regulate and coordinate social exchanges. For primates, the emission of vocal displays is the most conspicuous and frequently the principal modality employed to sound an alarm, attract a mate, define a territory, threaten a predator, appease a dominant conspecific, chastise a subordinate, or soothe a dependent. Nonhuman primates have dispersed from the rain forest arboreal canopy to savanna, alpine grassland, and numerous different forest habitats. Differences in the acoustic properties of these habitats, such as background noise, humidity levels, wind, and reverberation, may constrain the structure of primate utterances. The principal scope of this chapter is to review the acoustic properties of the distinct habitat types primates occupy, to evaluate the data in light of new technological advancements in comparative bioacoustics, and to highlight possible research opportunities that future investigators may wish to pursue.

The acoustic adaptation hypothesis is the proposition that ecological constraints for aural communication have acted as a source of selection on both the production of acoustic signals and the perceptual systems dedicated to their identification and classification (Gish and Morton 1981; Brown et al. 1995). The strongest evidence that acoustic communication in humans is sensitive to environmental impediments is given by the Lombard effect, an involuntary increase in vocal amplitude observed when humans strive to communicate within a noisy soundscape (Lombard 1911). The Lombard effect is not unique to humans. Evidence of its existence has been observed in birds (Ey and Fischer 2009), amphibians (Shen and Xu 2016), and a variety of mammals, including nonhuman primates and whales (Sinnott et al. 1975; Scheifele et al. 2005; Snowdon, Chap. 6). There are many avenues by which communication systems can be influenced by ecological factors beyond the parameters reflected by the Lombard effect. Researchers have proposed that selection may have led to changes in the pitch, duration, and temporal-spectral composition of utterances to heighten their audibility in the natural habitat. Even aspects of human language have been discussed in this context (Fought et al. 2004). Furthermore, the orientation, location and elevation of broadcast sites, the spacing of vocalizers, and the timing of broadcasts all impact audibility, and selection may have acted to optimize these parameters as well.

4.2 Geographic Range and Distribution of Nonhuman Primates

Genetic data suggest that the order Primates arose from arboreal mammals during the middle of the Cretaceous period around 85 million years ago (Tavaré et al. 2002), and the oldest primate fossils date from the late Paleocene in Africa (Williams et al. 2010). Genetic data from living primates has led to a considerable revision of the taxonomy, including increasing the number of species recognized. In 2001, about 350 species of primates were recognized (Groves 2001). By 2014 the number was raised to 488 (Rylands and Mittermeier 2014), and it is likely that additional genetic analyses will continue to increase these numbers.

Most nonhuman primates inhabit tropical or subtropical regions of Africa, Asia, or the Americas. Only humans are found on every continent and in all climatic regions. Primates exhibit a four orders of magnitude variation in size from the 30 g mouse lemur (*Microcebus murinus*) to the 200 kg lowland gorilla (*Gorilla gorilla*). Most primates, however, are relatively small bodied and live in or remain in proximity to trees. Most are arboreal; only humans are fully terrestrial. Semiterrestrial species include those living in open country, such as baboons (genus *Papio*), geladas (*Theropithecus gelada*), and patas monkeys (*Erythrocebus patas*), but also some species like l'Hoest's monkeys (*Cercopithecus l'hoesti*) and mandrills (*Mandrillus sphinx*) that are denizens of dense forests. Primates inhabit a range of altitudes. Mountain gorillas (*Gorilla beringei*) have been found at elevations of 4,200 m (Schaller 1963), and geladas have been found at elevations approaching 5,000 m (Stammbach 1987).

The ancestral habitats of most modern primates are best represented by three habitat types: rain forest, riverine forest, and savanna. A large majority of extant primates still use these habitats. Over the past 5,000–10,000 years, human population growth, deforestation, and climate change have progressively altered the natural world to the extent that the ancestral habitats in which primates evolved, flourished, and dispersed have shrunk dramatically, existing today principally as isolated patches (Haddad et al. 2015).

4.3 Forest and Savanna Acoustics

Very few studies have sought to quantify the acoustic properties of primate habitats despite the significance for primate communication. Three studies (Waser and Waser 1977; Waser and Brown 1986; Brown et al. 1995) have provided an in-depth investigation of primate habitat acoustics in East Africa (Fig. 4.1), and one has

Fig. 4.1 Savanna (*left panel*), riverine forest (*middle panel*), and rain forest (*right panel*) differ prominently in terms of vegetation, humidity, wind, and their impacts on sound propagation, attenuation, and reverberation

investigated primate habitat acoustics in South America (de la Torre and Snowdon 2002). Other studies have investigated habitat acoustics in environments where primates are present but have focused on birds, anurans, or other taxa (reviewed in Ey and Fischer 2009; Wiley 2015). This chapter focuses on the Brown and Waser studies, as they provide the most comprehensive overview of relevant concepts, and because East Africa is the location of the earliest primates and, therefore, an important environment for the evolution of both the earliest monkeys and humans.

4.3.1 Habitat Characteristics

The early Waser and Brown studies were carried out in two African rain forest sites in Uganda and Kenya. Both study sites were in mature forests with an understory of semiwoody plants shaded by a continuous canopy at 30 m with an occasional emergent reaching 50 m in elevation. The rainy and dry seasons occur twice annually in both forests, but seasonality is not pronounced, and there is considerable year-to-year variation.

Riverine forests are vegetationally simple relative to rain forests. In contrast to rain forest sites, the riverine forest canopy is incomplete, with only the occasional tree reaching 30 m in elevation. With more light penetrating the canopy, the understory can become very dense. Measurements were conducted in eastern Kenya where the forest comprises a strip of vegetation a few hundred meters wide that is subject to regular flooding and rainfall is highly seasonal.

Savanna measurements were conducted in central Kenya. The study sites consist of grassland with scattered thickets and small trees consistent with the acacia-commiphora thornbush characteristic of most of eastern and northern Kenya (Lind and Morrison 1974).

4.3.2 Studying Habitat Acoustics: Ambient Noise and Sound Transmission

Studies of habitat acoustics have varied considerably in methodology and, as discussed in Sect. 4.5, methodological variation often limits our ability to generalize results across habitats, seasons, and species. Michelsen (1978) pointed out some of the pitfalls plaguing sound propagation experiments, and many of the considerations discussed by Fischer et al. (2013) with regard to primate sound playback experiments are also relevant to studies of sound attenuation and degradation.

The Brown and Waser experiments attempted to deal with some common issues. For example, ambient-noise samples were designed to systematically sample the background noise generated by biotic and abiotic sources of sound as a function of time of day, season, and location within a habitat type. Sound transmission tests were broadcast through calibrated speakers and rerecorded at multiple distances, with tests replicated along orthogonal tracks both upwind and downwind. The initial inventory of broadcast sounds comprised pure tones, tone pulses, and tone pulse trains transmitted at octave intervals up to 4 kHz, spanning the frequencies and pulse rates observed in primate calls. Pure tone broadcasts were designed to measure attenuation and amplitude fluctuations, while the tone pulse and pulse train broadcasts measured reverberation time and modulation depth as functions of carrier frequency. Subsequently, broadcast stimuli consisted of representative rain forest and savanna monkey calls. Where possible, tests used a balanced design incorporating homologous calls from species living in the different habitats (Waser 1982; Gautier 1988) with all calls broadcast through both appropriate and inappropriate habitats. Call types were included only if it was possible to obtain exemplars recorded close to the vocalizer with very good signal-to-noise ratios and no apparent distortion, thereby ensuring the fidelity of the signal. In each site, ambient-noise measurements and sound transmission broadcasts were conducted at elevations appropriate to the primate inhabitants.

4.3.3 Primate Habitats: How They Differ Acoustically

4.3.3.1 Ambient Noise

Exposure to elevated levels of noise can produce four outcomes: permanent hearing loss, temporary hearing loss, masking of biologically important signals, and physiological and psychological stress (Miller 1974). Ambient noise generated from biotic sources can approach levels sufficient to produce permanent or temporary hearing loss. The noise of some chorusing insects (Young 1990) may exceed 110 dB (A scale), a level sufficient over time to produce hearing loss in most, if not all, species of birds and mammals (Miller 1974; Dooling et al. 2009). Primates presumably distance themselves from sources of dangerously high levels of ambient noise, and under most conditions the chief impediment posed by ambient noise is the masking of biologically important signals. Habitat noise will compromise the capacity of listeners to detect the presence of predators and prey, to hear alarm calls, to warn conspecifics, and to acoustically identify individuals and recognize species. These tasks differ in their complexity and require different signal-to-noise ratios to be executed accurately. Therefore, masking is a potential problem for all species of primates in all habitat types.

Ambient-noise surveys show that noise levels differ as a function of habitat type. Overall, ambient-noise levels are lowest in the savanna and highest in the riverine forest. Sound pressure levels are greatest for low-frequency sounds, decrease to a minimum in the vicinity of 1 kHz, and then increase for higher frequency bands in all three habitats. However, the details of the frequency spectrum for ambient noise differ significantly among habitats (Fig. 4.2).

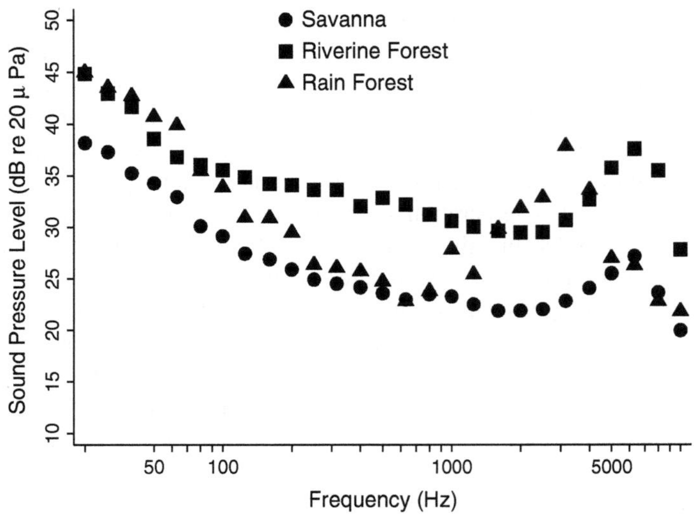

Fig. 4.2 Ambient noise averaged across time of day in savanna (*circles*), riverine (*squares*), and rain forest (*triangles*) habitats. (Data after Waser and Brown 1986)

Biotic sound sources, wind noise, and wind-induced vegetation motion are the principal sources of ambient noise in all habitats, but their relative contributions differ among habitat types. In both savanna and riverine forest, the interaction of the wind with vegetation is the primary source of noise. Even in the riverine forest there is little shelter from the wind. As the sun heats the landscape, the air temperature near the surface exceeds that at higher elevations, and this temperature difference produces refraction, eddies, turbulence, and a sound shadow near the ground (Wahlberg and Larsen 2017). Hence, as surface temperatures rise in open habitats, wind-induced noise increases and is prominent from midmorning until late afternoon.

In the savanna, the ambient noise is largely governed by the wind, and only at 0600 h are biotic noise sources prominent. Later in the day, biotic sources of sound are overwhelmed by wind-induced noise. In the riverine forest, biotic sound sources, primarily insects, are conspicuous in the midmornings, producing a 4–8 kHz peak. In rain forest, noise levels are consistently high below 100 Hz and between 2–4 kHz. The 2–4 kHz noise levels in the rain forest are due to birds and insects and consistently exceed those in the savanna and riverine forest. Though the rain forest is noisy in the 2–4 kHz band, it is comparatively quiet between 200–1,000 Hz. Unlike the case for savanna and riverine forest, wind noise is rarely prominent in the rain forest, but dripping condensation and rain are significant sources of ambient-noise. Ambient-noise levels change as a function of time of day in the rain forest, as they do in both the savanna and riverine forest, but the change in ambient-noise levels is primarily due to changes in the acoustic behavior of biotic sources and not due to wind.

Sound level recordings indicate that ambient noise should most adversely impact acoustic communication in the riverine forest. To a human listener, however, the sound of wind in the ears makes hearing seem most challenging in the savanna. Sound level recordings do not adequately capture *receptive wind noise* produced by the interaction between wind and the listener's head. Receptive wind noise is presumably governed by the shape of the pinna, the distribution of hair around the head and the ears, and other factors influencing turbulence around the head and in the ear canal (Shaw 1974). In fact, measuring background noise in the afternoon is often difficult due to wind-induced microphone noise, raising the possibility that some primates may have evolved specializations to counter this problem.

4.3.3.2 Attenuation

In an idealized anechoic environment, sound waves would be emitted from a point source and radiated in all directions equally. Sound attenuation would be governed by only one factor: spherical spreading. According to the inverse square law, the surface area of a sphere increases geometrically with increments of the radius, so the sound pressure level is reduced by (2r) or approximately 6 dB in all directions for each doubling of the distance. Excess attenuation is the decrement in sound amplitude greater than that expected by spherical spreading. In East African primate habitats, excess attenuation (measured 1 m above the ground) is more prominent in both the savanna and riverine forest habitats than in the rain forest (measured 7–15 m above the ground). The principal source of this difference is elevation above the ground; most rain forest primates are arboreal, while savanna and riverine forest species are often semiterrestrial. In addition, habitat differences in refraction, humidity, thermal gradients, and turbulence strongly influence sound propagation.

In both rain forest sites, attenuation for broadcasts at 200 Hz is less than that expected by the inverse square law (Fig. 4.3). These findings are consistent with the idea of a *sound window* for long-distance signaling in the arboreal environment (Morton 1975). The sound window, a range of frequencies within which sound attenuation is minimal, is consistently apparent at all source-receiver distances in rain forest habitats (Waser and Waser 1977; Waser and Brown 1984).

In contrast, in both the riverine forest and savanna environments, excess attenuation generally increases with increments in broadcast frequency, and a sound window for long-range signaling is not apparent. Excess attenuation measurements are much more variable in the savanna and riverine forest sites, especially at longer source-receiver distances, than rain forest measurements. This difference is likely due to refraction, wind, and turbulence. Even relatively low wind velocity muted upwind recordings of white noise bursts by 15 dB or more relative to downwind recordings (Brown, unpublished observations).

The increase in excess attenuation as a function of propagation distance is approximately logarithmic in all three habitats (Fig. 4.4). That is, averaged across broadcast frequency, excess attenuation increases by a similar value for each doubling of propagation distance from 12.5 m to 25 m to 50 m. The growth in excess attenuation with propagation distance is modest in the rain forest habitat and steeper for both the savanna and riverine forest environments.

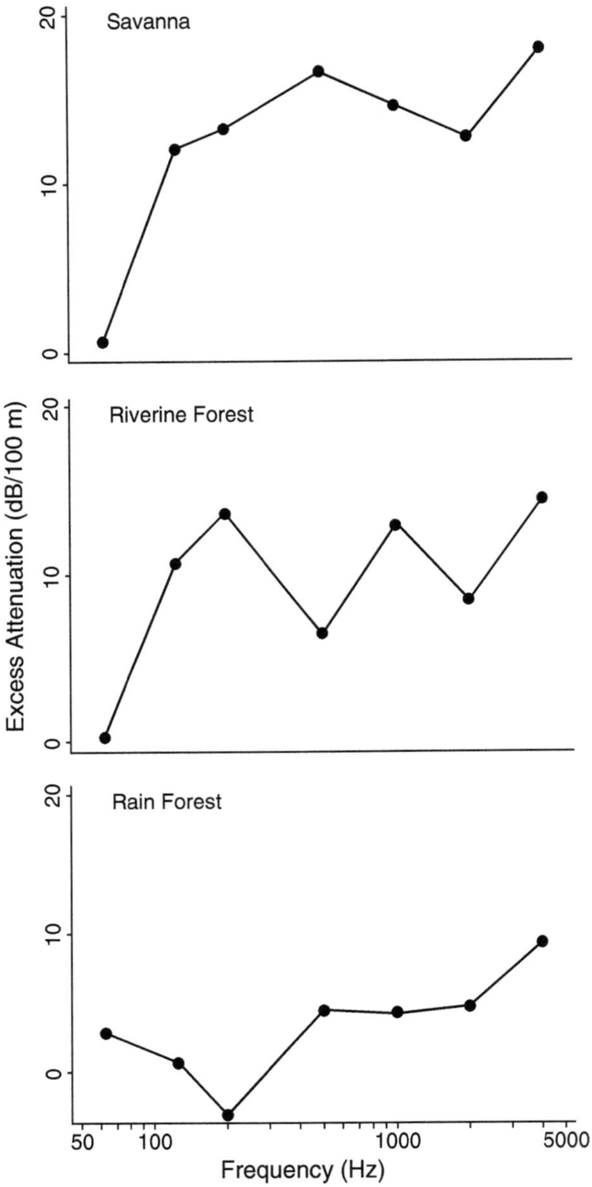

Fig. 4.3 Excess attenuation of broadcast tones as a function of frequency averaged across propagation distance in three primate habitats. (Data after Waser and Brown 1986)

4.3.3.3 Amplitude Fluctuations

Fluctuations in amplitude within 20-s tone broadcasts are presumably induced by moment-to-moment instabilities in thermal gradients, creating fluctuations in wind speed and wind direction, for example, within bubbles of turbulent air (Wiley 2015). Amplitude fluctuations are generally similar in all three habitats. Tones broadcast

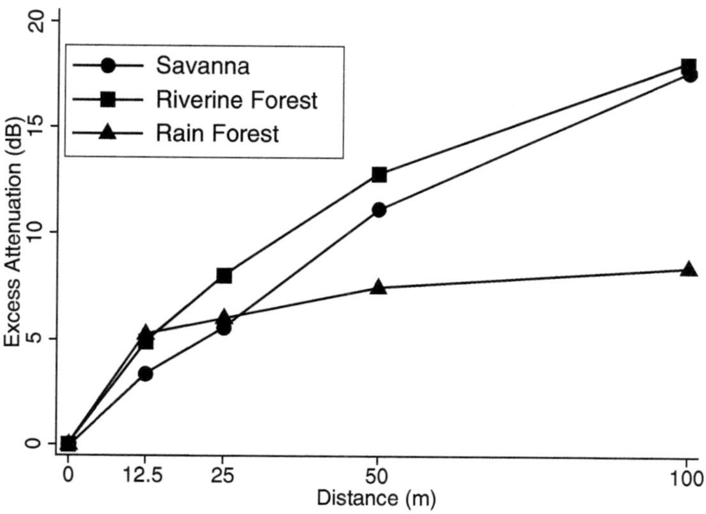

Fig. 4.4 Excess attenuation as a function of propagation distance averaged across broadcast frequency in three different primate habitats. (Data after Waser and Brown 1986)

over short propagation distances (e.g., 12.5 m) do not fluctuate in amplitude, while at long propagation distances (e.g., 100 m), tones can warble as much as 10–15 dB. Amplitude fluctuations are also influenced by carrier frequency, and tones from 100 to 200 Hz are less susceptible to fluctuations in level. This suggests that, as a rule, signals in the 100 to 200 Hz region are more favorably propagated and more stable in amplitude than signals at other frequencies, especially in the rain forest habitat.

4.3.3.4 Reverberation

Reverberation is the persistence of sound after the emission of the signal is terminated. Reverberation is created when the signal is reflected from various surfaces in the environment: eventually the sound is absorbed and dissipated and the echoes decay to zero amplitude. The RT_{60}, a measurement commonly used in architectural acoustics, is the time required for the amplitude of the reflections of the broadcast signal to decay 60 dB. In East African measurements, reverberation in the savanna is low (190 ms) for all pulse frequencies and propagation distances (Fig. 4.5). Discrete echoes are often audible in both forest habitats but are never apparent in the savanna. In both the riverine forest and rain forest sites, reverberation increases as a function of propagation distance up to approximately 600 ms averaged across carrier frequency. Some carrier frequencies are more prone to reverberation than others, and 500-Hz tone pulses in the riverine forest exhibit the longest reverberation times with RT_{60} values approaching 750 ms. The reverberation time for pulses broadcast at 125-Hz are short in all three habitats, confirming that sounds with longer wavelengths are relatively resistant to scattering and reflection in all environments.

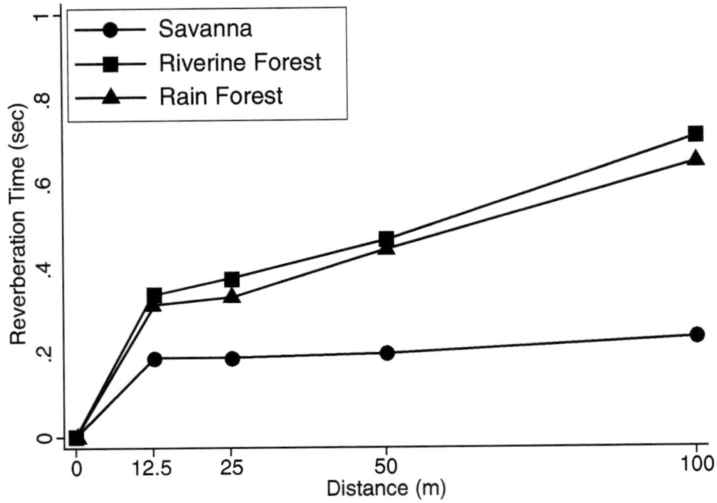

Fig. 4.5 Reverberation time (RT_{60}) as a function of propagation distance averaged across pulse carrier frequency in three different primate habitats. (Data after Waser and Brown 1986)

4.3.3.5 Modulation Depth

The persistence of sound due to reverberation poses an impediment for the perception of trilled and pulsed utterances. That is, the silent interval between each successive trill or pulse is not silent in a reverberant environment but is overlaid by the echo of the preceding pulse. *Modulation depth* is then the amplitude difference between the pulse and the interpulse interval. If reverberation is prominent, modulation depth is low; if reverberation is minimal, modulation depth is high. Pulsed signals retain greater modulation depth in savanna broadcasts than in forest broadcasts. Resistance to pulse train degradation in the savanna is most prominent at short propagation distances and for carrier frequencies of 200 Hz or more. In both forest habitats, modulation depth is significantly degraded for all carrier frequencies and propagation distances (Fig. 4.6). Modulation depth measurements are virtually identical for the two forest habitats.

4.3.3.6 Call Degradation

In general, primate utterances are composed of a variety of notes or frequency elements emitted in a specific sequence or temporal pattern, and not all of these components will be equally susceptible to modification by absorption, reflection, scattering, and refraction. Scattering may disturb the apparent temporal pattern of a call by masking the reception of low-amplitude or silent elements, overlaying them with indirectly transmitted reflections of the previously emitted, loud components. In the case of the baboon *wahoo* call, for example, the reflection of the initially

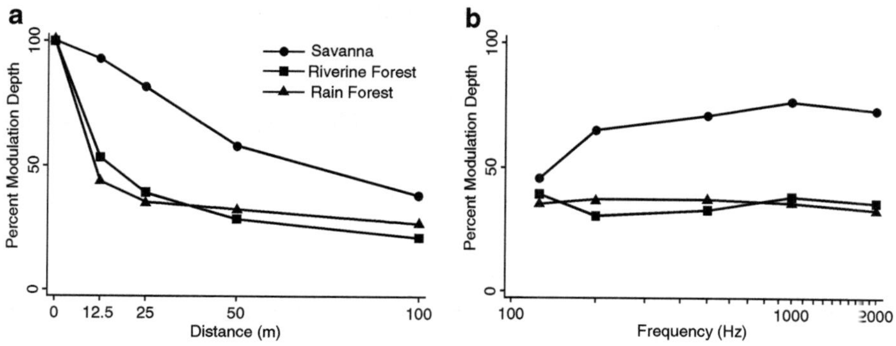

Fig. 4.6 Percent modulation depth: (**a**) averaged across pulse carrier frequency and pulse repetition rate as a function of propagation distance; (**b**) averaged across pulse repetition rate and propagation distance as a function of pulse carrier frequency. The percent modulation depth was calculated as the peak-to-trough sound level difference within a pulse train relative to that broadcast. (Data after Waser and Brown 1986)

emitted *wa* might mask the reception of the subsequently issued *hoo*, especially in forested environments. As another example, forest-living gray-cheeked mangabey males code individual identity in the number of rapidly repeated pulses in the *gobble* of their *whoop-gobble* calls (Waser 1982); reverberation presumably obscures this information for listeners far from the source.

Similarly, attenuation is typically frequency specific. Therefore, at a distance from the vocalizer, the *amplitude ratio* of different elements is disturbed, changing the apparent emphasis of different frequencies in the utterance. Because scattering and attenuation are largely independent processes, signal degradation may occur independently in the frequency and time domains. That is, the frequency components of a signal may be differentially attenuated or absorbed, while the temporal envelope of the signal is preserved. Alternatively, scattering may disturb the temporal patterning of a signal without the call spectrum being changed by differential attenuation. Hence, metrics of degradation have been developed for both frequency and time domains (Brown and Waser 1988; Brown et al. 1995).

4.3.3.7 Time Domain Distortion Analysis

To characterize the magnitude of distortion in the time domain, the broadcast signal (recorded 1 m from the source in an anechoic chamber) can be cross-correlated with the corresponding waveform recorded at 12.5 m or 100 m. The height of the ordinate of a normalized cross-correlogram would be 1.0 if the two waveforms were identical and would incrementally become smaller as the two waveforms became progressively dissimilar. As an example, average exemplars for the blue monkey *boom* call at propagation distances of 1, 12.5, and 100 m are shown in Fig. 4.7. The waveform for exemplars recorded at 12.5 m or 100 m does not change much relative

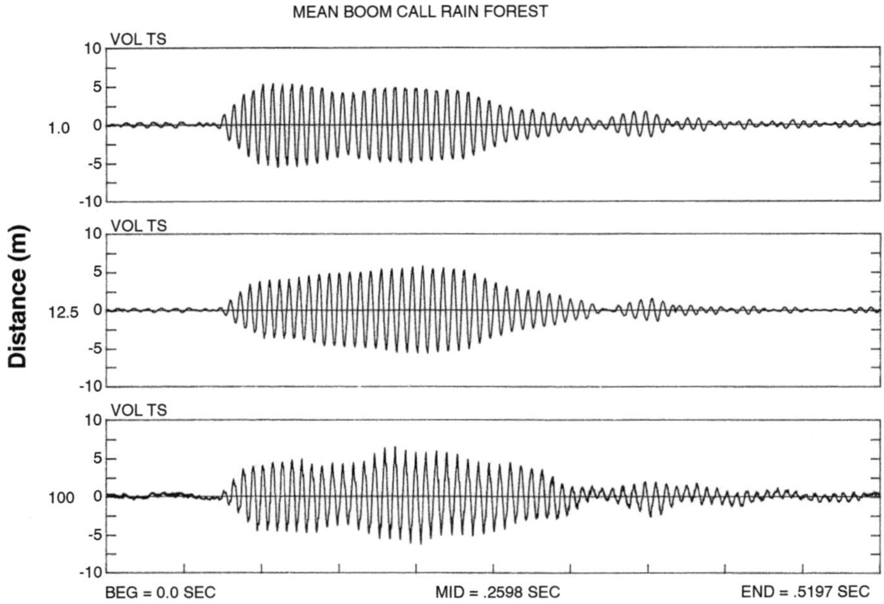

Fig. 4.7 Mean waveforms of the blue monkey *boom* call at propagation distances of 1.0 m, 12.5 m, and 100 m. (Data after Brown and Waser 1988)

to the waveform of the broadcast signal, and the cross-correlations for the boom at 12.5 and 100 m are 0.96 and 0.94, respectively. Hence, the *boom* is quite resistant to distortion in the time domain.

4.3.3.8 Frequency Domain Distortion Analysis

The frequency spectrum of the broadcast and propagated waveforms is provided by the Fourier transform, and distortion in the frequency domain may be measured by generating a difference spectrum derived by subtracting the propagated frequency spectrum from the broadcast frequency spectrum. If the spectrum of the broadcast signal is identical to that propagated and subsequently recorded in the natural habitat, the resulting difference spectrum would be represented by a horizontal line displaced below the original value by the mean attenuation of the signal as a whole. For example, Fig. 4.8 shows the frequency spectrum for the broadcast blue monkey *trill* (measured 1 m from the source in an anechoic chamber), the spectrum for the *trill* measured at a propagation distance of 12.5 m in the rain forest, and the corresponding difference spectrum. This difference spectrum shows both positive and negative values. Negative values indicate that the amplitude of the signal at 12.5 m, relative to the mean for the signal as a whole, exceeds that at 1 m, whereas

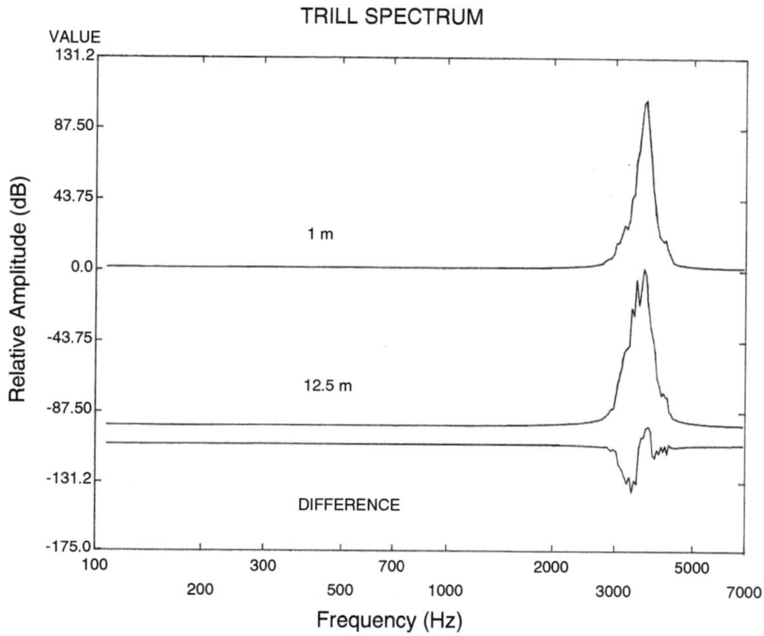

Fig. 4.8 Blue monkey *trill* call spectra at propagation distances of 1 m and 12.5 m shown in comparison with the difference spectrum for the 1 m and 12.5 m signals. (Data after Brown and Waser 1988)

positive values indicate that the relative amplitude of the signal at 1 m exceeds that at 12.5 m. The difference spectrum shows that the blue monkey *trill* is distorted in the frequency domain by the acoustics of the rain forest after only 12.5 m of transmission.

4.4 The Acoustic Adaptation Hypothesis

The *acoustic adaptation hypothesis* can be tested by analyzing the degradation of calls broadcast in appropriate and inappropriate habitats (Brown et al. 1995). If the acoustic properties of the environment in which the signal is normally produced have acted as a source of selection then, for example, distortion should be lower for calls of rain forest monkeys broadcast in rain forest than for those same calls broadcast in savanna.

Overall the results showed that distortion scores were greater for broadcasts conducted in the savanna relative to those conducted in the rain forest, as shown in Fig. 4.9 where the majority of scores fell below the diagonal line. This effect was particularly strong for the values derived from the frequency domain analysis.

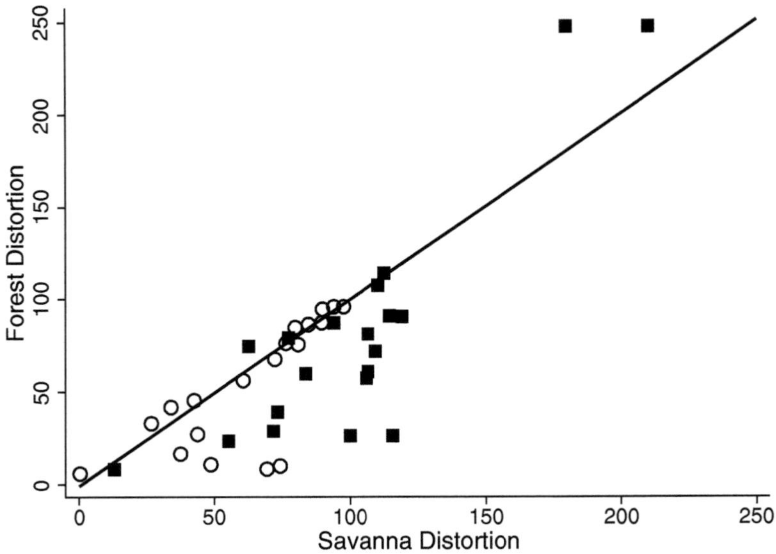

Fig. 4.9 Distortion as a function of habitat type. The results are pooled across species and transmission distance. The frequency-domain analysis data are plotted as *solid squares*, and the time-domain analysis data are plotted as *open circles*. Most scores fall below the *diagonal line*, indicating that distortion was greater for broadcasts conducted in the savanna relative to those conducted in the rain forest, particularly for the frequency-domain analysis. The *x*- and *y*-axes have no units because distortion scores are dimensionless integer numbers. Larger values indicate greater distortion that can be ranked in severity within a domain. (Data after Brown et al. 1995)

As shown in Fig. 4.10, many calls of rain forest monkeys (blue monkeys and grey-cheeked mangabeys) exhibited low distortion scores for broadcasts in the appropriate habitat (rain forest) relative to those conducted in the inappropriate habitat (savanna). However, the calls of savanna monkeys (vervet monkeys and baboons) exhibited similar distortion values for broadcasts conducted in both appropriate (savanna) and inappropriate (rain forest) habitats. The forest monkey calls that were most strongly distorted when broadcast in the savanna, the inappropriate habitat, included the blue monkey *grunt, pyow, trill, ka-train, chirp*, and the mangabey *chorused grunt, scream*, and *loud grunt*.

These findings are consistent with the idea that the structure of calls emitted by rain forest primates has been modified by natural selection to be relatively resistant to distortion, while in the savanna, with the availability of unobstructed visual signaling, frequent wind noise, and the absence of sound windows, the intensity of selection along this dimension has been relaxed.

The ratio of a call's distortion scores in appropriate compared to inappropriate habitats would seem to be a reasonable index of *acoustic adaptation*. Some forest monkey calls, including the blue monkey *ka-train, trill*, and *grunt*, indeed had impressively low distortion ratios. But the details raise numerous puzzles. For example, the mangabey *whoop-gobble*, used in long-distance interactions between

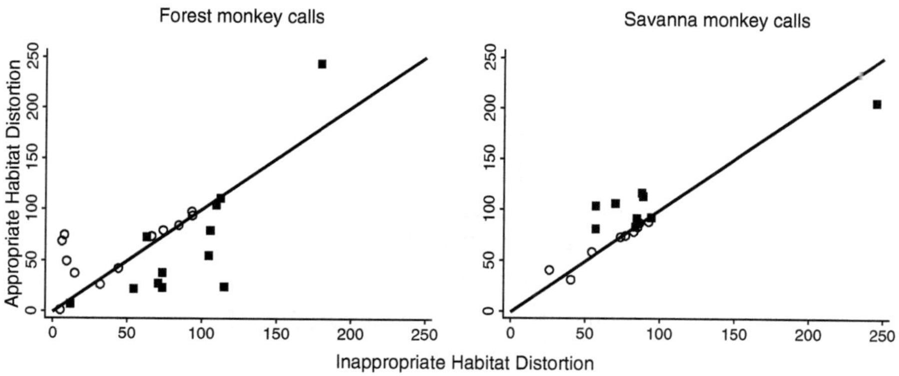

Fig. 4.10 Distortion in "appropriate" and "inappropriate" habitats. The *solid squares* show the results for the frequency-domain analysis; the results for the time-domain analysis are given by the *open circles*. The data points positioned below the *diagonal line* denote cases where distortion in the inappropriate habitat exceeded that in the appropriate habitat. Conversely, data points plotted above the diagonal line indicate the reverse. The *x*- and *y*-axes have no units because distortion scores are dimensionless integer numbers. Larger values indicate greater distortion that can be ranked in severity within a domain. (Data after Brown et al. 1995)

groups, is apparently one of the most distorted mangabey calls in both time and frequency domains and shows no obvious resistance to degradation in the appropriate habitat. Time domain distortion of the blue monkey *boom* call appears to be greater in appropriate than in inappropriate habitat. The blue monkey *ka-train*, usually a response to avian predators, has the lowest measured distortion ratio of any vocalization in the study, but other forest monkey alarm calls (blue monkey *chirps* and mangabey *staccato barks*) show no particular evidence of adaptation to minimize distortion. These interesting results highlight the need for additional studies, more sophisticated analyses, and more refined hypotheses (see Sect. 4.5.1).

4.4.1 The Audible Range of Short- and Long-Distance Calls

The audible range of primate calls is governed by habitat acoustics, signal parameters, and the auditory sensitivity of conspecific recipients. Calculations of audible range, therefore, should include five factors: (1) the spectrum and amplitude of the signal at the source; (2) propagation characteristics of signals broadcast in the natural habitat; (3) degradation and distortion of the signal as a result of habitat acoustics; (4) the spectra and amplitude of ambient noise in the natural habitat; and (5) masked auditory thresholds of conspecific listeners for signals embedded in environmental background noise. Masked auditory thresholds have been measured in blue monkeys and grey-cheeked mangabeys for eleven vocalizations using positive-reinforcement operant-conditioning procedures in a laboratory setting (Brown and Waser 1984). For the sake of comparison, audible distance was also calculated for

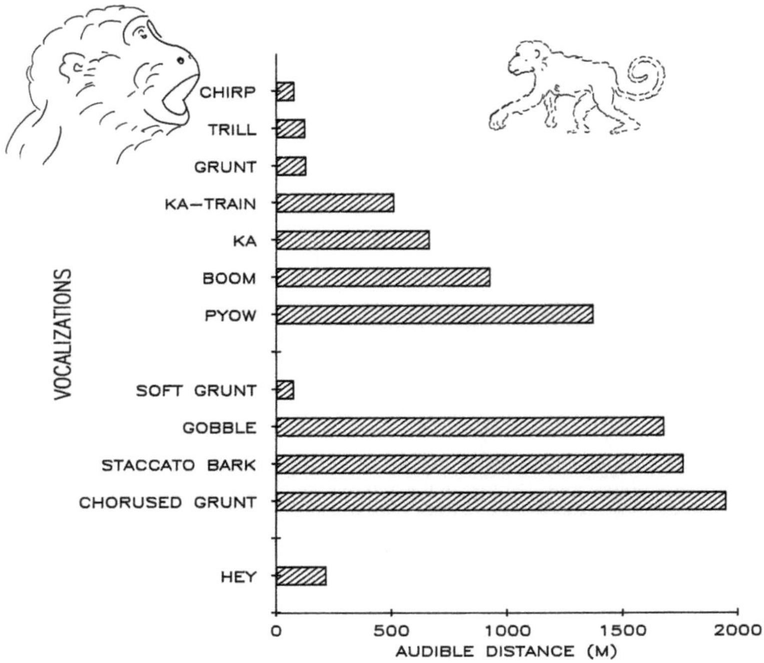

Fig. 4.11 Audible distance estimates (in meters) for eleven monkey calls and the human utterance *hey*. (Data after Brown 1989)

the utterance *hey* shouted by twelve undergraduate students. Audible range was calculated for vocalizers and recipients at an elevation of 1.5 m (humans) and 8 m (blue monkeys and mangabeys).

The audible distance for these twelve primate calls ranged from 79 m for the blue monkey *chirp* to nearly 2 km for several mangabey vocalizations (Fig. 4.11). The loudest *hey* recorded produced an audible distance of 434 m in the rain forest habitat, while the modal distance for this utterance was just 217 m. These results suggest that the audible distance of some forest monkey calls is nearly an order of magnitude greater than that for shouting humans. Differences in the elevation of the source and receiver are the principal parameters that restrict the audible range of humans relative to forest monkeys. Excess attenuation is inversely associated with elevation. Because humans are not arboreal, their loudest utterances will become inaudible after modest propagation distances. The same factor restricts the audible range in the savanna for baboons and vervet monkeys.

Primates analyze sound to detect the movements of conspecifics, predators, and prey; to identify by voice the age, gender, and social rank of troop members; and to acoustically recognize both individuals and species. These tasks differ in their perceptual complexity and require different signal-to-noise ratios to be calculated with precision. Call discrimination requires more favorable signal-to-noise levels than detection does; call recognition requires a more favorable signal-to-noise level than discrimination does (Dooling et al. 2009). The audible distance calculation is based on masked thresholds, and it would be interesting to see how the communicative range shrinks as the task shifts from detection to discrimination to recognition.

4.5 Evolutionary Aspects of Primate Vocal Signals

The notion that primate vocal signals evolve within constraints set by the acoustic environment, which motivated early experiments, has since expanded into a variety of more specific questions. Four of these fall under the general rubric of the acoustic adaptation hypothesis.

4.5.1 Have Some Calls Evolved to Maximize Active Space?

Sound propagation experiments focused on primates (e.g., Sugiura et al. 2006; Bezerra et al. 2012) have largely reinforced early generalizations that sound attenuates less at lower frequencies, but if the vocalizer is near the ground, sounds propagate poorly at all frequencies (Wiley 2015). Consistent with these generalizations, primate vocalizations thought to function in territorial defense are significantly lower in frequency than other calls in the repertoire, and species with larger home ranges generally produce calls with a lower dominant frequency (Mitani and Stuht 1998). Most extant primates are arboreal, and arboreal adaptations were central to the evolutionary emergence of primates. Phylogenetic analyses confirm that the presence of vocalizations that exploit the arboreal habitat's advantage for long-distance acoustic signaling is an ancestral trait (Wich and Nunn 2002). Nevertheless, this trait has been lost by most terrestrial and semiterrestrial primates, perhaps because such species have few opportunities to call (or listen) from elevated sites.

A number of primate calls utilize formant-like frequency bands in parts of the spectrum that are relatively free of ambient noise or exhibit other structural characteristics thought to increase audible range (reviewed by Ey and Fischer 2009). For example, Schneider et al. (2008) showed that the dominant frequencies used by four sympatric Indonesian primates are substantially lower than those used by most birds and insects. On the other hand, Morrill et al. (2013) documented clear daily and seasonal patterns in the spectral distribution of ambient noise in Brazilian forest but found no indication that common marmoset (*Callithrix jacchus*) *phee* calls exploit quiet regions.

By giving a listener's auditory system a well-defined "target" to listen for, calls with a fixed, rigid structure might allow listeners to better pull signal out of noise, and some primate long-distance calls are notably stereotyped (Mitani and Stuht 1998; Zuberbühler, Chap. 7). Little attention has been paid to the listener's side of the equation, and the impact of habitat acoustics on the coevolution of signal form and perceptual specializations remains a promising but largely unexplored field. For example, the exceptional low-frequency hearing of blue monkeys expands the active space for low-frequency calls that also exploit the low-attenuation sound window (Brown and Waser 1984). Could more specialized auditory processing mechanisms be tuned to detect signal attributes that are resistant to attenuation?

Comprehensive assessments of active space (including background noise, source intensity, propagation characteristics, and receiver thresholds) exist for only a few primate vocalizations. Future work should focus on cases where there is a clear relationship between greater audible distance and increased fitness, such as vocalizations involved in marking territorial boundaries or attracting mates. For example, in the *wahoo* calls of male baboons, the frequency of the *wa* and the duration of the *hoo* are indicators of rank (Fischer et al. 2004) and thus might influence female mate choice. These attributes are readily interpreted as honest indicators of male stamina, but might they also influence the distance over which a dominant male can be heard?

In addition, experimental work could compare the propagation of existing calls to variants that might theoretically represent the ancestral state. For example, the methodology of phylogenetic reconstruction that has been applied to the calls of anurans (Ryan and Rand 1999) infers the call structure of common ancestors from those of contemporary species. Might it be applied to primates that have radiated into a variety of habitats, like macaques or *Cercopithecus* species?

Even if it is not possible to reconstruct and synthesize putative ancestral primate calls, the propagation of the target species' call can be compared with that of homologous calls from close relatives living in habitats with different acoustic properties. Another promising approach is that of Morrill et al. (2013), who broadcast calls of common marmosets along with call variants whose frequency or temporal parameters were artificially modified in ways consistent with the range of variation that common marmosets are physiologically capable of emitting.

4.5.2 Do Primates Modify Call Structure or Use Based on Habitat Acoustics?

A fascinating variant of the idea that habitat acoustics constrain the evolution of call structure is that some aspects of primate vocal behavior are labile. For example, individuals modify aspects of call production in ways that increase audible distance as they move through acoustically heterogeneous environments. In many species, long-distance calls are primarily given at or near dawn, when wind noise is minimal and temperature or humidity gradients may improve transmission (reviewed by Ey and Fischer 2009). Golden lion tamarins (*Leontopithecus rosalia*) give loud calls from elevated sites, which should increase the distance over which they are audible (Sabatini and Ruiz-Miranda 2010), and common marmosets time their vocalizations to avoid interference from experimentally delivered noise bursts (Roy et al. 2011).

Olive baboons (*Papio anubis)* that live in relatively forested habitats in Uganda give contact grunts that last longer than those produced by baboons that live in more open habitats in Nigeria. Members of both populations of baboons increase the rate and duration of their grunts when they enter forest (Ey et al. 2009). Japanese macaque (*Macaca fuscata*) *coos* are longer when individuals are more widely separated (Koda and Sugiura 2010). Longer and more frequent calls presumably increase a listener's chances to detect the call where the environment is less acoustically transparent. On the other hand, Hedwig et al. (2015) found that mountain gorillas modified the form of their contact calls as a function of listener distance and vegetation density but not in directions expected to improve signaling distance.

Primate responses to habitat acoustics primarily involve modifications in call use rather than call structure. However, the contact *coos* of Japanese macaques are lower pitched in a population living in more forested habitat where low-frequency tones attenuate less; pitch differences between populations apparently develop during ontogeny (Koda and Sugiura 2010).

4.5.3 Might Selection Favor Calls that Degrade in Predictable Ways?

In studies that have not found evidence for optimal transmission, some researchers have suggested that call structure might instead have evolved to degrade in ways that provide reliable cues to source distance (e.g., de la Torre and Snowdon 2002; Sabatini et al. 2011). For example, degradation due to sound scattering in forests and to differential attenuation of high frequencies in all habitats could provide "ranging" cues. This idea has been extensively developed and tested for avian vocalizations (Naguib and Wiley 2001; Wiley 2015), but data for primates are more limited. Intriguingly, common marmoset *phee* calls appear to attenuate in a markedly predictable way compared to tones and clicks broadcast in the same location (Morrill et al. 2013).

In an elegant experimental approach that deserves wider application, Whitehead (1987) examined the idea that howler monkeys (*Alouatta palliata palliata*) use the degree of reverberation associated with another individual's *bark* vocalizations to assess its distance. He played a sequence of two call series separated by approximately 15 min, the first composed of recorded *barks*, followed by the same *barks* subjected to artificial reverberation. This sequence simulates withdrawal or an increase in spacing between groups, and howler monkey groups tended to respond by moving away from the playback site. In contrast, the monkeys responded with approach to the same stimuli played in the opposite order, a playback sequence that acoustically simulates an approaching group.

4.5.4 Are Evolutionarily Important Cues Coded in Attributes Resistant to Degradation?

The idea that call structure can be used to judge source distance highlights the fact that minimizing attenuation may not be the only, or even the most interesting, effect of natural selection on primate vocal communication. Do habitat acoustics constrain the form of vocal signals even when the audience is nearby? Degradation, rather than attenuation, might be particularly serious in closed environments where redundant visual signals may be ineffective, background noise may be high, and scattering and atmospheric turbulence are substantial.

In an early investigation of sound propagation in Ecuadorean forest, de la Torre and Snowdon (2002) showed that *J calls* and *long calls* of pygmy marmosets (*Callithrix pygmaea*), which are used when listeners are >20 m away, were less adversely affected by reverberation than *trills*, which were used when conspecifics were closer (also see Snowdon, Chap. 6). Investigating the degradation of particular acoustic parameters in the natural environment is an approach that holds great promise.

Consider a mammalian but nonprimate example. During the breeding season, male koalas (*Phascolarctos cinereus*) give loud *bellows*. Charlton et al. (2012a) inferred that males would benefit by transmitting reliable cues to their size and individual identity. Analysis of calls recorded from captive males showed that formant spacing during the later inhalation phases of the *bellow* was the most reliable predictor of the male's body size (Charlton et al. 2011). Playback of *bellows*, along with synthesized variants in which formant spacing was artificially shifted, in a two-speaker choice paradigm showed that estrous female koalas looked longer and more often at speakers broadcasting the variants whose formants indicated the larger male (Charlton et al. 2012b). In addition, playback to wild males showed that they responded differentially to synthetic call variants differing only in formant spacing (Charlton et al. 2013). Finally, having found a set of acoustic parameters that listeners apparently respond to in ways that might benefit the vocalizer, Charlton et al. (2012a) examined the propagation of those calls in eucalypt forest and found that formant spacing was less degraded than several other acoustic attributes.

A study of orangutan (*Pongo pygmaeus*) *long calls* illustrates a variant of this approach. Lameira and Wich (2008) broadcast these calls through as much as 300 m of Indonesian forest and showed that fundamental frequency and several other acoustic parameters (e.g., pulse duration) previously shown to be important in individual discrimination were not changed by transmission over these distances. Using discriminant function analysis, they asked how well individually identified males could be differentiated based on those parameters. Finally, they performed the same discriminant function analysis on the same calls rerecorded after transmission through the forest. Statistically, individual identity could be distinguished robustly as far as 300 m away from the source. Pulse duration and fundamental frequency were the most useful parameters for distinguishing the individual identity of vocalizers.

4.6 Future Research Directions

During the 1980s, computer power and software limitations were serious impediments for measuring the degradation of acoustic signals. These limitations meant that researchers had to digitize analog tape recordings, develop their own algorithms for quantifying degradation in the time and frequency domains and, for large datasets, conduct the signal processing analyses on a supercomputer (Brown et al. 1995). Computer advances have overcome these limitations in processing speed, and the widespread availability of sound analysis software has facilitated progress in primate bioacoustics. Where, then, are the limitations and opportunities for future research?

4.6.1 Thinking About the Physics

If primate habitats differ in the constraints they pose for acoustic communication, there must be identifiable physical principles that underlie those differences. Research in the physics of sound propagation has advanced significantly (Larsen and Wahlberg 2017; Wahlberg and Larsen 2017), but our understanding of acoustic habitat differences remains largely empirical and, at present, is not well articulated with advancements in physics and the measurement of relevant parameters.

In open habitats where wind and turbulence impede sound propagation, researchers should consider conducting wind speed measurements and track temperature changes in the ground and the air. As the surface temperature heats up, sound is refracted upward, creating a shadow zone near the ground, and future research could explore the impact of this process on savanna primate communication. Particularly for semiterrestrial primates, the impedance of the ground surface may influence sound propagation; where the investigator suspects this possibility, measurements of such factors as soil porosity, hydration, grass, and leaf litter characteristics should be considered. In forested habitats, air temperature and humidity may be stratified between the forest floor and canopy, and stratification can have a strong impact on sound propagation. Future researchers may consider systematically tracking temperature and humidity as a function of elevation and time of day (Wahlberg and Larsen 2017). The rain forest attributes that result in the presence of sound windows are still not robustly linked to the physics of sound propagation, and future researchers should address this problem.

In addition to advancing the understanding of sound attenuation, future research could examine more closely the issue of degradation and its measurement. There is no standard metric for measuring habitat-induced degradation for broadcast calls. Researchers have relied on cross-correlation between the emitted signal and recordings of the broadcast signal at various propagation distances (Brown et al. 1995; Nicholls and Goldizen 2006; Morrill et al. 2013), discriminant analysis of particular parameters (Lameira and Wich 2008; Charlton et al. 2012b), and tail/signal or blur ratios (Sabatini et al. 2011; Sandoval et al. 2015).

Ideally, call degradation studies would specifically target those parameters most likely to have been selected to promote species and individual recognition in the acoustic habitat. Because primates exhibit a four orders of magnitude variation in size, communicatively relevant parameters that do not show a simple allometric relationship with body size are obvious candidates for consideration.

For sound production, the prime candidates include the presence, morphology, or absence of accessory air sacs; the length and shape of the supraglottic airway; and anatomy of the nasal cavities. The acoustic impact of these parameters is malleable as air sacs are inflated or deflated, the larynx is raised or lowered in the trachea, the jaw is opened or closed, lips are protruded or retracted, and so forth (Brown and Cannito 1995). Collectively, changes in these parameters change the resonance of the airway, and resonance variation, like formant changes in speech perception, must be perceptible in the natural environment in order to play a communicatively relevant role. In parallel with variations in the morphology of the airway are species differences in the layered composition, shape, length, and viscoelastic properties of the vocal folds (Riede and Brown 2013). Furthermore, there is a good possibility that tissues other than the vocal folds oscillate during call production in some species. Primate vocal folds exhibit several different patterns or regimens of oscillation. It is probable that many distinctive primate utterances are produced by abrupt bifurcations between these regimens, and this phenomenon is likely to be relevant for species and individual recognition (Brown et al. 2003).

For sound perception, the parameters most relevant for habitat acoustics include the size, shape, and mobility of the pinna; the length of the auditory canal; and the volume of the middle ear (Quam, Martínez, Rosa, and Arsuaga, Chap. 8). Changes in these parameters would produce corresponding changes in the resonances of the ear, and this would impact the relative audibility of high-frequency versus low-frequency sounds (Webster and Webster 1980; Hemilä et al. 1995). Species differences in auditory sensitivity and the range of hearing have been observed in human and nonhuman primates (Brown and Waser 1984; Stebbins and Moody 1994; Ramsier and Rauschecker, Chap. 3), and it is possible that the patterns of contraction and relaxation of the outer hair cells in the cochlea differ between species in ways that influence the relative audibility of species-specific vocal gestures. The point here is to encourage the development of research strategies along with sound analysis algorithms tuned to the parameters expected to most greatly impact signal transmission and reception in natural habitats.

Naguib and Wiley (2001) noted that "few studies of sound transmission have measured the components of degradation in ways that reveal how it might be perceived by a receiver." This observation remains true, and future research should strive to develop a sound degradation algorithm that objectively scores degradation in perceptually relevant ways. In this respect, algorithms developed for time-series pattern recognition may hold promise. Music, bird song, speech, and primate utterances are all patterns of acoustic events over time. Machine recognition of speech has been successfully addressed by Hidden Markov Model algorithms, and this approach has been successfully adapted to animal bioacoustic research (Johnson and Clemins 2017). Auditory perception by nonhuman primates closely resembles human perception (Stebbins and Moody 1994; Sinnott and Brown 1997; Sinnott et al. 1997; but see Ramsier and Rauschecker, Chap. 3; Quam, Martínez, Rosa, and Arsuaga, Chap. 8), and it may be possible to develop algorithms that score degradation in ways that are relevant for general primate auditory perception.

Primate vocal signals are regarded as being emitted by an isolated vocalizer and the signals are subject to propagation constraints as defined by spherical spreading. As noted in Sect. 3.3.2, the amplitude of the wave front sampled at progressively greater propagation distances decreases by approximately 6 dB for each doubling of the distance from the source. However, many primate vocalizations are not emitted from an isolated vocalizer, and investigators have tended to overlook communally emitted calls (Campbell and Snowdon 2007). Duetting and dawn choruses occur in many primate species, while group encounters that induce counter-calling exchanges and anti-predator displays are virtually always emitted by multiple vocalizers. This raises the possibility that the propagation of some calls might be better described by cylindrical spreading. If sound sources are arrayed in a row, the sound energy propagates along the surface of an imaginary cylinder with a surface area proportional to $2\pi r$. For cylindrical spreading, sound amplitude is expected to decrease 3 dB for each doubling of propagation distance (Embleton 1996). In many instances in nature, sound transmission may be neither truly spherical nor truly cylindrical, but something in between, and researchers should consider the full spectrum of possibilities in evaluating sound emission in primates. The recruitment of vocal participants in a dawn chorus, for example, may not only amplify the level of the signal. More importantly, it may shift the geometry of the wave front from spherical spreading toward cylindrical spreading. In this respect, Wahlberg and Larsen (2017) note that a signal from a spherical source that would attenuate 26 dB in 20 m, would propagate 400 m from a cylindrical source before attenuating the same degree.

4.6.2 Moving Beyond First-Order Generalizations About Habitat Types

First-order generalizations have considered habitat types for African primates as represented by rain forest, riverine forest, and savanna, but clearly primates occupy other habitats, such as the temperate forests of Japan and China or the high elevation grasslands of Ethiopia. How acoustically uniform are rain forests? Are the East African forests comparable to the coastal forests of Brazil, the flooded forests of the Amazon, or the seasonally dry forests of south India? How similar are the acoustics of Kenyan savanna sites to those of South African grassland or fynbos? How much acoustic heterogeneity is present within each of these habitat types, for example, as a consequence of the seasonality of leaves dropping or flooding, the vocalizer's elevation above the ground, or spatial variation in vegetation structure on a local scale?

Only a handful of studies approach such questions. For example, Slabbekoorn (2004) investigated background noise using a common methodology in different forests in Cameroon. Bormpoudakis et al. (2013) investigated background noise patterns across multiple temperate sites and seasons. Such studies may eventually generate a database that is large enough to determine the heterogeneity of habitat types. Comparable, systematic analyses of local variation in sound propagation as well as "soundscapes" are needed. Where predictions from the acoustic adaptation hypothesis are not fulfilled (Ey and Fischer 2009), is this because the hypothesis is inappropriate or because our predictions are based on naïve assumptions about the details of habitat acoustics? Is between-habitat variation large enough that we should really expect to find evolutionarily fixed differences in signaling, or does most variation occur at the scale of individual populations or home ranges, so that we might more likely expect ontogenetic or learned variation in call use and structure?

4.6.3 Archiving Data

Relatively few studies of habitat acoustics have been carried out by primatologists. In principle, this limitation might be mitigated by the numerous studies carried out by investigators focused on birds (Barker 2008), frogs (Kime et al. 2000; Malone et al. 2014), or insects (Couldridge and van Staaden 2004) in habitats also occupied by primates. There are serious problems, however, in comparing studies of habitat acoustics that differ in design and analysis. The absence of a standard rubric adopted by multiple independent investigators is an enduring weakness for investigations of the acoustic adaptation hypothesis.

Different questions drive different investigators and this will inevitably lead to differences in methodology. As an alternative to standardizing methodology, background noise recordings and data from sound propagation tests need to be archived. For example, much of the sound propagation data collected by Brown and Waser have recently been archived at The Cornell Laboratory of Ornithology's Macaulay Library of Natural Sounds (macaulaylibrary.org). Archived data are then available for analysis by methodologies developed by other investigators. Archiving sound propagation data sets will ultimately strengthen our understanding of habitat acoustics and the insights derived by different algorithms and methodologies.

In addition, archived datasets will allow documentation of changes in the acoustic soundscape resulting from the passage of time and from human encroachment. The riverine forest site sampled by Waser and Brown (1986) is probably very different today. Hydroelectric dams have altered patterns of flooding, and the forest understory may have been dramatically changed by three decades of elephant poaching. The acoustic impact of these changes and those to follow can be monitored by the availability of archived data. Several groups have amassed large amounts of background noise data (Pijanowski et al. 2011), but such collections become far more valuable if they are curated systematically, enriched by the contribution of data from outside investigators, and have protocols that give access to investigators who wish to analyze data with comparable methodology across studies (Webster and Budney 2017).

4.6.4 Thinking Synthetically About Sound Production, Background Noise, Propagation and Perception

Most studies of primate bioacoustics concentrate primarily on one part of the system—they compare call frequencies to background noise frequencies, or they focus on attenuation or reverberation or auditory sensitivity. But animals face these factors in combination. If the goal is to assess the degree to which habitat acoustics shape call structure, what investigators really need to know is how strongly the signal contrasts with noise and how readily conspecific listeners can detect and classify the signal at progressively more distant locations in the natural habitat.

This raises a related issue: Nearly all studies of sound propagation understandably measure conditions when it's quiet and when there is little wind. However, animals also have to deal with communication under suboptimal conditions. Wiley (2015) has argued persuasively that communication must be viewed as a problem of signal detection in noise, whereby noise includes not only what is commonly thought of as background noise but also errors introduced by scattering, turbulence, and other acoustic phenomena and even the auditory capabilities of the listener. This approach is consistent with those of Lameira and Wich (2008) and Charlton et al. (2012a), who ask how readily specific types of information can be extracted from signals after their degradation by the local environment. The gold standard would be to ask the animals themselves how well they can extract the relevant information from degraded signals.

4.7 Conclusion

Ancestral primates dispersed from the rain forest canopy to savanna, fynbos, alpine meadows, flooded forests, and dry forests. Primates are a conspicuously vocal order, and they emit an assortment of calls to attract mates, define territories, signal alarm, scold subordinates, solicit aid, and soothe offspring. In each habitat the emitted call is subjected to reflection, constructive and destructive interference, reverberation, attenuation, and frequency-dependent distortion, and the call may be emitted when wind and a noisy background make hearing difficult. Measurements of ambient noise and sound propagation in the most common primate habitats of rain forest, savanna, and riverine forest show that these habitat types pose different obstacles for vocal communication. Wind noise and atmospheric turbulence impede sound propagation and hearing in the savanna but not in the rain forest. Vibrating and rattling vegetation noise is significant in the riverine forest but not in the savanna. Biotic noise disturbances are prominent in the rain forest but not in the savanna or riverine forest.

Furthermore, habitat differences provide different opportunities for acoustic signaling. The most obvious of these is the ability to maneuver vertically in forested habitats, and calls emitted from elevated locations propagate more favorably relative to terrestrial broadcasts. Habitat differences have likely served as a source of selection for the structure of primate calls and the mechanisms dedicated to their perception. This proposition, the acoustic adaptation hypothesis, has received some study and support, but improved algorithms for analyses are needed, inconsistent puzzles in the data exist, and many additional questions remain to be explored. The vocalizations of different primates may readily be modified by the presence, configuration, or absence of vocal air sacs, in addition to variations in jaw opening, lip protrusion, vocal tract length, nasal cavity anatomy, and the viscoelastic properties of the vocal folds.

In addition, species differences in sound perception may be influenced by differences in the resonance of the ear canal, pinna morphology, middle ear volume, and the filtering properties of the cochlea's outer hair cells. Bioacousticians have just scratched the surface of the parameters selection may have tweaked to heighten the fidelity of communication in acoustically untidy habitats. Hopefully, this chapter will inspire a new generation of researchers to reevaluate, contemplate, and tackle unaddressed questions and assume stewardship for the soundscapes and inhabitants of the ancestral homes within which human and all nonhuman primates arose.

Compliance with Ethics Requirements
Charles Brown declares he has no conflicts of interest.
Peter Waser declares he has no conflicts of interest.

References

Barker, N. K. (2008). Bird song structure and transmission in the neotropics: Trends, methods and future directions. *Ornitologia Neotropical*, 19, 175–199.

Bezerra, B. M., Souto, A. S., & Jones, G. (2012). Propagation of the loud "tcho" call of golden-backed uakaris, *Cacajao melanocephalus*, in the black-swamp forests of the upper Amazon. *Primates*, 53, 317–325.

Bormpoudakis, D. J., Sueur, J., & Pantis, J. D. (2013). Spatial heterogeneity of ambient sound at the habitat type level: Ecological implications and applications. *Landscape Ecology*, 28, 495–506.

Brown, C. H. (1989). The acoustic ecology of East African primates and the perception of vocal signals by grey-cheeked mangabeys and blue monkeys. In R. J. Dooling & S. H. Hulse (Eds.), *Comparative psychology of audition: Perceiving complex sounds* (pp. 201–239). Hillsdale, NJ: Lawrence Earlbaum Associates.

Brown, C. H., Alipour, F., Berry, D. A., & Montequin, D. (2003). Laryngeal biomechanics and vocal communication in the squirrel monkey (*Saimiri boliviensis*). *The Journal of the Acoustical Society of America*, 113, 2114–2126.

Brown, C. H., & Cannito, M. P. (1995). Articulated and inflected primate vocalizations: Developing animal models of speech. In D. Robbin & D. Beukleman (Eds.), *Disorders of motor speech: Assessment, treatment, and clinical considerations* (pp. 43–63). Baltimore, MD: Brooks Publishing Company.

Brown, C. H., Gomez, R., & Waser, P. M. (1995). Old world monkey vocalizations: Adaptations to the local habitat? *Animal Behaviour*, 50, 945–961.

Brown, C. H., & Waser, P. M. (1984). Hearing and communication in blue monkeys. *Animal Behaviour*, 32, 66–75.

Brown, C. H., & Waser, P. M. (1988). Environmental influences on the structure of primate vocalizations. In D. Todt, P. Goedeking, & D. Symmes (Eds.), *Primate vocal communication* (pp. 51–68). New York: Springer-Verlag.

Campbell, M. W., & Snowdon, C. T. (2007). Vocal response of captive-reared *Saguinus oedipus* during mobbing. *International Journal of Primatology*, 28, 257–270.

Charlton, B. D., Ellis, W. A., McKinnon, A. J., Cowin, G. J., et al. (2011). Cues to body size in the formant spacing of male koala (*Phascolarctos cinereus*) bellows: Honesty in an exaggerated trait. *Journal of Experimental Biology*, 214, 3414–3422.

Charlton, B. D., Ellis, W. A., Brumm, J., Nilsson, K., & Fitch, W. T. (2012a). Female koalas prefer bellows in which lower formants indicate larger males. *Animal Behaviour*, 84, 1565–1571.

Charlton, B. D., Reby, D., Ellis, W. A., Brumm, J., & Fitch, W. T. (2012b). Estimating the active space of male koala bellows: Propagation of cues to size and identity in a Eucalyptus forest. *PLoS One*, 7(9), e45420. doi:10.1371/journal.pone.0045420.

Charlton, B. D., Whisson, D. A., & Reby, D. (2013). Free-ranging male koalas use size-related variation in formant frequencies to assess rival males. *PLoS One*, 8(7), e70279. doi:10.1371/journal.pone.0070279.

Couldridge, V. C. K., & van Staaden, M. J. (2004). Habitat-dependent transmission of male advertisement calls in bladder grasshoppers (Orthoptera: Pneumoridae). *Journal of Experimental Biology*, 207, 2777–2786.

de la Torre, S., & Snowdon, T. (2002). Environmental correlates of vocal communication of wild pygmy marmosets, *Cebuella pygmaea*. *Animal Behaviour*, 63, 847–856.

Dooling, R. J., West, E. W., & Leek, M. R. (2009). Conceptual and computational models of the effects of anthropogenic noise on birds. *Proceedings of the Institute of Physics*, 31, 99–106.

Embleton, T. F. W. (1996). Tutorial on sound propagation outdoors. *The Journal of the Acoustical Society of America*, 100, 31–48.

Ey, E., & Fischer, J. (2009). The "Acoustic Adaptation Hypothesis"—A review of the evidence from birds, anurans, and mammals. *Bioacoustics*, 19, 21–48.

Ey, E., Rahn, C., Hammerschmidt, K., & Fischer, J. (2009). Wild female olive baboons adapt their grunt vocalizations to environmental conditions. *Ethology*, 115, 493–503.

Fischer, J., Kitchen, D. M., Seyfarth, R. M., & Cheney, D. L. (2004). Baboon loud calls advertise male quality: Acoustic features and their relation to rank, age and exhaustion. *Behavioral Ecology and Sociobiology*, 56, 140–148.

Fischer, J., Noser, R., & Hammerschmidt, K. (2013). Bioacoustic field research: A primer for acoustic analyses and playback experiments with primates. *American Journal of Primatology*, 75, 643–663.

Fought, J. G., Munroe, R. L., Fought, C. R., & Good, E. M. (2004). Sonority and climate in a world sample of languages: Findings and prospects. *Cross Cultural. Research*, 38, 27–51.

Gautier, J.-P. (1988). Interspecific affinities among guenons as deduced from vocalizations. In A. Gautier-Hion, F. Bourliere, J.-P. Gautier, & J. Kingdon (Eds.), *A primate radiation* (pp. 194–226). Cambridge: Cambridge University Press.

Gish, S. L., & Morton, E. S. (1981). Structural adaptations to local habitat acoustics in Carolina wren songs. *Zeitschrift für Tierpsychologie*, 56, 74–84.

Groves, C. (2001). *Primate taxonomy*. Washington, DC: Smithsonian Institution Press.

Haddad, N. M., Brudvig, L. A., Clobert, J., Davies, K. F., et al. (2015). Habitat fragmentation and its lasting impact on Earth's ecosystems. *Science Advances*, 1(2), e1500052. doi:10.1126/sciadv.1500052.

Hedwig, D. R., Mundry, R., Robbins, M. M., & Boesch, C. (2015). Audience effects, but not environmental influences, explain variation in gorilla close distance vocalizations—A test of the acoustic adaptation hypothesis. *American Journal of Primatology*, 77, 1239–1252.

Hemilä, S., Nummela, S., & Reuter, T. (1995). What middle ear parameters tell about impedance matching and high frequency hearing. *Hearing Research*, 85, 31–44.

Johnson, M. T., & Clemins, P. J. (2017). Hidden Markov model signal classification. In C. H. Brown & T. Riede (Eds.), *Comparative bioacoustics: An overview* (pp. 371–428). Oak Park, IL: Bentham Science.
Kime, N. M., Turner, W. R., & Ryan, M. J. (2000). The transmission of advertisement calls in Central American frogs. *Behavioral Ecology*, 11, 71–83.
Koda, H., & Sugiura, H. (2010). The ecological design of the affiliative vocal communication style in wild Japanese macaques: Behavioral adjustments to social contexts and environments. In N. Nakagawa (Ed.), *The Japanese macaques* (pp. 167–188). New York: Springer-Verlag.
Lameira, A. R., & Wich, S. A. (2008). Orangutan long call degradation and individuality over distance: A playback approach. *International Journal of Primatology*, 29, 615–625.
Larsen, O. N., & Wahlberg, M. (2017). Sound and sound sources. In C. H. Brown & T. Riede (Eds.), *Comparative bioacoustics: An overview* (pp. 3–60). Oak Park, IL: Bentham Scientific.
Lind, E. M., & Morrison, M. E. S. (1974). *East African vegetation*. London: Longman.
Lombard, É. (1911). Le signe de l'élévation de la voix. *Annales des Maladies de l'Oreille, du Larynx, du Nez et du Pharynx*, 37(2), 101–109.
Malone, J. H., Ribado, J., & Lemmon, E. M. (2014). Sensory drive does not explain reproductive character displacement of male acoustic signals in the upland chorus frog (*Pseudacris feriarum*). *Evolution*, 68, 1306–1319.
Michelsen, A. (1978). Sound reception in different environments. In M. Ali (Ed.), *Sensory ecology* (pp. 345–373). New York: Plenum Press.
Miller, J. D. (1974). Effects of noise on people. *The Journal of the Acoustical Society of America*, 56, 729–763.
Mitani, J. C., & Stuht, J. (1998). The evolution of primate loud calls: Acoustic adaptation for long-distance propagation. *Primates*, 39, 171–182.
Morrill, R. J., Thomas, A. W., Schiel, N., Souto, A., & Miller, C. T. (2013). The effect of habitat acoustics on common marmoset vocal signal transmission. *American Journal of Primatology*, 75, 904–916.
Morton, E. S. (1975). Ecological sources of selection on avian sounds. *American Naturalist*, 108, 17–34.
Naguib, M., & Wiley, R. H. (2001). Estimating the distance to a source of sound: Mechanisms and adaptations for long-range communication. *Animal Behaviour*, 62, 825–837.
Nicholls, J. A., & Goldizen, A. W. (2006). Habitat type and density influence vocal signal design in satin bowerbirds. *Journal of Animal Ecology*, 75, 549–558.
Pijanowski, B. C., Villanueva-Rivera, L. F., Dumyahn, A., Krause, B. L., et al. (2011). Soundscape ecology: The science of sound on the landscape. *BioScience*, 61, 203–216.
Riede, T., & Brown, C. H. (2013). Body size, vocal fold length, and fundamental frequency—implications for mammalian vocal communication. *Nova Acta Leopoldina N.F.*, 111(380), 295–314.
Roy, S., Miller, C. T., Gottsch, D., & Wang, X. (2011). Vocal control by the common marmoset in the presence of interfering noise. *Journal of Experimental Biology*, 214, 3619–3629.
Ryan, M. J., & Rand, A. S. (1999). Phylogenetic influence on mating call preferences in female tungara frogs, *Physalaemus pustulosus*. *Animal Behaviour*, **57**, 945–956.
Rylands, A. B., & Mittermeier, R. A. (2014). Primate taxonomy: Species and conservation. *Evolutionary Anthropology*, 23, 8–10.
Sabatini, V., & Ruiz-Miranda, C. R. (2010). Does the golden lion tamarin, *Leontopithecus rosalia* (Primates: *Callitrichidae*) select a location within the forest strata for long distance communication? *Zoologica*, 27, 179–183.
Sabatini, V., Ruiz-Miranda, C. R., & Dabelsteen, T. (2011). Degradation characteristics of golden lion tamarin *Leontopithecus rosalia* two-phrase long calls: Implications for call detection and ranging in the evergreen forest. *Bioacoustics*, 20, 137–158.
Sandoval, L., Dabelsteen, T., & Mennill, D. J. (2015). Transmission characteristics of solo songs and duets in a neotropical thicket habitat specialist bird. *Bioacoustics*, 24, 289–306.
Schaller, G. (1963). *The mountain gorilla*. Chicago, IL: University of Chicago Press.
Scheifele, P., Andrew, S., Cooper, R. A., Darre, M., et al. (2005). Indication of a Lombard vocal response in the St. Lawrence River beluga. *The Journal of the Acoustical Society of America*, 117, 1468–1492. http://asa.scitation.org/doi/pdf/10.1121/1.1835508.

Schneider, C., Hodges, K., Fischer, J., & Hammerschmidt, K. (2008). Acoustic niches of Siberut primates. *International Journal of Primatology*, 29, 601–613.
Shaw, E. (1974). Transformation of sound pressure level from the free field to the eardrum in the horizontal plane. *The Journal of the Acoustical Society of America*, 56, 1848–1861.
Shen, J.-X., & Xu, Z.-M. (2016). The Lombard effect in male ultrasonic frogs: Regulating antiphonal signal frequency and amplitude in noise. *Scientific Reports*, 6, 27103. doi:10.1038/srep27103.
Sinnott, J. M., & Brown, C. H. (1997). Perception of the English liquid /ra-la/ contrast by humans and monkeys. *The Journal of the Acoustical Society of America*, 102, 588–602.
Sinnott, J. M., Brown, C. H., Malik, W., & Kressley, R. A. (1997). A multidimensional scaling analysis of vowel discrimination in humans and monkeys. *Perception and Psychophysics*, 59, 1214–1224.
Sinnott, J. M., Stebbins, W. C., & Moody, D. B. (1975). Regulation of voice amplitude by the monkey. *The Journal of the Acoustical Society of America*, 58, 412–414.
Slabbekoorn, H. (2004). Habitat-dependent ambient noise: Consistent spectral profiles in two African forest types. *The Journal of the Acoustical Society of America*, ***116***, 3727–3733.
Stammbach, E. (1987). Desert, forest, and montane baboons: Multi-level societies. In B. Smuts, D. Cheney, R. Seyfarth, R. Wrangham, & T. T. Struhsaker (Eds.), *Primate societies* (pp. 112–120). Chicago, IL: The University of Chicago Press.
Stebbins, W. C., & Moody, D. B. (1994). How monkeys hear the world: Auditory perception in nonhuman primates. In R. R. Fay & A. N. Popper (Eds.), *Comparative hearing: Mammals* (pp. 97–133). New York: Springer-Verlag.
Sugiura, H., Tanaka, T., & Masataka, N. (2006). Sound transmission in the habitats of Japanese macaques and its possible effect on population differences in coo calls. *Behaviour*, 143, 993–1012.
Tavaré, S., Marshall, C. R., Will, O., Soligo, C., & Martin, R. D. (2002). Using the fossil record to estimate the age of the last common ancestor of extant primates. *Nature*, 416, 726–729.
Wahlberg, M., & Larsen, O. N. (2017). Propagation of sound. In C. H. Brown & T. Riede (Eds.), *Comparative bioacoustics: An overview* (pp. 61–120). Oak Park, IL: Bentham Science.
Waser, P. M. (1982). The evolution of male loud calls among mangabeys and baboons. In C. T. Snowdon, C. H. Brown, & M. R. Petersen (Eds.), *Primate communication* (pp. 117–143). Cambridge: Cambridge University Press.
Waser, P. M., & Brown, C. H. (1984). Is there a "sound window" for primate communication? *Behavioral Ecology and Sociobiology*, 15, 73–76.
Waser, P. M., & Brown, C. H. (1986). Habitat acoustics and primate communication. *American Journal of Primatology*, 10, 135–154.
Waser, P. M., & Waser, M. S. (1977). Experimental studies of primate vocalizations: Specializations for long-distance propagation. *Zeitschrift für Tierpsychologie*, 43, 239–263.
Webster, M. S., & Budney, G. F. (2017). Sound archives and media specimens in the 21st century. In C. H. Brown & T. Riede (Eds.), *Comparative bioacoustics: An overview* (pp. 479–503). Oak Park, IL: Bentham Science Publishers.
Webster, D. B., & Webster, M. (1980). Morphological adaptations of the ear in the rodent family Heteromyidae. *American Zoologist*, 20, 247–254.
Whitehead, J. M. (1987). Vocally mediated reciprocity between neighbouring groups of mantled howling monkeys, *Alouatta palliata palliata*. *Animal Behaviour*, 35, 1615–1627.
Wich, S. A., & Nunn, C. L. (2002). Do male "long-distance calls" function in mate defense? A comparative study of long-distance calls in primates. *Behavioral Ecology and Sociobiology*, 52, 474–484.
Wiley, R. H. (2015). *Noise matters: The evolution of communication*. Cambridge, MA: Harvard University Press.
Williams, B. A., Kay, R. F., & Kirk, C. (2010). New perspectives on anthropoid origins. *Proceedings of the National Academy of Sciences of the United States of America*, 107, 4797–4804.
Young, D. (1990). Do cicadas radiate sound through their ear drums? *Journal of Experimental Biology*, 151, 41–56.

Chapter 5
Evolutionary Origins of Primate Vocal Communication: Diversity, Flexibility, and Complexity of Vocalizations in Basal Primates

Elke Zimmermann

Abstract This chapter explores the importance of vocalizations in basal primates for modeling the evolutionary origins of primate social communication. A rich diversity in acoustic structures and signals, extending into the ultrasonic range, is used to govern various ecological and social challenges in their networks with varying degrees of social cohesiveness. Vocalizations convey information on specific emotions and on species, population, group, kin, and individual identity. Comparisons within the same phylogenetic group put forth the notion that natural selection limits cross-taxa vocal flexibility and favors universals in acoustic structure, whereas sexual and kin selection drive divergence. Bioacoustic research on vocal ontogeny depicted a babbling period with high vocal plasticity and an unexpected capability in adults to adapt their vocalizations to fluctuating background noise. Using a comparative bioacoustic approach, this chapter illuminates extraordinary cross-taxa variation in the acoustic spaces open for signal evolution and reveals that predation risk specifically shapes the acoustic space used. Comparison of vocal repertoire size, taken as an indicator of vocal complexity, reveals wide cross-taxa variation in call types and a striking variation in both solitary-ranging nocturnal and group-living diurnal species. The *phylogenetic hypothesis* and the *social complexity hypothesis* cannot fully explain this variation. Altogether, basal primates exhibit unique diversity, complexity, and flexibility of vocalizations for social communication, providing promising new avenues to trace the evolutionary origins of primate communication.

Keywords Bioacoustics • Comparative approach • Cross-taxa comparison • Emotional contagiousness • Evolution • Hearing • Phylogenetic hypothesis • Prosimian • Social complexity hypothesis • Vocal ontogeny

---

E. Zimmermann (✉)
Institute of Zoology, University of Veterinary Medicine,
Bünteweg 17, Hannover 30559, Germany
e-mail: elke.zimmermann@tiho-hannover.de

R.M. Quam et al. (eds.), *Primate Hearing and Communication*, Springer Handbook of Auditory Research 63, DOI 10.1007/978-3-319-59478-1_5

5.1 Introduction

5.1.1 Basal Primates and the Evolutionary Origins of Primate Vocalizations

Basal primates are important for understanding the evolutionary origins of primate vocalizations because they share features in their functional morphologies, brains, and life histories that are considered basal, ancestral, or plesiomorphic to those of monkeys and apes (Fig. 5.1) (Zimmermann and Radespiel 2013). The basal primates are mostly of the wet-nosed strepsirrhine clade, which makes up about one-third of all living primates (Kappeler 2012) and has two main groups: the Malagasy Lemuriformes (Yoder 1997; Mittermeier et al. 2010) with about 101 species and the African and Asian Lorisiformes with about twenty-eight species (Butynski et al. 2013). Also included in the basal primates is one group of the dry-nosed haplorhine clade—the Tarsiiformes—that includes about ten species (Groves and Shekelle 2010). Before recent molecular advances revealed that tarsiers are linked to the Haplorhini (e.g., Hartig et al. 2013), traditional primate taxonomic schemes grouped them with lemurs and lorises under the taxon Prosimii, based on phenotypic similarities (Fleagle 2013). Current phylogeographical and paleontological evidence point to the fact that the two major phylogenetic groups of the strepsirrhines, the Lorisiformes and the Lemuriformes, as well as the phylogenetic group of the Tarsiiformes (as the first representative of the haplorhines), emerged around the mid-Palaeocene, about 65–75 million years ago (Horvath et al. 2008; Matsui et al. 2009; Perelman et al. 2011).

This chapter gives a brief historical summary of research on basal primate vocalizations and then focuses on selected taxa for insight into the diversity, flexibility, and complexity of their acoustic communication systems. Cross-taxa variation in acoustic space is further explored with a comparative bioacoustic approach to assess the acoustic parameter space that is open for signal evolution. Likewise, a comparative approach is used to explore two of the most influential hypotheses explaining the evolution of primate acoustic communication: the *phylogenetic hypothesis* and the *social complexity hypothesis*. Finally, perspectives for future research will be addressed, particularly as relevant to modeling the evolutionary origins of primate vocal communication, including speech and language in humans.

5.1.2 Ecological and Social Challenges Shaping Vocal Communication in Basal Primates

Strepsirrhines and tarsiers were neglected groups of primates for a long time. Over the last few decades, however, empirically driven research on these groups has increased and revealed a unique species richness in different phylogenetic groups. Fascinating adaptations to various ecological and social challenges have been discovered, in particular, for nocturnal Malagasy lemurs (Mittermeier et al. 2010; Zimmermann and Radespiel 2014; Lehman et al. 2016), African galagos (Butynski et al. 2013), and Asian tarsiers (Groves and Shekelle 2010).

Fig. 5.1 Diversity of morphotypes in basal primates. Displayed are representatives of the Strepsirrhini: (**A**) the nocturnal Aye-aye (*Daubentonia madagascariensis*, Daubentoniidae); (**B**) the diurnal Coquerel's sifaka *(Propithecus coquereli*, Indriidae); (**C**) the nocturnal western woolly lemur (*Avahi occidentalis*, Indriidae); (**D**) the diurnal ring-tailed lemur *(Lemur catta*, Lemuridae); (**E**) the nocturnal gray mouse lemur (*Microcebus murinus*, Cheirogaleidae); (**F**) the nocturnal Milne-Edwards' sportive lemur (*Lepilemur edwardsi*, Lepilemuridae); (**G**) the nocturnal greater slow lori (*Nycticebus coucang*, Lorisidae); (**H**) the nocturnal northern lesser galago (*Galago senegalensis*, Galagidae); and as a representative of the basal Haplorhini, (**I**) the Lariang tarsier (*Tarsius lariang*, Tarsiidae). (Photographs: **A**, **C**, **D**, **E**, **G** by E. Zimmermann; **B**, **F** by M. Hokan; **H** by H. Zimmermann; **I** by S. Merker)

All tarsiers and most strepsirrhine species are arboreal and are adapted to life in complex, three-dimensional environments such as tropical rainforest, dry deciduous forest, and gallery/savannah forest (Table 5.1). Ring-tailed lemurs (*Lemur catta*) and some true lemurs (*Eulemur* spp.) represent the few basal primates that are semiterrestrial (Ward and Sussman 1979). With regard to moving and foraging in a three-dimensional world, Malagasy lemurs exhibit greater diversity in activity

Table 5.1 Vocal repertoire, acoustic spaces, and some eco-ethological features in basal primates[a,b]

Common and Latin name	VRS	AS (Hz)	Clade	Family	FU	BM (g)	PR	AC	SOS	HAB	LOC	IIC	EC	REF
Aye-Aye *Daubentonia madagascariensis*	6	174–9,200	Le	D	1	2775	l	N	s	DDF, RF	QRL	creee /screech		1–3
Western Woolly Lemur *Avahi occidentalis*	5	415–4,659	Le	I	3	912	m	N	p	DDF	VCL	whistle	AC	1, 4–7
Verreaux's Sifaka *Propithecus verrauxi*	6		Le	I	5,5	3250	m	D	g	DDF, HF, SF	VCL		chorus	1, 8–10
Silky Sifaka *Propithecus candidus*	10		Le	I	5,5		m	D	g	RF	VCL		chorus	1, 11
Indri *Indri indri*	9	106–1,237	Le	I	4	9,500	l	D	p	RF	VCL	contact-seeking whistle	duet/ chorus	1, 12–15
Milne-Edwards's Sportive Lemur *Lepilemur edwardsi*	9	633–5,641	Le	Le	1	931	m	N	p	DDF	VCL	screech	duet	1, 9, 16–18
Grey Mouse Lemur *Microcebus murinus*	10	8,000–40,000	Le	Ch	1	54	h	N	s	DDF, SF	QRL	whistle	AC	1, 19, 20
Coquerel's Giant Mouse Lemur *Mirza coquereli*	10	282–10,000	Le	Ch	1	310	m	N	s	DDF	QRL	whistle	AC	1, 21, 22
Western Fat-tailed Dwarf Lemur *Cheirogaleus medius*	8	4,000–34,000	Le	Ch	1	195	h	N	p	DDF	QRL	whistle	AC	1, 22, 23

Red-fronted Brown Lemur *Eulemur rufifrons*	8		Le	L	8,7	2250	m	C	g	DDF, RF	QRL			1, 24
Black Lemur *Eulemur macaco*	13		Le	L	10	2050	m	c	g	RF	QRL			1, 25
Crowned Lemur *Eulemur coronatus*	10	900–5,352	Le	L	6,2	1650	m	c	g	RF	QRL			1, 26
Black-and-white Ruffed Lemur *Varecia variegata*	12		Le	L	4,7	3350	l	d	g	RF	QRL	mew	chorus	1, 12, 27
Ring-tailed Lemur *Lemur catta*	22	240–8,320	le	L	14,5	2200	m	d	g	DDF	QRL, ST	infant contact call; mew	chorus	1, 28
Greater Slow Loris *Nycticebus coucang*	8	1,100–7,400	Lo	Lo	1	687		n	p	RF	QR	click, tsic	AC	29–31
Southern Lesser Galago *Galago moholi*	18	300–2,800	Lo	G	1	153		n	s	DF, SF	VCL	click	AC	32, 33
Northern Lesser Galago *Galago senegalensis*	17	100–8,100	Lo	G	1	313		n	s	DF, SF	VCL	zek	AC	32–34
Demidoff's Dwarf Galago *Galagoides demidovii*	8	400–1,600	Lo	G	1	58		n	s	DF, SF, RF	VCL	click		23, 32, 33, 35

(continued)

Table 5.1 (continued)

Common and Latin name	VRS	AS (Hz)	Clade	Family	FU	BM (g)	PR	AC	SOS	HAB	LOC	IIC	EC	REF
Thick-tailed Greater Galago *Otolemur crassicaudatus*	15		Lo	G	1	1175		n	s	DF, SF	QRL	click		32, 36–38
Northern Greater Galago *Otolemur garnetti*	11	215–1,539	Lo	G	1	836		n	s	DF, SF	QRL	click		32, 37, 39, 40
Spectral Tarsier *Tarsius tarsier*	8	3,000–12,000	T	T	1	112		n	p	RF	VCL	whistle	duet	41–43
Philippine Tarsier *Carlito syrichta*	7	9,752–22,000	T	T	1	128		n	s	RF	VCL			44–47

[a]Data References (*REF*): 1. Mittermeier et al. 2010; 2. Ramsier and Dominy 2012; 3. Stanger and Macedonia 1994; 4. Ramanankirahina et al. 2015; 5. Ramanankirahina et al. 2011; 6. Ramanankirahina et al. 2012; 7. Thalmann 2003; 8. Macedonia and Stanger 1994; 9. Petter and Charles-Dominique 1979; 10. Trillmich et al. 2004; 11. Patel and Owren 2012; 12. Geissmann and Mutschler 2006; 13. Maretti et al. 2010; 14. Pollock 1986; 15. Torti et al. 2013; 16. Méndez-Cárdenas and Zimmermann 2009; 17. Rasoloharijaona et al. 2003; 18. Rasoloharijaona et al. 2006; 19. Braune et al. 2005; 20. Zimmermann 2010; 21. Pages 1980; 22. Stanger 1993; 23. Stanger, unpublished data; 24. Pflüger and Fichtel 2012; 25. Gosset et al. 2003; 26. Gamba and Giacoma 2007; 27. Pereira et al. 1988; 28. Macedonia 1993; 29. Wiens and Zitzmann 2003; 30. Zimmermann 1985; 31. Zimmermann 1989b; 32. Butynski et al. 2013; 33. Zimmermann et al. 1988; 34. Zimmermann 1989a; 35. Charles-Dominique and Martin 1972; 36. Bearder 2007; 37. Clark 1988; 38. Tandy 1976; 39. Becker et al. 2003a; 40. Becker et al. 2003b; 41. Merker and Groves 2006; 42. Niemitz 1984; 43. Nietsch 2003; 44. Dagosto et al. 2003; 45. Neri-Arboleda et al. 2002; 46. Řeháková-Petrů et al. 2012; 47. Wright and Simons 1984

[b]Column headings: Vocal repertoire sizes (VRS, number of call types), acoustic space (AS) in basal primates in relation to *Clade*, taxonomic family (*Family*), foraging unit size (FU, median number of members in foraging unit), body mass (BM), predation risk (*PR*, according to Scheumann et al. 2007a; l = low, m = medium, h = high), activity pattern (AC: n, nocturnal; c, cathermal; d, diurnal), social system (SOS: s, solitary; p, pair; g, group; according to Kappeler 2012), habitat (HAB: DDF, dry deciduous forest; DF, dry forest; RF, rain forest; HF, humid forest; SF, savannah forest/gallery forest), locomotion (LOC: QRL, quadrupedal running and leaping; VCL, vertical clinging and leaping; QR, quadrupedal running; ST, semiterrestrial). IIC, infant isolation call; EC, emotional contagiousness (AC, antiphonal calling). Foraging unit size is according to the given references. The taxonomy is according to Groves and Shekelle (2010); Mittermeier et al. (2010); Butynski et al. (2013)

and locomotion patterns, diets, body sizes, and social life compared to African and Asian lorisiforms and tarsiers. Members of eight phylogenetic groups (genera) of the Malagasy lemurs and all members of the Lorisiformes (Table 5.1) are adapted to a nocturnal life, and all tarsiers are suggested to be secondarily nocturnal (descendants of a diurnal ancestor) (Martin and Ross 2005). Members of four phylogenetic groups of the Malagasy lemurs are diurnal (active solely during the day), and members of two groups became cathemeral (activity distributed approximately evenly throughout the 24 h of the daily cycle, or significant amounts of activity, particularly feeding and/or traveling, occur within both the light and dark portions of the 24-h cycle) (see Tattersall 1979).

Strepsirrhines show a rich variety of adaptations for locomotion (Fleagle 2013), from climbing and running quadrupedally (with no adaptations for leaping) to quadrupedal running, climbing, and leaping to specialized vertical clinging and leaping (Table 5.1). In contrast, all tarsiers are highly specialized for vertical clinging and leaping. Strepsirrhine primates, particularly lemurs, have developed a richly diverse diet, whereas all tarsiers are dietary specialists, feeding primarily on arthropods and small vertebrates. Habitat type, activity pattern, and the distribution of food are suggested to have major implications for foraging strategies, predator avoidance, sociality, infant care and, thereby, communication and cognition in primates (also see Ramsier and Rauschecker, Chap. 3; Snowdon, Chap. 6; Zuberbühler, Chap. 7).

Predation risk in basal primates is much higher than in monkeys and apes, in part because of their small to medium body size (Gursky and Nekaris 2007; Scheumann et al. 2007a; Fichtel 2012). Body sizes in lemurs vary about 300-fold from the smallest bodied extant primates, the mouse lemurs (30–60 g), to the largest bodied extant strepsirrhine, the indri (*Indri indri*, 9,500 g) (Table 5.1). In contrast, Lorisiformes (55–1,220 g) and Tarsiformes (110–140 g) exhibit much narrower body mass variations. Known predators of basal primates include carnivores, raptors, and reptiles such as snakes and monitor lizards. Since smaller bodied species have more predators than larger bodied ones, selection should favor crypsis, which also affects communication strategies.

In comparison to monkeys and apes, basal primates display a unique degree of flexibility in their patterns of social life (Table 5.1). All nocturnal strepsirrhines and tarsiers live in social networks with varying degrees of social cohesiveness, and they often have an individualized neighborhood system (as defined by Richard 1985; Radespiel 2000) in a transition state between solitariness and gregariousness (Müller and Thalmann 2000; Kappeler 2012; Lehman et al. 2016). Adults may travel or forage solitarily, but they often live in stable pairs, mixed-sex groups, or kin-related female groups in which members of social units are socially bonded and share unit-exclusive and unit-specific sleeping sites. If pair partners establish long-term pair bonds and share pair-exclusive home ranges and often sleeping sites but forage apart from each other, this social organization is termed a *dispersed pair*. Other nocturnal taxa have pair partners that also forage together, and that social organization is termed a *cohesive pair*. Cathemeral and diurnal lemurs are usually organized as cohesive pairs and families; only the diurnal ring-tailed lemurs are known to establish larger multi-male, multi-female groups, sometimes with more than one matriline. These varying degrees of sociality in basal primates pose specific social challenges that may shape the evolution of acoustic signaling systems.

Infant care patterns are more flexible in strepsirrhines and tarsiers than in monkeys and apes. In tarsiers and most nocturnal strepsirrhines, females leave their infants cached at safe shelters (e.g., tree holes, nests, or dense vegetation) for extended periods of time while foraging (Ross 2003; Zimmermann and Radespiel 2013). In contrast, two nocturnal genera of strepsirrhines, as well as most cathemeral and diurnal lemurs, maintain close body contact with clinging infants during foraging. These distinctions in infant care patterns are linked to specific challenges in the social and ecological world that these primates have overcome by evolving appropriate communication strategies in the acoustic domain. Taken as a whole, their basal phylogenetic status, substantial ecological diversity, and unique adaptations make basal primates excellent candidates for exploring the evolutionary origins of primate acoustic communication.

5.1.3 History of Research on Acoustic Communication in Basal Primates and the Discovery of Ultrasound in Vocalizations

The study of the origin and evolution of display behavior began more than 150 years ago with Darwin's groundbreaking work on the *Expression of the Emotions in Man and Animals* (Darwin 1872). However, it was not until the middle of the 20th century that systematic research on the vocal expressions of basal primates began. In his influential work on the calls and facial expressions among primates, Andrew (1963) was the first to compare the vocalizations of primates from different clades, including some strepsirrhines and tarsiers, relying on new technologies that enabled researchers to record, visualize, measure, and quantify vocalizations.

More than ten years later, Klopfer (1977) stated in his review that although several descriptions of the vocalizations of basal primates were available, systematic studies on their roles in communication were not yet published. About the same time, research teams around the world were starting the first field projects on the ecology and behavior of free-ranging strepsirrhines in western, eastern, and southern Africa (Bearder and Doyle 1974; Petter et al. 1977; Nash and Harcourt 1986) and on the island of Madagascar (Jolly 1966; Charles-Dominique and Martin 1972), as well as projects studying tarsiers in the Philippines (Niemitz 1984), with the goal of illuminating the biology of these hitherto neglected primates. Petter and Charles-Dominique (1979) were the first to provide a catalogue of spectrographically presented sounds of all major groups of strepsirrhines categorized according to their presumed functions. Shortly afterwards, Zimmermann (1981) presented the first empirical evidence for ultrasound (>20 kHz) in primate vocalizations in the infant isolation calls of the Lorisiformes. Cherry et al. (1987) then documented ultrasound production by Lemuriformes, and recently, Ramsier et al. (2012b) described ultrasound production by Tarsiiformes.

5.2 Diversity and Flexibility of Vocalizations

In recent decades, field and laboratory research has revealed that basal primates have evolved a remarkable diversity of vocalizations with regard to acoustic structure, reflecting a much broader frequency range that was available for signal evolution than in monkeys and apes.

5.2.1 General Remarks on the Cross-Taxa Diversity of Vocalizations

As outlined in Sect. 5.1.2, basal primates differ more in activity patterns and social cohesiveness within their social networks than monkeys and apes; nevertheless, they may establish individualized and permanent social bonds. Consequently, these primates face fundamental challenges to survive and reproduce, such as avoiding predators, limiting the potential for costly conflicts between competitors within and, if in sympatry, between species, coordinating movements in time and space while searching for common sleeping and/or feeding sites and reunions of mothers and infants, mates, pairs or groups, in particular when members are dispersed during foraging. In response to these challenges, a rich diversity of vocalizations evolved for acoustic signaling in basal primates (Table 5.1 and Sect. 5.2.2).

To the human ear, the vocalizations of basal primates sound fairly mysterious and even atypical for primates. The series of short, broadband, and high-frequency *click*, *tsic*, or *zek* noises (Fig. 5.2) used as infant isolation calls by bushbabies and lorises sound like high-frequency human tongue clicks, whereas the series of short, broadband, and low-frequency clicks given during grooming in lemurs (Scheumann et al. 2007b) sound like a purring cat. A series of low-frequency,

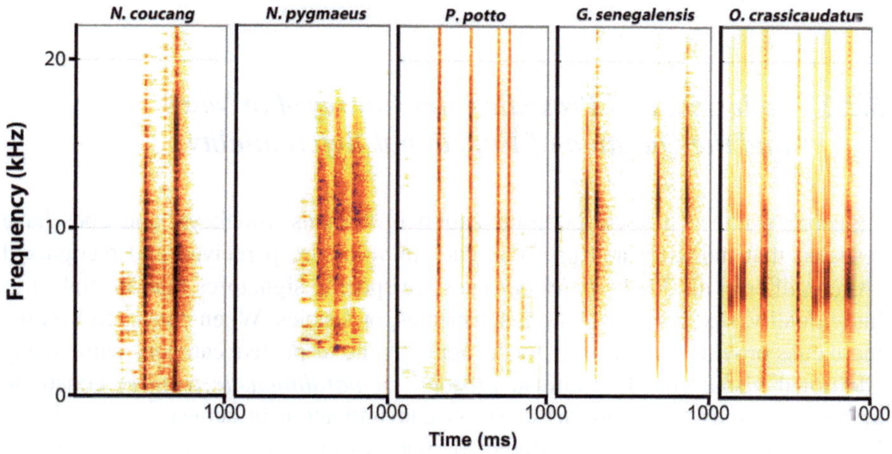

Fig. 5.2 Infant isolation calls of basal primates. Displayed are the major similarities in acoustic structure within the phylogenetic group of the Lorisiformes

broadband, and noisy *grunts* and *growls* (e.g., Zimmermann 1995a), which sound like hissing cats or growling/snarling dogs and acoustically mimic large and dangerous carnivores, occur across all nocturnal strepsirrhines and are associated with threatening situations. The *cries*, *whines*, or *barks* of the bushbabies (*Otolemur, Galago*), which are used during advertisement or in alarm situations, resemble human infant crying, whining, or a dog barking. The *mews* of ring-tailed lemurs or ruffed lemurs (*Varecia variegata*), which govern reunions, resemble the moaning meowing of cats, whereas the melodious low-frequency *songs* and *duets* in indris and sportive lemurs (*Lepilemur*), which are given in a territorial defense context, resemble howling in coyotes or wolves. The high-frequency/ultrasonic, multi-syllabic, and rapid frequency-modulated *trills* and *chirps* of mouse lemurs (*Microcebus*) and the *songs* of tarsiers (*Tarsius, Carlito*), which are produced in a reunion or mating context, sound like chirps in mating songs of crickets and grasshoppers. These sounds are barely perceptible or imperceptible to the human ear, but they resemble singing in birds when transferred to the human hearing range. The tonal, high-frequency, and almost constant-frequency *whistles* characteristic of cohesion or disturbance/alarm contexts in lorises (*Perodicticus, Nycticebus, Loris*), woolly lemurs (*Avahi*), tarsiers (*Tarsius*), and small-bodied lemurs (*Cheirogaleus, Microcebus*) sound quite similar to whistling sounds in humans. Note, however, that sounding like these other sound sources does not mean that the sound production mechanisms are comparable (also see Fitch 2010).

As in monkeys and apes, the high diversity of vocalizations in basal primates provides the potential for recognition even across distance and, consequently, for the gathering of dispersed members of social units at a particular site and a distinct time. A recent study in group-living lemurs pointed to the fact that vocal exchanges indicate the strength in social bonds between group members and allow grooming at a distance through vocal signaling networks (Kulahci et al. 2015). Furthermore, specific differences in acoustic signaling and recognition systems may represent an efficient premating isolation mechanism, contributing to species cohesiveness when living in sympatry (Zimmermann 2016).

5.2.2 Indexical and Prosodic Cues Conveyed in Voice and the Question of Functional Referentiality

Research studies on vocal communication in humans, monkeys, and apes have revealed that indexical and prosodic cues in voice are perceived and recognized (Altenmüller et al. 2013). *Indexical cues* are specific signatures that are linked to individuality, sex, kin, population, subspecies, or species. When conveyed vocally, these cues provide the substrate for recognizing the respective category without any additional visual cue. In contrast, *prosodic* or *paralinguistic features* constitute *affective prosody* (i.e., patterns of stress and intonation in acoustic expressions). Humans make use of these paralinguistic features in affective prosody not only to express what and how they feel but also to evoke or interpret these feelings in others, as well as to think about their own and others' feelings as they make decisions. Although humans are unique in the goal-directed usage of prosodic cues in voice for decision making in groups and social manipulation, the acoustic communication of emotions and emotional contagiousness evoked by voice also are an important and potentially universal feature of monkey and ape societies (e.g., for governing group movements and reunions) (Altenmüller et al. 2013). Furthermore, monkeys and apes often vocally express emotions linked to specific individualized contexts, potentially leading to context-specific calls. When these calls are then responded to in a context-specific way without any additional visual cue, they are often termed *functionally referential* (see Wheeler and Fischer 2015; Snowdon, Chap. 6; Zuberbühler, Chap. 7). To what extent do basal primates share these building blocks in acoustic communication with monkeys and apes?

Current research on basal primates has revealed that these primates also exhibit a rich diversity of vocalizations in which indexical cues are conveyed, as is known for monkeys and apes. Vocal indexical cues are widespread across both groups of basal primates (the strepsirrhines and the tarsiers) and thus are present irrespective of social system and phylogeny. Individual-specific acoustic signatures are described from calls of all major phylogenetic groups of the strepsirrhines, such as the nocturnal, solitary-ranging mouse lemurs (Leliveld et al. 2011) and bushbabies (Kessler et al. 2014), the nocturnal, dispersed pair-living sportive lemurs (Rasoloharijaona et al. 2006), the cohesive pair-living woolly lemurs (Ramanankirahina et al. 2015), the cathemeral group-living crowned lemurs (*Eulemur coronatus*, Gamba and Giacoma 2007) and red-bellied lemurs (*Eulemur rubiventer*) (Gamba et al. 2011; Gamba et al. 2012b), the diurnal pair-living indris (Giacoma et al. 2010; Torti et al. 2013), and the group-living silky sifakas (*Propithecus candidus*) (Patel and Owren 2012) and ring-tailed lemurs (Macedonia 1986; Oda 2002). Individual-specific variation of long-distance calls was also revealed in Philippine tarsiers (*Carlito*, previously *Tarsius, syrichta*) (Řeháková-Petrů et al. 2012).

Sex-specific variation in acoustic dimensions also occurs in both groups of the basal primates, such as in sex-specific syllables of duetting songs in the nocturnal Milne Edwards' sportive lemur (*Lepilemur edwardsi*) (Rasoloharijaona et al. 2006), the cathemeral crowned lemur (Gamba and Giacoma 2007), the diurnal indri (Giacoma et al. 2010; Torti et al. 2013), and nocturnal tarsiers (Niemitz et al. 1991; Nietsch 1999; Merker et al. 2010). The first evidence for vocal kin signatures and vocal kin recognition in basal primates was recently reported in gray mouse lemurs (*Microcebus murinus*) (Kessler et al. 2012, 2014).

Group, population, subspecies, and species-specific signatures in vocalizations are present in both groups of the basal primates. They are documented for strepsirrhines, such as the nocturnal, solitary-ranging lesser (e.g., Zimmermann 1990; Ambrose 2003) and larger bushbabies (Clark 1988; Masters 1991); the nocturnal, solitary-ranging lemurs (Hafen et al. 1998; Braune et al. 2008); the nocturnal, dispersed pair-living lemurs (Méndez-Cárdenas et al. 2008); the cathemeral group-living lemurs (Gamba et al. 2012a); the diurnal pair-living lemurs (Gamba et al. 2011) and group-living lemurs (Macedonia and Taylor 1985; Gamba et al. 2012b); and tarsiers (Nietsch 1999; Merker et al. 2009; Burton and Nietsch 2010).

In sum, vocal indexical cues are present in monkeys, apes, and humans (see Snowdon, Chap. 6; Zuberbühler, Chap. 7) but also in basal primates. Since these vocal cues are described from different phylogenetic clades (Zimmermann et al. 2013), such as tree shrews, bats, rodents, pigs, elephants, carnivores, dolphins and whales, vocal indexical cues likely represent universal features across all mammals, providing an important basis for sensory exploration, such as for the acoustic recognition of the respective categories. To what extent these acoustic cues for species, subspecies, population, group, sex, or kin recognition are used in the respective social system of the different phylogenetic groups of basal primates is an interesting but rather unexplored area in bioacoustic research.

Prosodic cues (referring to emotional/motivation states or potentially to particular categories in the external world, such as type of predator) are not only present in monkeys and apes but are also described for strepsirrhines, whereas nothing is known so far for tarsiers. As put forth by Morton (1977) and refined by August and Anderson (1987), motivation-structural rules should govern vocal production and favor the emergence of cross-taxa universals in vocal acoustics based on physiological constraints. Cross-taxa universals, conveying varying degrees of short-term stress or response urgency, were shown in the temporal pattern of different strepsirrhine vocalizations in which higher affect intensity, arousal, or response urgency was expressed by higher calling rates (Zimmermann et al. 2013). Likewise, a first cross-taxa comparison of the most commonly used calls in the vocal repertoire of lesser mouse lemurs of the genus *Microcebus* (Zimmermann 2016) pointed to cross-taxa universals in the acoustic contour of calls used either in the agonistic, pain, or alarm/disturbance context, whereas major species-specific divergences occurred in the acoustic contour of vocalizations used in the affiliation context (mating, group reunion, mother-infant reunion). These findings provide some support for the refined Morton motivation-structural rules and suggest that natural selection (predation, physiological constraints linked to short-term stress) limits cross-taxa vocal flexibility and favors universals in acoustic structure, whereas sexual and kin selection drives cross-taxa flexibility and divergence specifically.

In contrast to monkeys and apes, for basal primates there is only sparse empirical information on the significance of context-specific variation in vocalizations, which has been discussed as a prerequisite for the evolution of functional referentiality in the acoustic domain. Experimental studies displaying dummies of ground and aerial predators to groups of diurnal and semiterrestrial group-living ring-tailed lemurs revealed that predator dummies elicited acoustically distinct alarm calls linked to distinct predator classes but not response urgency (Pereira and Macedonia 1991), a result that was further confirmed by playback experiments within the respective alarm call category. In the diurnal, arboreal ruffed lemurs, however, comparable experiments failed to support a referential signaling or a response urgency system and pointed instead to a generalized alarm call system (Macedonia 1990). Playback studies in two other diurnal, arboreal group-living lemurs (red-fronted lemurs, *Eulemur rufifrons*; Verraux's sifakas, *Propithecus verrauxi*) provided evidence for urgency-based changes in the acoustic structure of alarm calls and mixed evidence for a referential alarm call system characterized by functionally referential calls for diurnal raptors but not for carnivores (Fichtel and Hammerschmidt 2002; Fichtel and Kappeler 2002). In contrast, nocturnal strepsirrhines and tarsiers are thought to behave either cryptically or produce generalized alarm calls that may serve to communicate with predators (Scheumann et al. 2007a; Fichtel 2012). Thus, in basal primates a combination of activity pattern, phylogeny, and degree of sociality seems to shape the evolution of functional referentiality in acoustic signaling systems linked to predation.

Outside of the alarm and disturbance context, there is limited empirical information on context specificity in basal primates. In the grey mouse lemur, an acoustic and contextual analysis revealed context specificity in infant's calls, conveying the infant's need and response urgency (Scheumann et al. 2007b), which is potentially used for decision making in mothers. Cross-taxa comparisons of the acoustic structure of infant isolation calls of strepsirrhines (e.g., given when young infants suddenly get isolated from their foraging mothers) point to an interesting phenomenon. In this context, all infants of the Lorisiformes (i.e., lorises and galagos) produce isolation calls with a universal acoustic structure (loud series of short, broadband *clicks*, *tsics*, or *zeks*) irrespective of body size or mode of locomotion (see Sect. 5.2.1, Table 5.1, and Fig. 5.2). These transient broadband sounds, lasting only for a few milliseconds, may provide useful localization cues for mothers trying to locate their lost offspring (Newman et al. 2013; Brown and Waser, Chap. 4). In contrast, infants of the Lemuriformes emit isolation calls with a quite distinct universal structure: a series of much longer, tonal, and frequency-modulated *whistles*, *screeches*, or *mews* (Table 5.1; Fig. 5.3) (Newman et al. 2013). Comparable phylogenetic constraints within the same phylogenetic group, most likely related to the fact that infants have to be recognized at some distance by their mothers while not being detected by predators, may have shaped the evolution of these two acoustically divergent, but potentially equally beneficial, signaling strategies in mother-infant communication.

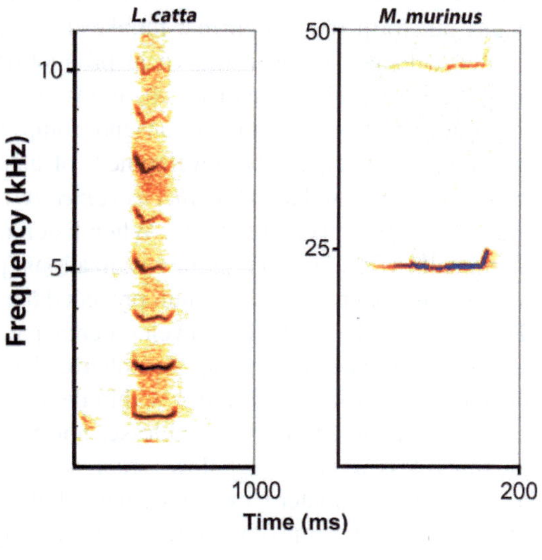

Fig. 5.3 Infant isolation calls of basal primates. Displayed are the major similarities in acoustic structure within the phylogenetic group of the Lemuriformes

Antiphonal calling, duetting, and chorusing in the context of intragroup cohesion and/or intergroup spacing is present across both groups of basal primates, suggesting that not only vocalizations of monkeys and apes but also those of basal primates convey emotional contagiousness. For example, nocturnal solitary-ranging females of the golden brown mouse lemurs (*Microcebus ravelobensis*) use antiphonal calling for governing group reunion (Braune et al. 2005), as do pair partners of nocturnal pair-living western woolly lemurs (*Avahi occidentalis*) when visually isolated at short distance (Ramanankirahina et al. 2015). Nocturnal, dispersed pair-living Milne-Edwards' sportive lemurs (Méndez-Cárdenas and Zimmermann 2009) as well as diurnal pair-living indris and group-living ruffed lemurs (Pollock 1986; Geissmann and Mutschler 2006), silky sifakas (Patel and Owren 2012), and ring-tailed lemurs (Bolt 2013) display loud duetting or chorusing (simultaneous calling of group members), often at resource sites at the beginning of their activity period to advertise pair/group strength and limit intergroup conflicts. Some syllables in these complex, emotionally contagious calls may also function as loud alarm calls for advertising the detection of disturbances or predators. The presence of acoustically induced emotional contagiousness in basal primates, including variations in vocal signaling behavior that are used to synchronize activities within a social unit and used to address conspecifics or predators that are currently out-of-sight, suggests that the evolution of acoustically induced emotional contagiousness in primates may represent a universal trait across different phylogenetic groups, originating deep in primate or even mammalian phylogeny.

5.2.3 Flexibility of Vocalizations Within a Species During Development and Adulthood

Previous research on the ontogeny of acoustic communication in nonhuman primates put forth the notion that monkeys and apes, in contrast to humans, show limited evidence for learning vocal production, whereas learning influences the usage and comprehension of vocalizations (e.g., Fitch 2010). Empirical research on the ontogeny of vocal production and comprehension for basal primates is just beginning to emerge. The development of alarm call recognition in free-ranging infant Verraux's sifakas (*Propithecus verreauxi*) was studied using playback experiments in the field (Fichtel 2008). When listening to conspecific and heterospecific alarm calls as well as non-alarm vocalizations (parrot song), the ability to discriminate alarm from non-alarm stimuli preceded the appearance of adult-like responses. These findings coincide with developmental patterns described for various monkeys and apes (see Snowdon, Chap. 6; Zuberbühler, Chap. 7). Comparing the development of vocal patterns in infants within the Lorisiformes revealed prominent divergences between the highly vocal lesser northern galago (*Galago senegalensis*) and the much less vocal greater slow loris (*Nycticebus coucang*) (Zimmermann 1995a, b). Calls of the greater slow loris emerged almost fully developed from the beginning; in contrast, some of the infant calls of galagos underwent prominent structural

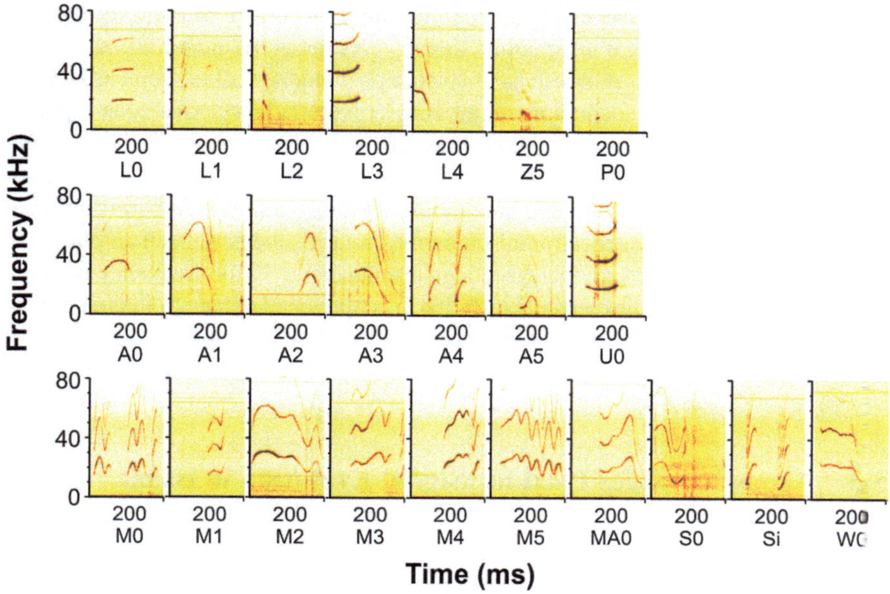

Fig. 5.4 Spectrograms of infants of the grey mouse lemur studied in the reunion context. Sounds produced suggest a babbling-like period early in development. An example of the high variation of syllable shapes present in vocal streams of sucklings is provided here by representative spectrograms (200 ms total duration). **Top row:** *L0 to P0*, quasi-constant frequency (*L0*) or single up- or downward frequency-modulated shape (*L1–L4*, *Z5*, *P0*); **Middle row:** *A0–A5*, *U0*, single up- and downward frequency modulated shape (inverted U or U shape); **Bottom row:** *M0–M5*, *MA0*, *S0*, *Si*, *W0*, multiple frequency-modulated shapes. Note that most vocalizations are in the ultrasonic range and inaudible to humans

changes similar to the pattern revealed for the grey mouse lemur during development (Zimmermann 1995b). The vocal ontogeny of the latter species was studied systematically during different phases of development in a reunion context in which the mother returned to her infants (Fig. 5.4). When the syllable structure within a call was analyzed, infants showed high variation of syllables within a vocal stream, which stabilized later in age (Zimmermann 1991; Linn 2013).

Similar phenomena were described for pygmy marmoset monkeys (*Cebuella pygmaea*) (see Snowdon, Chap. 6) and bats (Knörnschild 2014) and were taken as evidence for a babbling period. At present, it is unknown to what extent these phenomena in mouse lemurs are linked to maturational processes or to auditory-controlled vocal learning as postulated for bats. That calls in adult mouse lemurs are more flexible than previously assumed was shown by colony- and deme-specific variations in affiliation calls, which could not be explained merely by genetic distance (Hafen et al. 1998; Zimmermann and Hafen 2001), as well as a recent study on the effect of fluctuating ambient noise.

Fluctuating ambient noise often disturbs signal transmission and poses severe problems for primate acoustic communication. Under these conditions, humans are known to adapt their own voice by integrating auditory feedback with the vocal motor system (i.e., auditory-vocal integration), thereby enhancing call amplitude and modifying the spectrotemporal dimensions of voice, a phenomenon called Lombard speech (Junqua 1996). Similar adaptations improving signal efficiency have been reported for some monkeys (Brumm et al. 2004; Hage 2013; Ackermann et al. 2014). A study exploring whether components of Lombard speech emerge early in primate evolution examined calling behavior in male grey mouse lemurs in an induced courtship paradigm (Schopf 2013). Evidence for a Lombard effect (i.e., increased call amplitude in noise) was found, as was an increased signal redundancy in the frequency domain. More research on this unexpected capability to adapt vocalizations to fluctuations in the acoustic or socio-acoustic environment is needed to understand how the vocal capacity of basal primates fits within the evolution of primate communication.

5.3 Acoustic Spaces Used for Signal Evolution in Basal Primates

5.3.1 The Concept of the Acoustic Space

Like other arboreal mammals, strepsirrhines and tarsiers face major ecological challenges when communicating with sound (Zimmermann 2016). The message has to be loud and clear enough to be transmitted over some distance to conspecific recipients without too much distortion. Ambient noise from abiotic (e.g., wind, rain, thunderstorms) or biotic (e.g., cicadas, crickets, frogs, birds, mammals) sources and habitat structure (e.g., density of vegetation and resulting reflections and reverberations) may mask or modify the transmitted sound (see Brown and Waser, Chap. 4). Similarly, prey and predators can eavesdrop on the sounds emitted.

With these considerations in mind, it is generally postulated that natural selection drives acoustic signal evolution toward a multidimensional effective "acoustic window" or "acoustic space" to optimize transmission (Wilkins et al. 2013). This notion was also put forth by the *acoustic adaptation hypothesis* (Morton 1975) or the broader *sensory drive framework* (Endler 1992), describing the coevolution of signals, sensory systems, and microhabitat choice. An acoustic window, niche, or space thereby outlines the acoustic parameter space that is available for signal evolution for a given taxon in a given habitat (Wilkins et al. 2013).

5.3.2 Cross-Taxa Variation in Acoustic Spaces Used for Communication

When taking fundamental frequency ranges of the most commonly used tonal call types of adults as an indicator for acoustic space, basal primates use a much broader range of acoustic spaces for communication than monkeys and apes

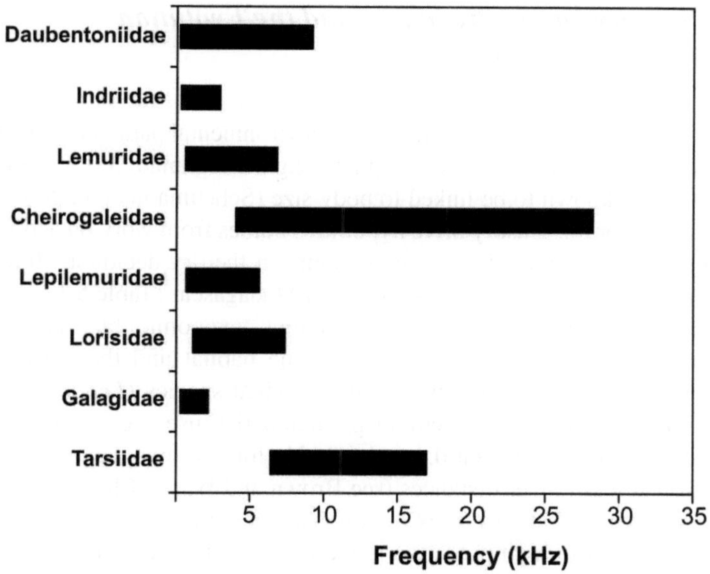

Fig. 5.5 Acoustic space used for communication in eight taxonomic families of basal primates. Acoustic spaces were determined by fundamental frequency ranges of tonal calls across the vocal repertoire for the species. A median value was calculated when there was more than one species studied in that taxonomic family (see Table 5.1)

(Table 5.1; Fig. 5.5). All bushbabies (Galagidae) communicate in a relatively low frequency space. This is true for small-bodied and medium-bodied taxa. Both the medium-bodied lorises (Lorisidae) and, in particular, the small-bodied tarsiers (Tarsiidae) use a much higher frequency acoustic space. The loris and tarsier species studied are nocturnal, live either solitarily or in pairs, inhabit both rain and dry deciduous forests, and show contrasting locomotion styles. Malagasy lemurs exhibit the broadest range of acoustic niches in basal primates. The medium-bodied and larger bodied lemurs (including Daubentonidae, Lepilemuridae, Indridae, and Lemuridae) use a low-frequency acoustic space for communication. The species studied in these groups are nocturnal, cathemeral, or diurnal; range either as solitary individuals, as a pair, or group; and inhabit a broad variety of habitats where they move by specific modes of locomotion.

In contrast, small-bodied lemurs of the family Cheirogaleidae, which are nocturnal, solitary, or pair-living and show a rather unspecialized mode of locomotion, use a medium- to high-frequency/ultrasonic space for communication. These findings suggest that the variation in acoustic space among basal primates is independent of activity pattern, ambient noise (reflected in habitat type), social system, and phylogenetic distance. The fact that small-bodied lemurs (belonging to the strepsirrhines) and small-bodied tarsiers (belonging to the haplorhines) show comparable high-frequency acoustic niches for communication points to comparable constraints across both groups.

5.3.3 Sensory Drive, Predation, and the Evolution of Acoustic Spaces

The *sensory drive hypothesis* postulates that environmental parameters, such as predation, drive the evolution of social signals (e.g., Boughman 2002). Furthermore, predation risk is known to be linked to body size (Scheumann et al. 2007a; Fichtel 2012). Support for the sensory drive hypothesis comes from work on acoustic niche partitioning in the nocturnal lemur community in the dry deciduous forest of the National Park Ankarafantsika in northwestern Madagascar (Table 5.1).

Five nocturnal species from three different taxonomic families (Indriidae, Lepilemuridae, Cheirogaleidae) share the same habitat and thus face the same ambient-noise conditions. The two medium-bodied species (*Lepilemur edwardsi* and *Avahi occidentalis*) with medium predation risk use a comparable low-frequency acoustic niche between 0.4 and 6.0 kHz for communication, favoring call transmission across longer distances (see Brown and Waser, Chap. 4). In contrast, the three smaller bodied species (*Cheirogaleus medius*, *Microcebus murinus*, and *M. ravelobensis*) with high predation risk communicate in a high-frequency/ultrasonic acoustic niche, between 4 and 40 kHz. Calls in this frequency range have more limited transmission potential, but they also have more limited detection potential by raptors, which are described as major predators for these species (Goodman et al. 1993).

Small-bodied lemurs, bushbabies, lorises, and tarsiers are subjected to two major categories of predators: raptors and carnivores. Of these, raptors cannot hear frequencies above 10 kHz, while the carnivores may have difficulties in locating their prey by the high-frequency signals that are barely perceived (for audiograms and auditory sensitivities of predators, see Fay 1988).

Acoustic spaces of small-bodied basal primates extend to the high-frequency/ultrasonic range with the exception of the smallest bodied galagids (*Galagoides*), which use a low-frequency acoustic space for communication that is comparable to medium- and larger bodied strepsirrhines. A potential explanation for this exception may be lower predation risk and/or different ecological settings in tropical African forests as compared to Malagasy and Asian forests. Indeed, predator pressure from owls appears to be less imposing in Africa than in Madagascar. In South African lesser bushbabies, for example, predation rate was estimated to be about 15% per year (Cheney and Wrangham 1987), whereas in the Malagasy mouse lemurs, predation by merely one raptor species, the barn owl, was assessed to be greater than 25% of the population per year (Goodman et al. 1993).

All in all, these findings support the sensory drive hypothesis (e.g., Boughman 2002) for the evolution of acoustic spaces in basal primates. Due to lower predation risk, medium- and larger bodied species are able to explore the low-frequency acoustic niche that favors long-distance communication; however, small-bodied lemurs and tarsiers are constrained for long-range communication because high predation pressure forces them to shift to the high-frequency and ultrasonic range. This high-frequency range is beneficial since it allows communication in a private communication channel without eavesdropping by predators (see

Zimmermann 2016). Comparative functional morphological and physiological research in basal primates may help to explore the extent to which there are specializations in the peripheral vocal motor system, the ear, or the central auditory system that facilitate high-frequency sound production and hearing, as is known for bats (Simmons et al. 2008; Veselka 2010).

5.4 Variation in Vocal Complexity across Basal Primates

5.4.1 Assessment of Vocal Complexity

In animal communication research, vocal repertoire size is measured by the number of call types that have been described and is often taken as a proxy for vocal complexity in a given species (e.g., McComb and Semple 2005). Vocal complexity can be assessed in different ways, such as by an information theoretical approach or by a typological approach in which the number of acoustically distinct signals is determined and their roles in regulating social interactions are assessed (e.g., Freeberg et al. 2012). As a first approach, the number of acoustically distinct calls within the repertoire of a species is taken here as a proxy for vocal complexity, which is then used to explore two of the most influential hypotheses proposed to explain the evolution of acoustic communication in mammals: *the phylogenetic hypothesis* (e.g., Harvey and Pagel 1991; Ord and Garcia-Porta 2012) and the *social complexity hypothesis* (e.g., McComb and Semple 2005; Freeberg et al. 2012; Pollard and Blumstein 2012). Data on the vocal repertoire size of strepsirrhines and tarsiers were taken from published information (Table 5.1). For the latter, only studies in which the adult vocal repertoire was systematically described, based on acoustic structure and context, were included.

5.4.2 The Phylogenetic Hypothesis and the Evolution of Vocal Complexity

The *phylogenetic hypothesis* argues that the vocal repertoire of a species is largely determined by its phylogeny and diversity between taxa is largely due to neutral mechanisms. Specifically, signal complexity between closely related members of a taxonomic group should be more similar to each other (based on their shared evolutionary history) than to signals of genetically more distant phylogenetic groups. There are empirical data on the vocal repertoires of basal primate taxa that can be used to explore the phylogenetic hypothesis.

Vocal complexity in the Strepsirrhines as a group varies from 5 to 22 call types: 8–18 in the Lorisiformes (six species) and 5–22 in the Lemuriformes (fourteen species) (see Table 5.1; Fig. 5.6). Furthermore, the variation in vocal repertoire size of the Lorisiformes is almost as high as in the Lemuriformes. Within the

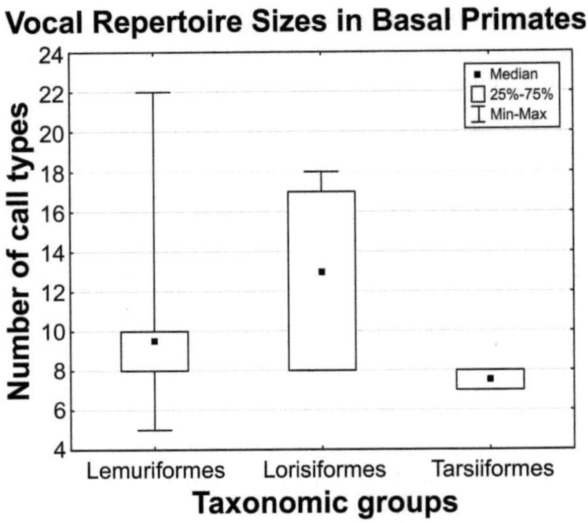

Fig. 5.6 Variation of vocal repertoire sizes for the three phylogenetic groups: the Lemuriformes (14 species), the Lorisiformes (6 species), and the Tarsiiformes (2 species). The size of the vocal repertoire of a species was taken as a proxy for vocal complexity. Findings revealed that the variation in vocal repertoire size is not significantly different (Mann-Whitney U test) between the Lemuriformes and the Lorisiformes, thus phylogenetic group does not predict repertoire size variation. However, it is remarkable that the Lorisiformes showed almost the same variation as the Lemuriformes. Findings suggest that vocal complexity in primates derives from a basal vocal repertoire of 7–8 call types in the last common ancestor of Strepsirrhini and Haplorhini

Lemuriformes, the variation of vocal complexity ranges between 5 and 10 call types in the Indriidae (four species), 8–10 in the Cheirogaleidae (three species), and 8–22 in the Lemuridae (five species).

Vocal repertoire size variation in the basal Haplorhines (Tarsiiformes) lies within the variation of the Strepsirrhines (7 and 8 call types in two species). Thus, at present, findings do not suggest that phylogenetic relatedness shapes the variation in vocal complexity of basal primates. Consequently, the currently available dataset on strepsirrhines and tarsiers, while admittedly small, does not support the phylogenetic hypothesis for the evolution of vocal complexity in the basal primates Furthermore, findings suggest very tentatively that vocal communication in primates derives from a common ancestor with a vocal complexity of 7 to 8 call types, the smallest shared vocal repertoire of both the Strepsirrhini and the Haplorhini.

5.4.3 The Social Complexity Hypothesis and the Evolution of Vocal Complexity

The *social complexity hypothesis* predicts that increases in social complexity may have driven the evolution of communication, including human language (e.g., McComb and Semple 2005; Freeberg et al. 2012), since more complex groups need more complex communication systems to govern interactions and relationships between group members. How social complexity can be reliably determined is debated. Some researchers rely on group size (e.g., Kappeler 2012) or foraging unit size (e.g., Ramsier et al. 2012a; Ramsier 2013). In contrast, others (e.g., Thierry et al. 2000; Freeberg et al. 2012) suggest that more complex distinctions need to be considered, and they characterize social complexity by social unit size (stable in space and over time), unit density, nature of member roles, egalitarian/despotic structure, and/or interunit density. Diurnal and cathemeral strepsirrhines (both belonging exclusively to the Lemuriformes) exhibit cohesive social systems, mostly in pairs or small family groups with long-term social bonds, and egalitarian to despotic social structures (Table 5.1) (Müller and Thalmann 2000; Kappeler 2012). While nocturnal strepsirrhines and tarsiers were previously often considered solitary, ongoing field research has revealed more flexibility with regard to forming groups and stable social relations within groups, making assessments of social complexity challenging (see Sect. 5.1.2) (Lehman et al. 2016). For the present purposes, group size (according to Kappeler 2012) and foraging unit size (according to Ramsier et al. 2012a; Ramsier 2013) were taken as a proxy of social complexity.

The vocal repertoire size in basal primates displays a broad variation, ranging from 5 to 22 call types. When vocal complexity is linked to social group size, the outcome is quite interesting (see Fig. 5.7). Variation ranges from 6 to 18 call types in nocturnal, solitary-ranging strepsirrhines and the solitary-ranging tarsier species, from 5 to 9 call types in pair-living nocturnal and diurnal strepsirrhines and one tarsier species, and from 6 to 22 call types in the group-living cathemeral and diurnal strepsirrhines (exclusively lemurs). Thus, it is remarkable that the variation in vocal repertoire size of nocturnal and diurnal pair-living lemurs is comparably low but that of nocturnal solitary-ranging strepsirrhines is nearly as large as that of the diurnal group-living lemurs. While social system does not seem to affect vocal repertoire size (see Fig. 5.7), foraging unit size is weakly correlated (see Fig. 5.8).

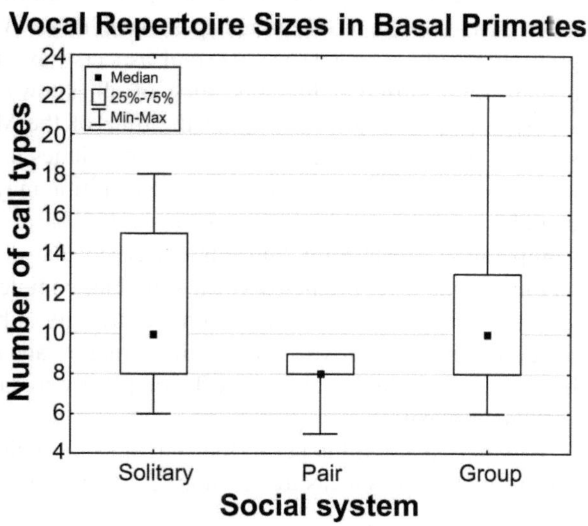

Fig. 5.7 Variation of vocal repertoire size in relation to social system (*solitary:* 9 solitary-ranging species; *pair:* 6 pair-living species; *group*: 7 group-living species). Kruskal-Wallis ANOVA for the effect of group size (df = 2; $F = 1.3543$; p not significant) showed that social system does not significantly affect vocal repertoire size. Note that size variation in the solitary-ranging group is almost as high as in the group-living basal primates

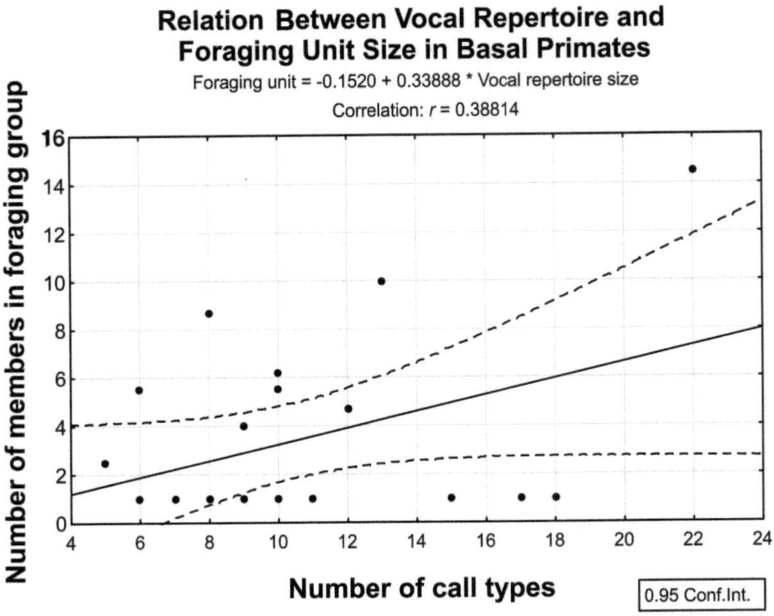

Fig. 5.8 Relationship between vocal repertoire size and foraging unit size in basal primates (22 species). A multiple regression analysis revealed that the number of call types is only weakly related to the number of members in foraging units (df = 1.2, $F = 3.5472$, $p < 0.0742$). *Conf. int.*, confidence interval shown as dashed lines

Thus, the currently available dataset for basal primates does not fully support the social complexity hypothesis either. These findings contrast with those of McComb and Semple (2005), who found that increases in the size of the vocal repertoire among forty-two nonhuman primate species (including some basal primates in addition to monkeys and apes) were linked to both the time spent on social grooming and on group size. Several aspects have to be taken into account to explain these different findings. First of all, knowledge of social group size in strepsirrhines was quite limited at the time of the McComb and Semple (2005) study so the values for group size in strepsirrhines used in their study differ from our current understanding. Another consideration may be that the proxy for social complexity for a comparative analysis for basal primates is too simplistic. Comparative research on squirrels discerned that demographic role complexity, and not group size, predicted vocal signaling complexity (Pollard and Blumstein 2012). Therefore, further research on the sociality of basal primates is needed with regard to demographic role complexity, structure and strength of social bonds, and interunit density and social network complexity to gain a better approximation of social complexity.

Another factor influencing the outcome of such a comparative approach is how primatologists measure the number of acoustically different calls. Variations in bioacoustic methods may lead to different estimates of repertoire size even in the same species (Zimmermann 1995a, b). However, it was not until recently that acoustic characterization of vocalizations in primates has become more standardized. Such a standardized and comparable metric is needed for a cross-taxa bioacoustic approach (also see Altenmüller et al. 2013; Semple and Higham 2013), perhaps analogous to the Facial Action Coding System (e.g., Parr et al. 2005; Caeiro et al. 2013) for facial signals in the visual domain or audiograms for sound perception in the auditory domain (Ramsier and Rauschecker, Chap. 3). Nevertheless, variation in vocal repertoire size across species is often used successfully as a first step in studies of the evolution of acoustic communication (Freeberg et al. 2012).

Two further aspects have to be considered when performing comparative approaches to explore the evolution of primate vocal complexity. Vocal repertoires in basal primates often do not merely consist of acoustically discrete call units. Rather, one call may grade insensibly into another, or different call types may be arranged in a highly complex temporal sequence (see Altenmüller et al. 2013), potentially encoding a much higher degree of complexity and signal efficiency (see Bregman 1990). Complexity cannot be fully considered if the analysis is limited to the number of acoustically distinct call types as done here and in previous research due to the limitations in the available datasets. Another aspect is that for most of the studied basal primates, it is not yet clear to what extent conspecific listeners pay attention to variations within a call type or differences between call types or homogeneous or heterogeneous call series as monkeys and apes do (see Snowdon, Chap. 6; Zuberbühler, Chap. 7). In line with this, vocal complexity in basal primates may be largely underestimated, since some species are able to produce subtle and perceptually meaningful variations within a call type and are able to combine such signals efficiently to convey more complex messages, increasing the expressive power of a limited repertoire of call types (Semple and Higham 2013; Zuberbühler, Chap. 7). These issues merit much more attention in the future, particularly in nocturnal basal primates.

All in all, this overview points to the fact that acoustic communication systems in extant basal primates may be derived from an acoustic signaling system with a vocal complexity of 7–8 call types, the shared vocal complexity of both the Strepsirrhini and the basal Haplorhini. This complexity most likely represents the lowest number of vocalizations needed to govern predator avoidance, group and individual spacing, social cohesion, and decision making in ancestral and dispersed primate societies. Likewise, findings indicate that the currently available datasets do not fully support either the phylogenetic hypothesis or the social complexity hypothesis. Findings discussed here highlight the need for a multifactorial approach and collection of standardized comparable datasets on a broader range of taxa of basal primates. Additional research is required to fully understand the link between phylogeny and social and communicative complexity and to better embed the vocal complexity of basal primates into the evolution of primate vocal communication.

5.5 Summary and Perspectives for Future Research

In this chapter it was argued that extant basal primate vocal signaling systems represent excellent candidates for exploring the evolutionary origins of acoustic communication in primates, given the basal phylogenetic status of these primates, their substantial ecological diversity, and their resulting unique adaptations. Indeed, a comparative bioacoustic approach on basal primates revealed that vocal signaling shows a remarkable diversity in acoustic structure, from narrowband to broadband, and from tonal, almost constant-frequency calls to complex frequency-modulated syllables that can be given as single units or arranged into complex vocal streams as in monkeys and apes. As in monkeys and apes, vocal signaling is used by basal primates as an important communication channel that governs specific challenges in the ecological and social world. Vocalizations also convey indexical cues and varying emotions, and they may induce vocal responses in conspecifics of the same type as produced by the caller, supporting emotional contagiousness. This trait may be universal across different phylogenetic groups, likely originating deep in primate or even mammalian phylogeny. Currently, it remains an open question to what extent basal primates are able to decode the rich information conveyed by vocal indexical and prosodic cues.

Cross-taxa comparative research within the same phylogenetic group of basal primates resulted in the hypothesis that natural selection (e.g., predation, physiological constraints linked to short-term stress) limits cross-taxa vocal flexibility and favors universals in acoustic structure that convey emotions, whereas sexual and kin selection drive cross-taxa flexibility and divergence. Research on vocal ontogeny and flexibility in the best-studied basal primate, the grey mouse lemur, found a babbling period with high vocal plasticity during ontogeny, and colony- and deme-specific variation in affiliation calls, as well as an unexpected capability of adults to adapt their own vocalizations to fluctuating background noise. Further research on vocal development in basal primates and the ability to adapt vocalizations to fluctuations in the acoustic or socio-acoustic environment will enable an understanding of how the vocal capacity of basal primates fits within the larger framework of the evolution of primate communication.

Using a comparative bioacoustic approach, the chapter further illuminated an extraordinary cross-taxa variation in the acoustic spaces open for signal evolution, extending from the hearing range of humans up to the ultrasonic range and reaching higher frequencies than in monkeys and apes. Findings provide support for the sensory drive hypothesis that predation risk linked to body mass shapes the acoustic space in the respective species. The proximate mechanisms for high-frequency and ultrasonic signaling in basal primates are largely unknown, and further examinations of the peripheral vocal motor system, the ear, and the central auditory system are needed. A comparison of vocal repertoire sizes among basal primates as an indicator of vocal complexity revealed wide cross-taxa variation in call types (5–22 types) and a striking variation in both solitary-ranging nocturnal and group-living diurnal taxa. Neither the phylogenetic hypothesis nor the social complexity hypothesis can fully explain that variation. Rather, a multifactorial approach and a standardized assessment of social and vocal complexity are needed, as is more research to increase sample sizes within and among the different phylogenetic groups of primates. Furthermore, the variation of vocal complexity in basal primates points to the idea that their acoustic communication systems may have originated from a signaling system with a vocal complexity of 7–8 call types, which is the shared vocal complexity of both the Strepsirrhini and the basal Haplorhini. That level of vocal complexity potentially represents the least number of vocalizations needed to govern predator avoidance, spacing, social cohesion, and decision making in ancestral and dispersed primate societies.

All in all, basal primates exhibit a unique diversity, flexibility, and complexity of vocalizations for communication in their social networks. Comparative bioacoustic approaches that examine the rich diversity of extant basal primates can provide promising new avenues to trace the evolutionary origins of primate vocal communication, including speech and language in humans.

Acknowledgments I thank Marissa Ramsier, Rolf Quam, and Art Popper for inviting me to contribute to this volume and for their constructive remarks. I further thank Sabine Schmidt, who commented on a previous version of this manuscript and helped to clarify some arguments, and Ute Radespiel and Marina Scheumann for their long-term collaboration. Sönke von den Berg and Tjard Bergmann are thanked for technical support. My research on acoustic communication in primates was supported by various resources (e.g., Deutscher Akademischer Austauschdienst [DAAD], Volkswagenstiftung [VW], Deutsche Forschungsgemeinschaft [DFG], and private sponsors).

Compliance with Ethics Requirements
Elke Zimmermann states that she has no conflicts of interest.

References

Ackermann, H., Hage, S. R., & Ziegler, W. (2014). Brain mechanisms of acoustic communication in humans and nonhuman primates: An evolutionary perspective. *Behavioral and Brain Sciences*, 37(6), 529–546.

Altenmüller, E., Schmidt, S., & Zimmermann, E. (2013). *Evolution of emotional communication: From sounds in nonhuman mammals to speech and music in man*. Oxford: Oxford University Press.

Ambrose, L. (2003). Three acoustic forms of Allen's galagos (Primates; Galagonidae) in the Central African region. *Primates*, 44, 25–39.

Andrew, R. J. (1963). The origin and evolution of the calls and facial expressions of the primates. *Behaviour, 20*, 1–109.

August, P. V., & Anderson, J. G. T. (1987). Mammal sounds and motivation-structural rules: A test of the hypothesis. *Journal of Mammalogy*, 68(1), 1–9.

Bearder, S. K. (2007). A comparison of calling patterns in two nocturnal primates, *Otolemur crassicaudatus* and *Galago moholi*, as a guide to predation risk. In S. L. Gursky & K. A. I. Nekaris (Eds.), *Primate anti-predator strategies* (pp. 206–221). New York: Springer Science+Business Media.

Bearder, S. K., & Doyle, G. A. (1974). Field and laboratory studies of social organization in bushbabies (*Galago senegalensis*). *Journal of Human Evolution*, 3(1), 37–50. doi:10.1016/0047-2484(74)90105-5.
Becker, M., Buder, E., Bakeman, R., Price, M., & Ward, J. (2003a). Infant response to mother call patterns in *Otolemur garnettii*. *Folia Primatologica*, 74, 301–311.
Becker, M. L., Buder, E. H., & Ward, J. P. (2003b). Spectrographic description of vocalizations in captive *Otolemur garnettii*. *International Journal of Primatology*, 24(2), 415–446.
Bolt, L. M. (2013). Squealing rate indicates dominance rank in the male ring-tailed lemur (*Lemur catta*). *American Journal of Primatology*, 75(12), 1174–1184. doi:10.1002/ajp.22179.
Boughman, J. W. (2002). How sensory drive can promote speciation. *Trends in Ecology and Evolution*, 17(12), 571–577.
Braune, P., Schmidt, S., & Zimmermann, E. (2005). Spacing and group coordination in a nocturnal primate, the golden brown mouse lemur (*Microcebus ravelobensis*): The role of olfactory and acoustic signals. *Behavioral Ecology and Sociobiology*, 58, 587–596.
Braune, P., Schmidt, S., & Zimmermann, E. (2008). Acoustic divergence in the communication of cryptic species of nocturnal primates (*Microcebus* spp.) *BMC Biology*, 6, 19.
Bregman, A. S. (1990). *Auditory scene analysis: The perceptual organization of sound*. Cambridge, MA: The MIT Press.
Brumm, H., Voss, K., Köllmer, I., & Todt, D. (2004). Acoustic communication in noise: Regulation of call characteristics in a New World monkey. *Journal of Experimental Biology*, 207, 443–448.
Burton, J. A., & Nietsch, A. (2010). Geographical variation in duet songs of Sulawesi tarsiers: Evidence for new cryptic species in south and southeast Sulawesi. *International Journal of Primatology*, 31(6), 1123–1146.
Butynski, T., Kingdon, J., & Kalina, J. (2013). *Mammals of Africa*. New York: Bloomsbury Publishing.
Caeiro, C. C., Waller, B. M., Zimmermann, E., Burrows, A. M., & Davila-Ross, M. (2013). OrangFACS: A muscle-based facial movement coding system for orangutans (*Pongo* spp.) *International Journal of Primatology*, 34(1), 115–129.
Charles-Dominique, P., & Martin, R. D. (1972). Behavior and ecology of nocturnal prosimians. *Zeitschrift fuer Tierpsychologie*, 9, 1–89.
Cheney, D. L., & Wrangham, R. W. (1987). Predation. In B. B. Smuts, D. L. Cheney, R. M. Seyfarth, R. W. Wrangham, & T. T. Struhsaker (Eds.), *Primate societies* (pp. 227–239). Chicago, IL: University of Chicago Press.
Cherry, J. A., Izard, M. K., & Simons, E. L. (1987). Description of ultrasonic vocalizations of the mouse lemur (*Microcebus murinus*) and the fat-tailed dwarf lemur (*Cheirogaleus medius*). *American Journal of Primatology*, 13, 181–185.
Clark, A. B. (1988). Interspecific differences and discrimination in auditory and olfactory signals of *Galago crassicaudatus* and *Galago garnettii*. *International Journal of Primatology*, 9(6), 557–571.
Dagosto, M., Gebo, D. L., & Dolino, C. N. (2003). The natural history of the Philippine tarsier (*Tarsius syrichta*). In P. C. Wright, E. L. Simons, & S. Gursky (Eds.), *Tarsiers: Past, present, and future* (pp. 237–259). New Brunswick, NJ: Rutgers University Press.
Darwin, C. (1872). *The expression of the emotions in man and animals*. New York: D. Appleton and Company.
Endler, J. A. (1992). Signals, signal conditions, and the direction of evolution. *American Naturalist*, 139, S125–S153.
Fay, R. R. (1988). *Hearing in vertebrates: A psychophysics databook*. Winnetka, IL: Hill-Fay Associates.
Fichtel, C. (2008). Ontogeny of conspecific and heterospecific alarm call recognition in wild Verreaux's sifakas (*Propithecus verreauxi verreauxi*). *American Journal of Primatology*, 70, 127–135.
Fichtel, C. (2012). Predation. In J. C. Mitani, J. Call, P. M. Kappeler, R. Palombit, & J. B. Silk (Eds.), *The evolution of primate societies* (pp. 169–194). Chicago, IL: University of Chicago Press.

Fichtel, C., & Hammerschmidt, K. (2002). Responses of redfronted lemurs to experimentally modified alarm calls: Evidence for urgency-based changes in call structure. *Ethology*, 108, 763–777.

Fichtel, C., & Kappeler, P. M. (2002). Anti-predator behavior of group-living Malagasy primates: Mixed evidence for a referential alarm call system. *Behavioral Ecology and Sociobiology*, 51, 262–275.

Fitch, W. T. (2010). *The evolution of language*. New York: Cambridge University Press.

Fleagle, J. G. (2013). *Primate adaptation and evolution* (3rd ed.). San Diego, CA: Academic Press.

Freeberg, T. M., Dunbar, R. I. M., & Ord, T. J. (2012). Social complexity as a proximate and ultimate factor in communicative complexity. *Philosophical Transactions of the Royal Society of London B: Biological Sciences*, 367(1597), 1785–1801. doi:10.1098/rstb.2011.0213.

Gamba, M., & Giacoma, C. (2007). Quantitative acoustic analysis of the vocal repertoire of the crowned lemur. *Ethology Ecology & Evolution*, 19(4), 323–343.

Gamba, M., Friard, O., & Giacoma, C. (2012a). Vocal tract morphology determines species-specific features in vocal signals of Lemurs (*Eulemur*). *International Journal of Primatology*, 33(6), 1453–1466.

Gamba, M., Colombo, C., & Giacoma, C. (2012b). Acoustic cues to caller identity in lemurs: A case study. *Journal of Ethology*, 30(1), 191–196. doi:10.1007/s10164-011-0291-z.

Gamba, M., Favaro, L., Torti, V., Sorrentino, V., & Giacoma, C. (2011). Vocal tract flexibility and variation in the vocal output in wild indris. *Bioacoustics, 20*(3), 251–265.

Geissmann, T., & Mutschler, T. (2006). Diurnal distribution of loud calls in sympatric wild indris (*Indri indri*) and ruffed lemurs (*Varecia variegata*): Implications for call functions. *Primates*, 47(4), 393–396. doi:10.1007/s10329-006-0189-5.

Giacoma, C., Sorrentino, V., Rabarivola, C., & Gamba, M. (2010). Sex differences in the song of *Indri indri*. *International Journal of Primatology*, 31(4), 539–551.

Goodman, S. M., O'Connor, S., & Langrand, O. (1993). A review of predation on lemurs: Implications for the evolution of social behavior in small, norturnal primates. In P. M. Kappeler & J. U. Ganzhorn (Eds.), *Lemur social systems and their ecological basis* (pp. 51–65). New York: Plenum Press.

Gosset, D., Fornasieri, I., & Roeder, J. J. (2003). Acoustic structure and contexts of emission of vocal signals by black lemurs. *Evolution of Communication*, 4(2), 225–251.

Groves, C., & Shekelle, M. (2010). The genera and species of Tarsiidae. *International Journal of Primatology*, 31(6), 1071–1082. doi:10.1007/s10764-010-9443-1.

Gursky, S. L., & Nekaris, K. A. I. (2007). *Primate anti-predator strategies*. New York: Springer Science+Business Media.

Hafen, T., Neveu, H., Rumpler, Y., Wilden, I., & Zimmermann, E. (1998). Acoustically dimorphic advertisement calls separate morphologically and genetically homogenous populations of the grey mouse lemur (*Microcebus murinus*). *Folia Primatologica*, 69(Suppl.1), 342–356.

Hage, S. R. (2013). Audio-vocal interactions during vocal communication in squirrel monkeys and their neurobiological implications. *Journal of Comparative Physiology A: Neuroethology, Sensory, Neural, and Behavioral Physiology*, 199(7), 663–668.

Hartig, G., Churakov, G., Warren, W. C., Brosius, J., et al. (2013). Retrophylogenomics place tarsiers on the evolutionary branch of anthropoids. *Scientific Reports*, 3. doi:10.1038/srep01756

Harvey, P. H., & Pagel, M. D. (1991). *The comparative method in evolutionary biology*. Oxford: Oxford University Press.

Horvath, J. E., Weisrock, D. W., Embry, S. L., Fiorentino, I., et al. (2008). Development and application of a phylogenomic toolkit: Resolving the evolutionary history of Madagascar's lemurs. *Genome Research*, 18(3), 489–499. doi:10.1101/Gr.7265208.

Jolly, A. (1966). Lemur social behavior and primate intelligence—Step from prosimian to monkey intelligence probably took place in a social context. *Science*, 153(3735), 501–506. doi:10.1126/science.153.3735.501.

Junqua, J. C. (1996). The influence of acoustics on speech production: A noise-induced stress phenomenon known as the Lombard reflex. *Speech Communication*, 20(1-2), 13–22. doi:10.1016/S0167-6393(96)00041-6.

Kappeler, P. M. (2012). The behavioral ecology of strepsirrhines and tarsiers. In J. C. Mitani, J. Call, P. M. Kappeler, R. A. Palombit, & J. B. Silk (Eds.), *The evolution of primate societies* (pp. 7–28). Chicago, IL: University of Chicago Press.

Kessler, S. E., Scheumann, M., Nash, L. T., & Zimmermann, E. (2012). Paternal kin recognition in the high frequency/ultrasonic range in a solitary foraging mammal. *BMC Ecology*, 12(1), 26. doi:10.1186/1472-6785-12-26.

Kessler, S. E., Radespiel, U., Hasiniaina, A. I. F., Leliveld, L. M. C., & Nash, L. T. (2014). Modeling the origins of mammalian sociality: Moderate evidence for matrilineal signatures in mouse lemur vocalizations. *Frontiers in Zoology*, 11, 14.

Klopfer, P. H. (1977). Communication in prosimians. In T. A. Sebeok (Ed.), *How animals communicate* (pp. 841–850). Bloomington, IN: Indiana University Press.

Knörnschild, M. (2014). Vocal production learning in bats. *Current Opinion in Neurobiology*, 28, 80–85. doi:10.1016/j.conb.2014.06.014.

Kulahci, I. G., Rubenstein, D. I., & Ghazanfar, A. A. (2015). Lemurs groom-at-a-distance through vocal networks. *Animal Behaviour*, 110, 179–186.

Lehman, S. M., Radespiel, U., & Zimmermann, E. (2016). *Dwarf and mouse lemurs of Madagascar: Biology, behavior, and conservation biogeography of the Cheirogaleidae*. Cambridge: Cambridge University Press.

Leliveld, L. M. C., Scheumann, M., & Zimmermann, E. (2011). Acoustic correlates of individuality in the vocal repertoire of a nocturnal primate (*Microcebus murinus*). *The Journal of the Acoustical Society of America*, 129(4), 2278–2288.

Linn, S. N. (2013). *Mother-offspring communication in grey mouse lemurs (Microcebus murinus) during reunions*. M.S. thesis, University of Veterinary Medicine Hannover, Hannover, Germany.

Macedonia, J. M. (1986). Individuality in a contact call of the ringtailed lemur (*Lemur catta*). *American Journal of Primatology*, 11, 163–179.

Macedonia, J. M. (1990). What is communicated in the antipredator calls of lemurs: Evidence from playback experiments with ringtailed and ruffed lemurs. *Ethology*, 86, 177–190.

Macedonia, J. M. (1993). The vocal repertoire of the ring-tailed lemur (*Lemur catta*). *Folia Primatologica*, 61(4), 186–217.

Macedonia, J. M., & Stanger, K. F. (1994). Phylogeny of the Lemuridae revisited—Evidence from communication signals. *Folia Primatologica*, 63(1), 1–43. doi:10.1159/000156787.

Macedonia, J. M., & Taylor, L. L. (1985). Subspecific divergence in a loud call of the ruffed lemur (*Varecia variegata*). *American Journal of Primatology*, 9, 295–304.

Maretti, G., Sorrentino, V., Finomana, A., Gamba, M., & Giacoma, C. (2010). Not just a pretty song: An overview of the vocal repertoire of *Indri indri*. *Journal of Anthropological Sciences*, 88, 151–165.

Martin, R. D., & Ross, C. F. (2005). The evolutionary and ecological context of primate vision. In J. J. Kremers (Ed.), *The primate visual system: A comparative approach* (pp. 1–36). New York: John Wiley & Sons..

Masters, J. C. (1991). Loud calls of *Galago crassicaudatus* and *G. garnettii* and their relation to habitat structure. *Primates*, 32(2), 153–167.

Matsui, A., Rakotondraparany, F., Munechika, I., Hasegawa, M., & Horai, S. (2009). Molecular phylogeny and evolution of prosimians based on complete sequences of mitochondrial DNAs. *Gene*, 441(1-2), 53–66. doi:10.1016/j.gene.2008.08.024.

McComb, K., & Semple, S. (2005). Coevolution of vocal communication and sociality in primates. *Biology Letters*, 1, 381–385.

Méndez-Cárdenas, M. G., & Zimmermann, E. (2009). Duetting—A mechanism to strengthen pair bonds in a dispersed pair-living primate (*Lepilemur edwardsi*)? *American Journal of Physical Anthropology*, 139(4), 523–532. doi:10.1002/ajpa.21017.

Méndez-Cárdenas, M., Randrianambinina, B., Rabesandratana, A., Rasoloharijaona, S., & Zimmermann, E. (2008). Geographic variation in loud calls of sportive lemurs (*Lepilemur* ssp.) and their implication for conservation. *American Journal of Primatology*, 70(9), 828–838.

Merker, S., & Groves, C. P. (2006). *Tarsius lariang*: A new primate species from western Central Sulawesi. *International Journal of Primatology*, 27(2), 465–485.

Merker, S., Driller, C., Perwitasari-Farajallah, D., Pamungkas, J., & Zischler, H. (2009). Elucidating geological and biological processes underlying the diversification of Sulawesi tarsiers. *Proceedings of the National Academy of Sciences of the United States of America*, 106(21), 8459–8464.

Merker, S., Driller, C., Dahruddin, H., Wirdateti, Sinaga, W., et al. (2010). *Tarsius wallacei*: A new tarsier species from Central Sulawesi occupies a discontinuous range. *International Journal of Primatology*, 31(6), 1107–1122.

Mittermeier, R. A., Louis Jr., E. E., Richardson, M., Schwitzer, C., et al. (2010). *Lemurs of Madagascar* (3rd ed.). Arlington, VA: Conservation International.

Morton, E. S. (1975). Ecological sources of selection on avian sounds. *American Naturalist*, 109, 17–34.

Morton, E. S. (1977). On the occurrence and significance of motivation-structural rules in some bird and mammal sounds. *American Naturalist*, 111, 855–869.

Müller, A. E., & Thalmann, U. (2000). Origin and evolution of primate social organisation: A reconstruction. *Biological Reviews*, 75, 405–435.

Nash, L. T., & Harcourt, C. S. (1986). Social organization of galagos in Kenyan coastal forests. 2. *Galago garnettii*. *American Journal of Primatology*, 10(4), 357–369. doi:10.1002/ajp.1350100407.

Neri-Arboleda, I., Stott, P., & Arboleda, N. P. (2002). Home ranges, spatial movements and habitat associations of the Philippine tarsier (*Tarsius syrichta*) in Corella, Bohol. *Journal of Zoology, 257*, 387–402.

Newman, J. D., Depeine, C. D., & Becker, M. L. (2013). Variation in strepsirrhine infant isolation calls and its evolutionary implications. In J. Masters, M. Gamba, & F. Génin (Eds.), *Leaping ahead: Advances in prosimian biology* (pp. 279–286). New York: Springer Science+Business Media.

Niemitz, C. (1984). Vocal communication of two tarsier species (*Tarsius bancanus* and *Tarsius spectrum*). In C. Niemitz (Ed.), *Biology of tarsiers* (pp. 129–141). Stuttgart: Gustav Fischer Verlag.

Niemitz, C., Nietsch, A., Warter, S., & Rumpler, Y. (1991). *Tarsius dianae*: A new primate species from central Sulawesi (Indonesia). *Folia Primatologica*, 56, 105–116.

Nietsch, A. (1999). Duet vocalizations among different populations of Sulawesi tarsiers. *International Journal of Primatology*, 20(4), 567–583.

Nietsch, A. (2003). Outline of the vocal behavior of *Tarsius spectrum*: Call features, associated behaviors, and biological functions. In P. C. Wright, E. L. Simons, & S. Gursky (Eds.), *Tarsiers: Past, present, and future* (pp. 196–220). New Brunswick, NJ: Rutgers University Press.

Oda, R. (2002). Individual distinctiveness of the contact calls of ring-tailed lemurs. *Folia Primatologica*, 73, 132–136.

Ord, T. J., & Garcia-Porta, J. (2012). Is sociality required for the evolution of communicative complexity? Evidence weighed against alternative hypotheses in diverse taxonomic groups. *Philosophical Transactions of the Royal Society of London B: Biological Sciences*, 367(1597), 1811–1828. doi:10.1098/rstb.2011.0215.

Pages, E. (1980). Ethoecology of *Microcebus coquereli* during the dry season. In P. Charles-Dominique (Ed.), *Nocturnal Malagasy primates: Ecology, physiology, and behavior* (pp. 97–116). New York: Academic Press.

Parr, L. A., Waller, B. M., & Fugate, J. (2005). Emotional communication in primates: Implications for neurobiology. *Current Opinion in Neurobiology*, 15, 716–720.

Patel, E. R., & Owren, M. J. (2012). Silky sifaka (*Propithecus candidus*) "zzuss" vocalizations: Sexual dimorphism, individuality, and function in the alarm call of a monomorphic lemur. *The Journal of the Acoustical Society of America*, 132(3), 1799–1810. doi:10.1121/1.4740475.

Pereira, M. E., & Macedonia, J. M. (1991). Ringtailed lemur anti-predator calls denote predator class, not response urgency. *Animal Behaviour*, 41(3), 543–544.

Pereira, M. E., Seeligson, M. L., & Macedonia, J. M. (1988). The behavioral repertoire of the black-and-white ruffed lemur, *Varecia variegata variegata* (Primates, Lemuridae). *Folia Primatologica*, 51(1), 1–32. doi:10.1159/000156353.

Perelman, P., Johnson, W. E., Roos, C., Seuánez, H. N., et al. (2011). A molecular phylogeny of living primates. *PLoS Genetics*, 7(3), e1001342. doi:10.1371/journal.pgen.1001342.

Petter, J. J., & Charles-Dominique, P. (1979). Vocal communication in prosimians. In G. A. Doyle & R. D. Martin (Eds.), *The study of prosimian behavior* (pp. 247–305). New York: Academic Press.

Petter, J. J., Albignac, R., & Rumpler, Y. (1977). *Faune de Madagascar 44: Mammiferes, Lemuriens (primates prosimiens)*. Paris: Ostrom CNRS.

Pflüger, F. J., & Fichtel, C. (2012). On the function of redfronted lemur's close calls. *Animal Cognition*, 15(5), 823–831.

Pollard, K. A., & Blumstein, D. T. (2012). Evolving communicative complexity: Insights from rodents and beyond. *Philosophical Transactions of the Royal Society of London B: Biological Sciences*, 367(1597), 1869–1878. doi:10.1098/rstb.2011.0221.

Pollock, J. I. (1986). The song in the indris (*Indri indri*; Primates: Lemuroidea): Natural history, form and function. *International Journal of Primatology*, 7(3), 225–264.

Radespiel, U. (2000). Sociality in the gray mouse lemur (*Microcebus murinus*) in northwestern Madagascar. *American Journal of Primatology*, 51, 21–40.

Ramanankirahina, R., Joly, M., & Zimmermann, E. (2011). Peaceful primates: Affiliation, aggression, and the question of female dominance in a nocturnal pair-living lemur (*Avahi occidentalis*). *American Journal of Primatology*, 73(12), 1261–1268. doi:10.1002/ajp.20998.

Ramanankirahina, R., Joly, M., & Zimmermann, E. (2012). Seasonal effects on sleeping site ecology in a nocturnal pair-living lemur (*Avahi occidentalis*). *International Journal of Primatology*, ***33***(2), 428–439.

Ramanankirahina, R., Joly, M., Scheumann, M., & Zimmermann, E. (2015). The role of acoustic signaling for spacing and group coordination in a nocturnal, pair-living primate, the western woolly lemur (*Avahi occidentalis*). *American Journal of Physical Anthropology*, 159(3), 466–477. doi:10.1002/ajpa.22898.

Ramsier, M. A. (2013). Evolution of auditory sensitivity among strepsirrhine primates. In J. Masters, M. Gamba, & F. Génin (Eds.), *Leaping ahead: Advances in prosimian biology* (pp. 265–270). New York: Springer Science+Business Media.

Ramsier, M. A., & Dominy, N. J. (2012). Receiver bias and the acoustic ecology of aye-ayes (*Daubentonia madagascariensis*). *Communicative and Integrative Biology*, 5(6), 637–640. doi:10.4161/cib.21509.

Ramsier, M. A., Cunningham, A. J., Finneran, J. J., & Dominy, N. J. (2012a). Social drive and the evolution of primate hearing. *Philosophical Transactions of the Royal Society of London B: Biological Sciences*, 367(1597), 1860–1868.

Ramsier, M. A., Cunningham, A. J., Moritz, G. L., Finneran, J. J., et al. (2012b). Primate communication in the pure ultrasound. *Biology Letters, **8***(4), 508–511.

Rasoloharijaona, S., Rakotosamimanana, B., Randrianambinina, B., & Zimmermann, E. (2003). Pair-specific usage of sleeping sites and their implications for social organization in a nocturnal Malagasy primate, the Milne Edwards' sportive lemur (*Lepilemur edwardsi*). *American Journal of Physical Anthropology*, 122(3), 251–258. doi:10.1002/ajpa.10281.

Rasoloharijaona, S., Randrianambinina, B., Braune, P., & Zimmermann, E. (2006). Loud calling, spacing, and cohesiveness in a nocturnal primate, the Milne Edwards' sportive lemur (*Lepilemur edwardsi*). *American Journal of Physical Anthropology*, 129, 591–600.

Řeháková-Petrů, M., Policht, R., & Peške, L. (2012). Acoustic repertoire of the Philippine tarsier (*Tarsius syrichta fraterculus*) and individual variation of long-distance calls. *International Journal of Zoology*, 2012, 10. Article ID 602401.

Richard, A. F. (1985). *Primates in nature*. New York: W.H. Freeman and Company.

Ross, C. (2003). Life history, infant care strategies, and brain size in primates. In P. M. Kappeler & M. E. Pereira (Eds.), *Primate life histories and socioecology* (pp. 266–284). Chicago, IL: University of Chicago Press.

Scheumann, M., Rabesandratana, A., & Zimmermann, E. (2007a). Predation, communication, and cognition in lemurs. In S. L. Gursky & K. A. I. Nekaris (Eds.), *Primate anti-predator strategies* (pp. 100–126). New York: Springer Science+Business Media.

Scheumann, M., Zimmermann, E., & Deichsel, G. (2007b). Context-specific calls signal infants' needs in a strepsirrhine primate, the gray mouse lemur (*Microcebus murinus*). *Developmental Psychobiology*, 49, 708–718.

Schopf, C. (2013). *Acoustic communication in noise: The gray mouse lemur as a primate model.* Ph.D. dissertation, Tierärztliche Hochschule, Institut für Zoologie, Hannover.
Semple, S., & Higham, J. P. (2013). Primate signals: Current issues and perspectives. *American Journal of Primatology*, 75(7), 613–620. doi:10.1002/ajp.22139.
Simmons, N. B., Seymour, K. L., Habersetzer, J., & Gunnell, G. F. (2008). Primitive early eocene bat from Wyoming and the evolution of flight and echolocation. *Nature*, 451(7180), 818–821. doi:10.1038/nature06549.
Stanger, K. F. (1993). *Structure and function of the vocalizations of nocturnal prosimians (Cheirogaleidae).* Ph.D. dissertation, Eberhard-Karls-Universität, Tübingen, Germany.
Stanger, K. F., & Macedonia, J. M. (1994). Vocalizations of aye-ayes (*Daubentonia madagascariensis*) in captivity. *Folia Primatologica*, 62(1-3), 160–169. doi:10.1159/000156773.
Tandy, J. M. (1976). Communication in *Galago crassicaudatus. Primates, 17*(4), 513–526.
Tattersall, I. (1979). Patterns of activity in the Mayotte lemur, *Lemur fulvus mayottensis. Journal of Mammalogy*, 60(2), 314–323. doi:10.2307/1379802.
Thalmann, U. (2003). Avahi, woolly lemurs, Avahy, Fotsy-fe, Ampongy, Tsarafangitra, Dadintsifaky. In S. M. Goodman & J. P. Benstead (Eds.), *The natural history of Madagascar* (pp. 1340–1342). Chicago, IL: University of Chicago Press.
Thierry, B., Iwaniuk, A. N., & Pellis, S. M. (2000). The influence of phylogeny on the social behaviour of macaques (Primates: Cercopithecidae, genus *Macaca*). *Ethology*, 106, 713–728.
Torti, V., Gamba, M., Rabemananjara, Z. H., & Giacoma, C. (2013). The songs of the indris (Mammalia: Primates: Indridae): Contextual variation in the long-distance calls of a lemur. *Italian Journal of Zoology*, 80(4), 596–607.
Trillmich, J., Fichtel, C., & Kappeler, P. M. (2004). Coordination of group movements in wild verreaux's sifakas (*Propithecus verreauxi*). *Behaviour*, 141, 1103–1120.
Veselka, N. (2010). A bony connection signals laryngeal echolocation in bats. *Nature*, 463, 939–942. doi:10.1038/nature08737.
Ward, S. C., & Sussman, R. W. (1979). Correlates between locomotor anatomy and behavior in two sympatric species of Lemur. *American Journal of Primatology*, 50(4), 575–590.
Wheeler, B. C., & Fischer, J. (2015). The blurred boundaries of functional reference: A response to Scarantino & Clay. *Animal Behaviour*, 100, E9–E13. doi:10.1016/j.anbehav.2014.11.007.
Wiens, F., & Zitzmann, A. (2003). Social structure of the solitary slow loris *Nycticebus coucang* (Lorisidae). *Journal of Zoology*, 261, 35–46. doi:10.1017/S0952836903003947.
Wilkins, M. R., Seddon, N., & Safran, R. J. (2013). Evolutionary divergence in acoustic signals: Causes and consequences. *Trends in Ecology & Evolution*, 28(3), 156–166.
Wright, P. C., & Simons, E. L. (1984). Calls of the Mindanao tarsier (*Tarsius syrichta*). *American Journal of Physical Anthropology*, 63(2), 236.
Yoder, A. D. (1997). Back to the future: A synthesis of strepsirrhine systematics. *Evolutionary Anthropology*, 6(1), 11–22.
Zimmermann, E. (1981). First record of ultrasound in two prosimian species. *Naturwissenschaften*, 68(10), 531–532.
Zimmermann, E. (1985). Vocalizations and associated behaviors in adult slow loris (*Nycticebus coucang*). *Folia Primatologica*, 44(1), 52–64. doi:10.1159/000156197.
Zimmermann, E. (1989a). Reproduction and behavioral development in slow loris (*Nycticebus coucang*). *Human Evolution*, 4, 171–179.
Zimmermann, E. (1989b). Aspects of reproduction, behavioral and vocal development of Senegal bushbabys (*Galago senegalensis*). *International Journal of Primatology*, 10, 1–16.
Zimmermann, E. (1990). Differentiation of vocalizations in bushbabies (Galaginae, Prosimiae, Primates) and the significance for assessing phylogenetic relationships. *Zeitschrift fuer zoologische Systematik und Evolutionsforschung*, 28, 217–239.
Zimmermann, E. (1991). Ontogeny of acoustic communication in prosimian primates. In A. Ehara, O. Takenaka, & M. Iwamoto (Eds.), *Primatology today* (pp. 337–340). Amsterdam: Elsevier.
Zimmermann, E. (1995a). Acoustic communication in nocturnal prosimians. In L. Alterman, G. A. Doyle, & M. K. Izard (Eds.), *Creatures of the dark: The nocturnal prosimians* (pp. 311–330). New York: Plenum Press.

Zimmermann, E. (1995b). Loud calls in nocturnal prosimians: Structure, evolution and ontogeny. In E. Zimmermann, J. D. Newman, & U. Jürgens (Eds.), *Current topics in primate vocal communication* (pp. 47–72). New York: Plenum Press.

Zimmermann, E. (2010). Vocal expression of emotion in a nocturnal prosimian primate group, mouse lemurs. In S. M. Brudzynski (Ed.), *Handbook of mammalian vocalization: An integrative neuroscience approach* (pp. 215–226). New York: Academic Press.

Zimmermann, E. (2016). Acoustic divergence in communication of cheirogaleids with special emphasis to mouse lemurs. In S. M. Lehman, U. Radespiel, & E. Zimmermann (Eds.), *Dwarf and mouse lemurs of Madagascar: Biology, behavior and conservation biogeography of the Cheirogaleidae* (pp. 405–421). Cambridge: Cambridge University Press.

Zimmermann, E., & Hafen, T. G. (2001). Colony specificity in a social call of mouse lemurs (*Microcebus* spp.). *American Journal of Primatology*, 54, 129–141.

Zimmermann, E., & Radespiel, U. (2013). Primate life histories. In W. Henke & I. Tattersall (Eds.), *Handbook of paleoanthropology* (Vol. 3, pp. 1527–1592). Berlin: Springer Science+Business Media.

Zimmermann, E., & Radespiel, U. (2014). Species concepts, diversity, and evolution in primates: Lessons to be learned from mouse lemurs. *Evolutionary Anthropology*, 23(1), 11–14. doi:10.1002/evan.21388.

Zimmermann, E., Bearder, S. K., Doyle, G. A., & Andersson, A. B. (1988). Variations in vocal patterns of Senegal and South African lesser bushbabies and their implications for taxonomic relationships. *Folia Primatologica (Basel)*, 51(2-3), 87–105.

Zimmermann, E., Leliveld, L. M. C., & Schehka, S. (2013). Toward the evolutionary roots of affective prosody in human acoustic communication: A comparative approach to mammalian voices. In E. Altenmüller, S. Schmidt, & E. Zimmermann (Eds.), *Evolution of emotional communication: From sounds in nonhuman mammals to speech and music in man* (pp. 116–132). Oxford: Oxford University Press.

Chapter 6
Vocal Communication in Family-Living and Pair-Bonded Primates

Charles T. Snowdon

Abstract Family-living and pair-bonded primates, such as gibbons, night monkeys, titi monkeys, marmosets, and tamarins, have some different social and ecological challenges than other primates and thus display some differences in vocal communication. Shared parental care, territory defense, pair-bond maintenance, and frequent exchange of roles throughout the day are common in family-living and pair-bonded primates. These species are usually sexually monomorphic, and they show relatively few sex differences in vocal output. Vocal communication is important in forming and maintaining pair bonds and in defending the pairs or family territory. In addition, these species appear to use vocal communication to a greater degree during social learning and putative teaching behavior, and adults appear to guide vocal development in young through reinforcement of vocal behavior. Adults of these species show great flexibility and plasticity in both vocal structure and usage in response to both social and environmental variation. They also adjust vocal output according to habitat acoustics to maximize audibility and minimize risk of predation. This chapter examines each of these areas of vocal communication to illustrate how family-living and pair-bonding primates use vocal communication.

Keywords Babbling • Dialects • Family-living primates • Habitat acoustics • Ontogeny • Pair bonds • Psychophysics • Sex differences • Social learning • Territory defense • Vocal communication

6.1 Introduction

The focus of a chapter on communication in family-living and pair-bonded primate species may seem at first blush to be unusual since most reviews are organized by phylogeny. However, there is merit in focusing a review based on the social system

---

C.T. Snowdon (✉)
Department of Psychology, University of Wisconsin,
1202 West Johnson Street, Madison, WI 53706, USA
e-mail: Snowdon@wisc.edu

R.M. Quam et al. (eds.), *Primate Hearing and Communication*, Springer Handbook of Auditory Research 63, DOI 10.1007/978-3-319-59478-1_6

rather than phylogeny. Much of the interest in studying primate vocal communication has been because of their phylogenetic closeness to human beings. Hence, research on our closest ancestors, the great apes, may have great importance for understanding the evolution of human communication, including language. However, none of the great apes shares a social system similar to that in which most humans today live, a system of pair-bonded adults raising offspring within the context of a family. Although the study of communication in great apes may illuminate some of the important cognitive underpinnings of the evolution of human communication, the divergence in social systems may obscure many other important social aspects of communication that may be specific to family-living and pair-bonded species. The goal of this chapter is to examine how vocal communication in family-living and pair-bonded primates might differ from that of species in other social systems.

What is different about family-living primates? These species are typically characterized by close social and emotional bonds between mates. Rather than simply mate and separate, an adult male and female will spend much of their time together; for example, pair-bonded common marmosets (*Callithrix jacchus*) may spend up to one-fifth of their daily activity (2–3 hours) grooming each other (Lazaro-Perea et al. 2004). Pair-bonded primates are known to show emotional responses when separated from each other, and they often prefer to associate with each other instead of other same-sex and opposite-sex members of their species or even their own infants (Mendoza and Mason 1986). In contrast to great apes, pair-bonded, family-living primates typically share parental care, foraging, and territorial defense duties, often exchanging roles throughout the day (Savage et al. 1996). Thus, careful and precise communication is important to share these different roles among group members, and one should expect to find greater attention to signals produced by group mates than might be the case in other species. In contrast to great apes, pair-bonded primates are generally sexually monomorphic (i.e., no obvious physical differences between males and females) and intra- and intersexual selection acts on both sexes equally. Thus, one might expect vocal communication to be less sexually dimorphic than that found in species with other social systems. Vocal communication also plays a role in developing and sustaining a pair bond and reducing stress in partners. Pair-bonded, family-living primates have exhibited the best evidence to date of cooperative foraging, rapid social learning, and even teaching of young (e.g., Rapaport 2006; Burkart et al. 2007; Humle and Snowdon 2008). Each of these activities is facilitated by vocal communication.

Family-living primates are noteworthy for the extensive energy that both parents invest in parental care. This joint investment is a major difference between family-living nonhuman primates and other nonhuman primates for which mothers are the major or sole providers of infant care. However, shared parental care leads to critical problems for each parent. Unlike female mammals where maternity is obvious due to pregnancy and parturition, a male mammal can never be certain of the paternity of his offspring. Consequently, a male needs to develop some trust that his mate will be unlikely to mate with anyone else, and at the same time, a female needs to have trust that her mate will still be with her when her infants are born to help her to care for them. Thus, it is important for both sexes to be selective about their mates and for both sexes to exhibit traits that can be used to evaluate mate quality. Both mates have a shared interest in defending their relationship from intruders of either sex, and territorial defense is often marked by vocal signaling.

Developmental processes, as well as plasticity of adult communication, appear to be different in these species, with evidence of babbling (highly variable versions of adult calls that become more stable with age), contingent actions by adults that shape vocal development, and the capacity to change vocal structure when an individual joins a new group. It is almost canonical in the literature that primate vocal development follows a fixed trajectory with no vocal learning or modification of call structure and usage (Hammerschmidt and Fischer 2008), yet social processes do play an important role in vocal development in many pair-bonded, family-living primates, with adults reinforcing vocal signals in the young and facilitating their development of adult structure and usage (Snowdon 2009). Evidence of population-specific dialects and vocal convergence when individuals change group membership illustrate vocal plasticity throughout the life span, not just in young animals. These primates also demonstrate remarkable flexibility in using vocal signals in novel contexts and in altering vocal structure in response to environmental noise and distance between conspecifics.

The pair-bonded, family-living species for which there is the most information about vocal communication include gibbons (*Hylobates* spp.) and siamangs (*Syndactylus* spp.), lesser apes found in Asia, and owl (or night) monkeys (*Aotus* spp.), titi monkeys (*Callicebus* spp.), Goeldi's monkeys (*Callimico goeldi*), marmosets (*Cebuella pygmaea*, *Callibella* spp., *Callithrix* spp.), and tamarins (*Saguinus* spp., *Leontopithecus* spp.) all from the South American tropics. Each of these species is almost exclusively arboreal where visual contact between group members is minimal except with animals in close proximity, which makes the use of vocal signals often of greater importance in communication than visual signals.

The remainder of this chapter reviews research on vocal communication in pair-bonded, family-living primates. The topics examined include sexual selection and sex differences (Sect. 6.2); formation and maintenance of pair bonds, including identification of mate (Sect. 6.3); cognitive aspects of vocalizations (Sect. 6.4); vocal signals used in social learning and teaching (Sect. 6.5); developmental processes, including babbling and adult contingent reinforcement of infant calls (Sect. 6.6); flexible adult vocal structure, including dialects and response to environmental noise (Sect. 6.7) and habitat acoustics, and use of vocalizations (Sect. 6.8).

6.2 Sexual Selection and Sex Differences

Most family-living primates display minimal sexual dimorphism (differences between males and females) in body structure and hair or skin coloration, and a logical prediction is that sexual dimorphism should be reduced in vocal communication as well. In many primate species living in multi-male, multi-female groups or in one male, multi-female groups, males may be up to twice the size of females, and in many species there are sex-specific vocal adaptations, including the flanges of mature male orangutans and the resonating throat pouches of male howling monkeys (*Alouatta palliata palliata*). Gauthier and Gauthier (1977) described several sex differences in vocalizations of Old World monkeys, including many species that have loud calls produced exclusively by males. In addition, males often had smaller vocal repertoires than females and called less frequently (Gauthier and Gauthier-Hion 1982).

Most family-living species are socially monogamous with males often playing an important role in infant care. This leads to different predictions from those usually made concerning sexual selection: males are said to compete with each other for mates, whereas females should be coy and choose carefully among many potential mates. In monogamous species, sexual selection should apply equally to both sexes. If a male is going to be more heavily involved in energetically costly parental care activities, then he should be selective about his partner, and one might expect greater competition among females to choose the best mate. At the same time, if a female relies on assistance from her mate for successful infant care, then she also needs to be selective, and males should compete among each other for the best mate. This should lead to similar displays indicating mate quality in both sexes, including vocal displays. In particular, both sexes should be similar in terms of pitches of vocalizations and might be expected to have similar vocal repertoires and to use them in similar contexts.

Although overall sex differences are predicted to be minimal in family-living nonhuman primates, some sex differences in vocalizations have been observed, such as differences in temporal parameters and usage, and some subtle differences in fundamental frequency rather than the large differences in fundamental frequency seen clearly in humans (e.g., Puts et al. 2012). Where fundamental frequency is sexually dimorphic, it is often males with higher frequencies. In the species of gibbons in which duetting between mates is common, males and females typically produce different sequences (Marshall and Marshall 1976) with females appearing to induce male singing in some species. (Deputte 1982). Lan (1993) reported that morning singing was dominated by males and that males and females produce different calls. Kloss's gibbons (*Hylobates klossii*) do not show the coordinated singing (duets) found in most other gibbon species (Dooley et al. 2013). Playbacks of male *H. klossii* solo songs elicited responses only from resident males, whereas playbacks of female songs elicited responses only from resident females (Raemakers and Raemakers 1985). Song structure may provide information relevant to mate choice in gibbons. Barelli et al. (2013) measured male song structure and fecal androgens, and they found males with higher androgen levels produced longer calls with higher pitch.

In common marmosets, males have higher frequency and greater variability in *phee calls* than females (Norcross and Newman 1993). In Weid's black-tufted-ear marmosets (*Callithrix kuhlii*) differences in frequency parameters distinguish male and female *phee calls*, and marmosets responded differently to playbacks of male and female calls (Smith et al. 2009). Miller et al. (2004) reported sex differences in the *combination long calls* of cotton-top tamarins (*Saguinus oedipus*) with males having shorter calls than females. Females were more attracted to male *long calls* with shorter notes, and males were more attracted to female calls with longer note duration, suggesting to Miller et al. (2004) that these *long calls* may play a role in sexual selection.

A natural playback experiment designed to see how cotton-top tamarins would respond to hearing calls of unfamiliar monkeys found a sex-specific response (McConnell and Snowdon 1986). Males gave *chirps* and females gave *long calls* in the early minutes, but both sexes converged on *chirp plus long call* vocalizations at the peak of arousal. However, a replication of the experiment on later generations of the same colony 20 years later (Scott et al. 2006) found a complete reversal, with males giving *long calls* and females giving *chirps* in the initial response to hearing an unfamiliar group. This replication provides a caution about attributing sex differences to tamarins. Although these studies have been done with captive animals, there is little reason to suspect different results from wild populations.

In summary, although some sex differences in vocalizations (and in response to vocalizations) have been reported in several family-living species, these are often quite subtle, requiring discriminant analyses of calls using multiple acoustic parameters to uncover sex differences. In comparison with species with other breeding systems, the sex differences in family-living species are relatively minor and, given the reversal of results in the same colony after 20 years, might be labile.

6.3 Formation and Defense of the Pair Bond

A strong pair bond is critical for socially monogamous species and in mammals is a necessary precursor to male parental care (Lukas and Clutton-Brock 2013). All male mammals face the problem of never being certain of paternity, and a monogamous relationship can provide some confidence that the infants the male is helping to rear are likely to be his own offspring. Vocalizations play an important role in forming and maintaining a pair bond and in keeping other individuals away. Most family-living species have *long calls* or songs that are often coordinated between mates and may serve to both reinforce the pair bonds and exclude others.

6.3.1 Duetting, Coordinated Songs, and Long Calls

Most gibbon species, as well as titi monkeys, show coordinated duetting or singing behavior, and there has been much interest in its coordination and function. Duetting is found among monogamous species of several Old World primates, including strepsirrhines [e.g., tarsiers (*Tarsius spectrum*), indris (*Indri indri*)], a langur species (Mentawai langur, *Presbytis potenziani*), and gibbons (*Hylobates* spp.). Haimoff (1986) has noted convergence in the structure and timing of duetting across these diverse species, including narrowband calls at dawn with a restricted frequency range and few harmonics, suggesting a convergence of duetting in monogamous Old World primates. Although duetting in gibbons and titi monkeys may have a role in pair-bond formation and strengthening the relationship between mates, most of the research, including playback experiments and naturalistic observations, suggests these calls primarily function to exclude intruders and maintain spacing.

6.3.2 Vocal Responses to Intruders

Several studies on gibbons have looked at responses to playbacks of vocalizations from familiar and unfamiliar animals. Raemakers and Raemakers (1985) found that male white-handed gibbons (*H. lar*) would respond as if to evict intruders if the songs were from solo males or pairs but not from solo females, whereas females reacted to the songs of solo females only. As with white-handed gibbons, female Bornean gibbons (*H. muelleri*) led group approaches and initiated singing to playback of female song, whereas males led group approaches and initiated singing to male songs. Studies on the Bornean gibbon found that playback location influenced responses (Mitani 1984, 1985). When the playback speaker was placed within the territory, mated males led approaches toward the songs, whereas songs played on the periphery led most commonly to singing behavior by the mated pair. Playbacks from deep within a neighbor's territory yielded neither approaches nor singing.

The role of duetting in territory maintenance has been studied in several species of titi monkeys. There are many curious species differences among titi monkeys that do not lead to any simple conclusions about the territorial functions of complex calls. In the red-bellied titi (*Callicebus moloch*), calling and counter calling led neighboring groups to approach each other and served to reinforce territory boundaries (Robinson 1979b, 1981), whereas in the collared titi (*Callicebus torquatus*), playbacks of solo male calls led to avoidance of the caller and playbacks of paired song led to counter calling but not approaching (Kinzey and Robinson 1983). However, in masked titi monkeys (*Callicebus personatus*), group encounters were rare and exclusively vocal with few signs of territorial behavior (Price and Piedade 2001). Caselli et al. (2014) found several variations of loud calls in the black-fronted titi (*Callicebus nigrifrons)* and argued that the calls were not used between groups and did not defend access to mates, but instead they regulated access to resources.

Marmosets and tamarins produce multi-syllabic whistle-like calls that appear to be used in multiple contexts. In some species, individual syllables are relatively flat in frequency, whereas in other species, syllables are frequency modulated. Three types of *long calls* were identified in cotton-top tamarins: one used in response to hearing calls of unfamiliar animals; another used when pair mates were separated or at a distance from each other; and a third form, the *combination long call* that includes both chirps and whistle notes, used mainly by nonreproductive individuals (Cleveland and Snowdon 1982). But Miller et al. (2005) found the *combination long call* to be common among reproductive adults in their colony. Playback studies found each of these three call types elicited different behavioral responses in cotton-top tamarins (Snowdon et al. 1983).

In a captive study that involved open doors between colony rooms to simulate the approach of unfamiliar animals, cotton-top tamarins increased their rates of *long calling*, suggesting that *long calls* play a role in territorial behavior (McConnell and Snowdon 1986; Scott et al. 2006). Playbacks of *long calls* of an unfamiliar cotton-top tamarin elicited antiphonal calling from residents and was used to census populations in the wild (Savage et al. 2010). These results imply a territorial function for *long calling*. Norcross and Newman (1993) found that *phee calls* from separated marmosets differed in structure from *phee calls* used in territorial contexts from the home cage. Furthermore, Norcross and Newman (1997) found that common marmosets rarely produced territorial *phee calls* when living in their natal group, but they began producing *phee calls* within four days after being paired with a mate. Golden lion tamarins (*Leontopithecus rosalia*) also have three distinct forms of long calls that are used for within-group cohesion, by animals separated from their group, and in territorial encounters, respectively (Halloy and Kleiman 1994). Thus, some *long calls* have a clear territorial function but other variants are used in other contexts.

6.3.3 Partner Separation

It is common for many species to call when separated from their group, but in marmosets and tamarins, calling is stronger when separated specifically from their mates. Playback of calls from a mate can reduce the stress of separation. Porter (1994) separated cotton-top tamarin mates into different rooms for 30 min and recorded a high rate of *long calls* from both sexes with males giving significantly more calls than females. Similarly, increased calling rates (and elevated cortisol levels) have been reported in marmosets and golden lion tamarins when housed alone or in novel social environments (Smith et al. 1998; Norcross and Newman 1999; Shepherd and French 1999). In a captive experiment, Ruckstalis and French (2005) played back vocalizations of mates to isolated marmosets and found that cortisol levels were significantly reduced compared with levels under control (no playback of mate calls) conditions. Thus, marmosets and tamarins display distress through increased *long calls* when separated from their mates, but this distress can be alleviated simply through playback of the mate's vocalizations. These results imply the ability to recognize specific individuals on the basis of call structure, and this has been shown explicitly in studies of pygmy marmosets (*Cebuella pygmaea*) (Snowdon and Cleveland 1980) and cotton-top tamarins (Snowdon et al. 1983).

6.3.4 Summary: Formation and Defense of Pair Bonds

Although coordinated duetting or singing behavior is often thought to be involved in indicating or maintaining a pair bond, there is little direct evidence of this except from marmosets and tamarins for whom the main function of coordinated calls is to indicate territory boundaries or maintain spacing between groups. However, in marmosets and tamarins there are also acoustic differences between the *long calls* used when bonded animals are separated from each other and the *long calls* used in territorial displays. Although separation induces *long calling* that is associated with increased stress hormone levels, playback of the mate's calls is sufficient to reduce cortisol levels, suggesting that the mate's voice has a stress-reducing effect.

6.4 Cognitive Aspects of Vocalizations

Many scientists have been interested in the cognitive components of communication. This research has mainly focused on Old World primates and great apes, but there has been increasing research on family-living primates. Among the topics that have been studied are whether signals are purely emotive or can also reference objects or events outside of the communicator, whether there is any syntactic structure to call sequences, the ordering of turn taking among individuals within a group, long-term memory for vocalizations, and perception of signals.

6.4.1 Referential Signals

Referential signals are calls that refer to a specific object or event in the environment. Some investigators (e.g., Zuberbühler 2000) have equated these signals with the prototypes of words in human language, but there is an emotional component in these calls as well. An animal communicating about food may also be communicating about its own desire for or interest in food. An animal that gives a predator-specific alarm call is not just identifying a predator but also is likely to be indicating some state of fear or arousal as well. Both food-associated calls and predator alarm calls have been studied in family-living primates.

6.4.1.1 Food Calls

Many nonhuman primates have specific calls that they give when they discover food. Elowson et al. (1991) measured individual food preferences for six foods in cotton-top tamarins and subsequently recorded calls associated with each of these foods. They reported two subtly different forms of calls: *C-chirps* were given as an animal approached the food and *D-chirps* were given after animals had taken the food. The rate of anticipatory calls (*C-chirps*) correlated directly with an individual's preference for foods. Benz (1993) replicated this study with golden lion tamarins and twelve different types of food and also found a correlation between an individual's preference and the rate of calling and specific call variants for protein, dried fruit, and grapes.

Caine et al. (1995) studied food calls in red-bellied tamarins and found more food calls with larger quantities and more palatable foods. However, they failed to find food calls during food exchanges between adults, similar to the results for cotton-top tamarins (Joyce and Snowdon 2007) but in contrast to adult lion tamarins (Brown and Mack 1978). Caine et al. (1995) also found that red-bellied tamarins called more often when they could see other group members than if they found food alone. In contrast, Roush and Snowdon (1994) failed to replicate the relationship between food preferences and rate of calling in cotton-top tamarins.

6.4.1.2 Predator Alarm Calls

Several nonhuman primates produce calls that are either specific to predator species or to the general context in which a predator operates (i.e., aerial/canopy versus ground). These predator specific calls, most famously among the vervet monkeys (*Chlorocebus aethiops*), have provoked considerable interest as a possible semantic signal parallel to words in speech (e.g., Seyfarth et al. 1980). Family-living primates are no exception. White-handed gibbons produce predator-specific calls to tigers and leopards and nonspecific alarm calls to eagles and pythons (Clarke et al. 2012). Black-fronted titi monkeys produce one type of alarm to raptors and to capuchin monkeys (*Cebus capucinus*) found in the canopy, and a different type of call is given to terrestrial threats (Cäsar et al. 2012a). In a study that played these calls back to groups of black-fronted titi monkeys, the monkeys looked up to the sky and canopy when the aerial alarm was played and looked at the caller when the terrestrial alarm was played, suggesting that the monkeys made inferences about the type or location of a predator based on call structure alone (Cäsar et al. 2012b). Sympatric saddle-back and moustached tamarins (*Saguinus fuscicollis* and *S. mystax*) also have predator-specific alarms for aerial and terrestrial predators, and they responded in a similar fashion when each was given (Kirchhof and Hammerschmidt 2006). Both species responded equally to the calls of their own as well as to those of the other species, illustrating cross-species recognition of alarm calls.

6.4.1.3 Other Signals

White-handed gibbons produce a seemingly similar *hoo call* (a moderate to soft call with a broad frequency range, given as a single call or in bouts of two to three calls in a variety of contexts). However, when these calls were analyzed in terms of structure, several subtle variants were identified that were consistently correlated with specific contexts: feeding, separation from group, encountering predators, interacting with neighbors, and duet songs (Clarke et al. 2015). Similar results were reported much earlier in cotton-top tamarins: eight different varieties of *chirps* were each associated with different contexts (alarm, mobbing, unfamiliar animal, approaching feeding, feeding, within group coordination; Cleveland and Snowdon 1982) (see Fig. 6.1). The differentiation of variants, in what initially sounds to human observers like a single call, indicates a greater complexity of vocal structure and contextual reference for these variants than previously appreciated.

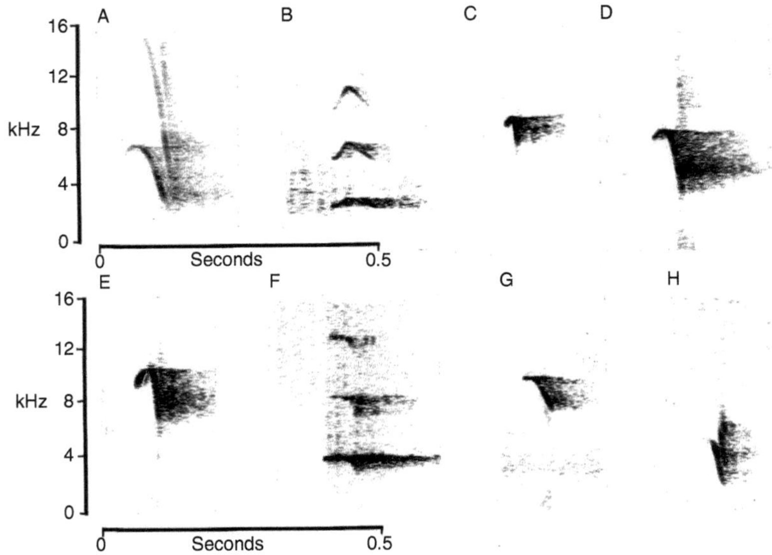

Fig. 6.1 Chirp variants in cotton-top tamarins: *type A* used in mobbing; *type B* used in investigation of novel objects; *type C* used during foraging, and *type D* used during eating. *Type E* chirps serve as alarm calls. *Type F* chirps are given in response to hearing calls of novel animals. *Type G* chirps are exchanged between calm animals within a group and *type H* chirps are used as mild alarms. (Modified from Snowdon 1982)

6.4.2 Syntax

Syntax in animal signals refers to the orderly sequencing of multiple calls or notes. Much of bird song is highly organized in terms of the structure and sequencing of different notes or themes, and there is also evidence of this in family-living primates. The songs of gibbons are highly structured with a series of notes produced in the duetting song and coordination of singing between the male and female (see Sect. 6.2). While the same notes are also found in songs that are given in response to predators, the structural organization of the notes differs. In white-handed gibbons, out-of-sight animals responded differently to the two types of songs, indicating that they were using the sequencing of notes rather than the notes themselves to discriminate between the two types (Clarke et al. 2006). In red-bellied titi monkeys, several calls are repeated, and these calls are organized into sequences involving different call types. These sequences were quite regular, and when playbacks of calls in altered sequences were presented to titi monkeys, they showed some ability to discriminate between normal and abnormal sequences (Robinson 1979a).

Tamarins and marmosets also show examples of syntax. Cleveland and Snowdon (1982) described several sequences in calls of cotton-top tamarins with a few general rules. Chirp-like calls always preceded longer constant-frequency calls within a sequence and, within a series of constant-frequency calls, each successive note was higher pitched than the previous one. In most cases, the sequence could not be decomposed into separate parts. That is, the sequenced call did not have the same function as each of the component parts did individually. This is phonological syntax, akin to the use in speech of different phonemes to create different meanings, such as "dog" versus "god." However, cotton-top tamarins showed a few examples of lexical syntax, wherein each component of the sequence has its own context and the sequence represents the combination of these contexts. For example, after an alarming event an animal will combine an alarm call with an affiliation call, and after this, other group members become active again. A second example is calling in response to the calls of novel animals: the male and female initially each use different calls but combine both types of calls at the peak of arousal (McConnell and Snowdon 1986). Miller et al. (2005) presented tamarins with manipulated *long calls* and found that recognition of call type and of caller occurred in separate stages of sensory processing.

6.4.3 Turn Taking

Duetting between mated pairs was discussed previously in Sect. 6.3.2 on pair bonding, but coordination of calling among group members is also seen outside of the calling between mates. In a group of three pygmy marmosets, Snowdon and Cleveland (1984) found that each animal within the group was more likely to call before another animal called a second time, and one possible order of turn taking (e.g., ABC, BCA, or CAB) was more common than the other order (CBA, BAC, or ACB). The development of turn taking is dependent upon the ability to recognize each individual based on voice alone.

Several studies have looked at antiphonal calling (the exchange of calls between two or more individuals or groups), which is common among marmosets and tamarins. The results included evidence of individual recognition within antiphonal calling (Miller and Thomas 2012), different structure in initial calls versus answering calls (Miller et al. 2010), and evidence of learning turn-taking behavior during development (Chow et al. 2015). Vocal turn taking by marmosets shows similar dynamics as vocal turn taking by humans, implying a converging evolution of cooperative vocal behavior in these two cooperatively breeding species (Takahashi et al. 2013).

6.4.4 Vocal Memory

Individual recognition by voice is critically important in any social group of primates, and recognition of voices of mates and of other family members is important in family-living species (also see Sect. 6.3). Little work has been done on long-term memory for vocalizations. However, in the natural environment where animals of both sexes disperse and form new family groups, recognition of the voices of relatives might be important in avoiding inbreeding. One study of cotton-top tamarins demonstrated that memories of calls of former family members last up to 5.5 years (Matthews and Snowdon 2011). To date, this is the longest duration of vocal memory in any nonhuman primate.

6.4.5 *Perception*

In human speech, phonemes are produced along a variety of continua, such as voice onset time or place of articulation, and human perceptual systems organize these vocal continua into discrete categories that allow the perception of distinct phonemes instead of multiple variations. Do similar processes exist in other species? Pygmy marmosets produce many variants of *trills*, which are sinusoidal, frequency-modulated calls varying in bandwidth and duration (see Fig. 6.2). Although several variants are used in similar contexts (see Sect. 6.8), two trill types are used in distinct contexts: the *closed mouth trill* is used as an affiliative contact call, whereas the *open mouth trill* is used in agonistic contexts. The main structural difference between these two calls in a captive population was duration with all *closed mouth trills* being shorter than 250 ms and all *open mouth trills* being longer. Snowdon and

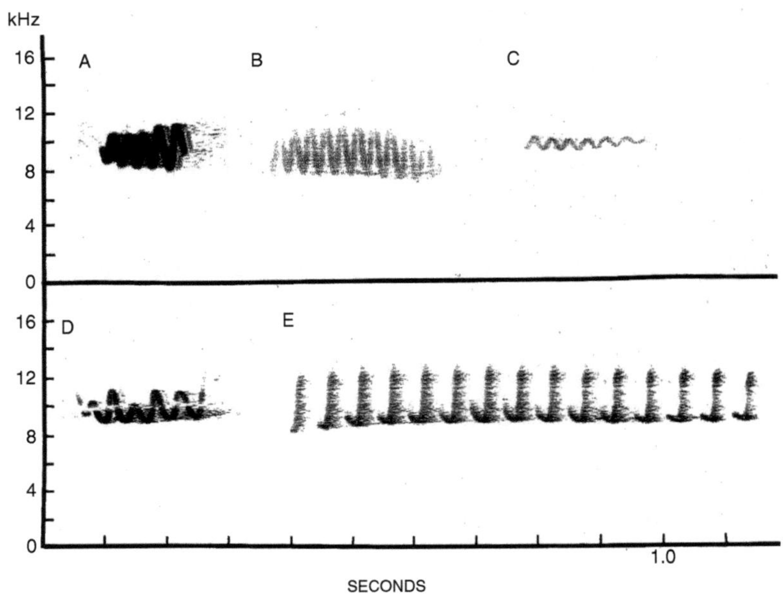

Fig. 6.2 Trill variants in pygmy marmosets: **(A)** *closed mouth trill*; **(B)** *open mouth trill*; **(C)** *quiet trill*; **(D)** *juvenile trill*; **(E)** *J-call*. (From Snowdon 1982, reprinted with permission of Cambridge University Press)

Pola (1978) synthesized trills and varied them along dimensions of bandwidth, rate of frequency modulation, and duration and played these synthesized trills to the marmosets. On the duration dimension, there was a clear category boundary at 250 ms with calls on either side of the boundary (varying by only 8 ms) eliciting different responses. *Closed mouth trills* elicited an immediate antiphonal response, whereas *open mouth trills* did not.

Masataka (1983) played synthesized alarm calls to Goeldi's monkeys (*Callimico goeldii*) and found that an increase of 0.2 kHz in the frequency range of the modulating sweep was sufficient to induce different behavioral responses, from a response appropriate to a mobbing call (i.e., approaching the caller to attack a predator) at a low-frequency range to a response appropriate to an alarm call (i.e., freezing) at a higher frequency range. Thus, both pygmy marmosets and Goeldi's monkeys show a human-like categorical perception of their own calls.

In a perceptual study of cotton-top tamarins, Ghazanfar et al. (2001) played back partial phrases or complete *combination long calls* and found that isolated tamarins responded significantly more to the entire call than to any component parts. They concluded that, from a tamarin's perspective, the entire *long call* forms the unit of perception. Bauers and Snowdon (1990) selected the two most acoustically similar of the eight *chirps* produced by cotton-top tamarins (*F* and *G chirps*, see Fig. 6.1) and found a clear difference in behavioral responses between the two playbacks.

6.4.6 Summary: Cognitive Aspects

There is considerable evidence for cognitive complexity in vocal communication in family-living primates. Referential signals communicate about food quality and predator types, and there is evidence of subtle variation in call structure that is correlated with specific contexts. Several species have call sequences that are consistent and predictable, and different sequences are used in different contexts. Many species show turn-taking behavior that indicates rule-based structures governing who will call as well as individual recognition of group members. There is some evidence of long-term vocal memory that may be important in avoiding inbreeding, and the perception of vocalizations has several parallels to the perception of speech sounds by humans.

6.5 Vocalizations in Social Learning and Teaching

Studies of social learning and teaching rarely mention the role of communication, yet vocal communication may play an important role. This section examines two sets of findings: one on how vocal communication might influence social learning and the other on putative teaching behavior in tamarins.

6.5.1 Social Learning

Although there is good evidence that rodents and birds can learn from others to avoid noxious foods (Galef and Giraldeau 2001), there has been little evidence among nonhuman primates. An illustrative example is on tufted capuchin monkeys (*Cebus apella*), which are not pair bonded or cooperatively breeding. When invisible white pepper was added to a familiar preferred food, mozzarella cheese, Visalberghi and Addessi (2000) found that capuchin monkeys learned to avoid the food individually. That is, there was no effect of watching other animals sample the adulterated food.

In a replication of the food avoidance study, in this case with cotton-top tamarins, Snowdon and Boe (2003) added white pepper to highly preferred tuna fish and found that only a third of the tamarins ever sampled the adulterated tuna, meaning that the other two-thirds of the animals avoided this previously preferred food. Furthermore, when tuna was later presented without any pepper, several animals continued to avoid eating tuna for more than a year after the initial experiment. What could account for the difference between these two studies? There was no evidence of any communication between the non-family-living capuchin monkeys that Visalberghi and Addessi studied, whereas cotton-top tamarins that sampled the adulterated tuna significantly reduced the number of food calls produced and increased the number of alarm calls (a novel use of alarm calls, see Sect. 6.7.4). The monkeys that first sampled the food also gave an increased frequency of visual disgust responses. Thus, the use of vocalizations (and visual signals) by the tamarins that first sampled the adulterated tuna may have facilitated the rapid and enduring social learning to avoid tuna.

6.5.2 Teaching

The existence of teaching in nonhuman animals has long been controversial. However, Caro and Hauser (1992) provided a simple operational definition. They have four criteria: (1) the teacher must alter its behavior only in the presence of a naïve animal; (2) the teacher must incur some cost or at least no immediate benefit; (3) the teacher's behavior encourages, punishes, or sets an example for the naïve animals; and (4) as a result, the naïve animal acquires a skill faster than it might otherwise. An additional criterion might be that the teacher is sensitive to the changes in the learner's behavior and alters its own behavior accordingly.

Tamarin and marmoset species are interesting because adults often share food with infants beginning at the time of weaning. This appears to modulate any weaning conflict and leads to young animals being able to feed on solid food at an earlier age than they might otherwise. Vocalizations play an important role in this process. Infants of many species beg for food, but adult tamarins who are prepared to share food with infants give distinct variations on normal food calls (see Sect. 6.4.1.1).

Adults produce not only more bouts of food calls but also produce many more calls within a bout at a much faster rate than they do with only adults present (Joyce and Snowdon 2007). The probability of an infant being able to obtain food from an adult is dependent on the adult producing the call (Roush and Snowdon 2001; Joyce and Snowdon 2007). Adults have modified their vocal behavior specifically for use in the food sharing context. Since these calls are energetically more costly than normal food calls and the adults are giving up some of their food, they are clearly incurring a cost. When twins are present (twinning is common among marmosets and tamarins), adults begin to give these rapid food calls and to share food almost a month earlier than when there is only a single infant present. Twins who receive food sharing at an earlier date also begin to forage on their own earlier than singletons, suggesting that the initially naïve animals are acquiring skills as a result of the adult behavior.

Food sharing begins at the end of the second month of life, peaks during the third month, and is rarely seen by five months of age. At this point all young tamarins are foraging successfully by themselves and giving food-associated calls similar to those of adults. Humle and Snowdon (2008) tested juvenile cotton-top tamarins seven months and older on a novel foraging task. Two opaque tubes with a food container suspended inside each tube were introduced first to the parents, and each parent was trained on a different method of solution. One solution was to walk along a branch and reach up into a tube to obtain food. The other solution was to hang suspended from the ceiling and pull up the food container hand over hand. Once the adults were well-trained, a juvenile was introduced. Even though food sharing and infant forms of food calling had not been observed for more than two months, the adults again began to give infant food calls and shared with the juveniles, but they only did this in the presence of the novel task and not on control days when food was present in a food dish. However, as soon as the juvenile was successful in obtaining food from the apparatus, the adult model stopped vocalizing and no longer engaged in food sharing. This is clear evidence that adult tamarins are sensitive to the changes in the learner's behavior and are adjusting their own behavior.

Parallel results have been reported in both captive and field studies of golden lion tamarins. Captive golden lion tamarins are more likely to share novel or difficult-to-process foods with infants (Rapaport 1999), and in the wild, where young tamarins have difficulty catching insect prey, adults successively withhold assistance from juveniles as their insect-catching skills improve (Rapaport 2006; Rapaport and Ruiz-Miranda 2006). In both golden lion tamarins and cotton-top top tamarins, adults have been observed calling near a prey source or assisting a young animal in obtaining food. This scaffolding behavior is a mark of human teaching, and its presence in tamarins contrasts sharply with the absence of any coaching or scaffolding behavior in chimpanzees, even when young individuals are feeding on potentially painful biting ants (Humle et al. 2009). However, despite the evidence for adults appearing to be sensitive to the abilities of young animals in cotton-top tamarins and lion tamarins, research on common marmosets did not show evidence of such sensitivity (Brown et al. 2005).

6.5.3 Summary: Vocalizations in Social Learning and Teaching

Vocal signals play an important role in both social learning and in teaching behavior in tamarins, and one is tempted to argue that such communication may be responsible for facilitating the rapid social learning seen in these species and absent in capuchin monkeys and chimpanzees. However, this is a hypothesis that needs to be tested closely in other family-living species as well as nonhuman primates with other forms of social organization. Most researchers on social learning have not been interested in the role of communication, but this may prove to be important.

6.6 Vocal Development

As noted in Section 6.1, it is commonly thought that vocal structures are innate in primates with little or no developmental modification. However, family-living primates appear to demonstrate a greater influence of social and environmental factors on vocal structure than has been seen in other nonhuman primates. This section first reviews various models and methods of studying vocal development followed by information about babbling and consideration of some naturalistic and experimental studies that suggest that vocal development of family-living primates is sensitive to social and environmental factors. Section 6.7 then examines plasticity in adult vocal structure and usage.

6.6.1 Models and Methods of Vocal Development

Three aspects of the development of vocal communication can be distinguished: (1) signal structure; (2) appropriate usage; and (3) comprehension of signals. Each of these may be subjected to different developmental processes. Four models can be used to explain developmental processes in vocal communication. These include (1) *innate or genetic determination*, whereby signal structure, usage, or comprehension are fixed at birth; (2) *maturation*, whereby signal structure, usage, or comprehension changes as a function of physical or social maturation but without any explicit learning process; (3) *limited learning*, whereby only certain aspects of signal structure, usage, or comprehension can be developed and only during a limited period in development; and (4*) open-ended learning* where structure, usage, or comprehension can be modified throughout an animal's life span.

It is generally accepted that nonhuman primates display developmental flexibility in the usage and comprehension of signals, but vocal structures are innate and not susceptible to modification by experience (Seyfarth and Cheney 1997). Janik and Slater (1997, 2000) have argued that evidence of vocal learning requires that an animal be able to acquire vocalizations from outside their natural species-specific repertoire. They further state that only songbirds and a few other genera of birds, cetaceans, bats, and humans show this ability, whereas no nonhuman primates do. This view has been reinforced by early studies of squirrel monkeys (*Saimiri sciureus*) and rhesus macaques (*Macaca mulatta*) that were reared in isolation. The isolate-reared squirrel monkeys had a normal adult vocal repertoire and responded with appropriate vocalizations in the proper contexts (i.e., giving alarm calls to predators never seen before) (Winter et al. 1973; Herzog and Hopf 1983, 1984). Similarly, isolate-reared rhesus macaques showed only minor perturbations in the structure of their *coo* vocalizations (Newman and Symmes 1974). When isolate-reared rhesus macaques were tested in a situation where one animal saw a stimulus that indicated a shock and a second animal could only see the facial expression of the monkey seeing the stimulus but had to respond to save both animals from getting shocked, the isolate-reared animals were effective communicators, but they could not "read" the signals of another monkey when they had to respond (Miller 1967). This suggests that, whereas the production of the signal and its use in an appropriate context were not affected by isolate rearing, the comprehension of the signal was impaired.

Isolate rearing of nonhuman primates is not ethically acceptable today, but cross-fostering and hybridization are two less invasive methods. In a study that cross-fostered rhesus and Japanese macaques with mothers of the opposite species, there was no evidence that the cross-fostered infants acquired the vocalizations of its foster species, but the foster mothers rapidly learned to respond appropriately to the calls of the foster infant (Owren et al. 1993). Hybridization between two species of squirrel monkeys found that the hybrid offspring tended to acquire the call characteristics of their mothers (Newman and Symmes 1982). However, in the wild, male squirrel monkeys are typically excluded from the group after mating, so it is possible that infant squirrel monkeys normally learn call structure from their mothers. Two studies on hybrid gibbon infants found that the calls of infants did not resemble those of either parent and, in some cases, contained aspects of the vocal structure of unrelated species. The mechanisms of vocal development in gibbons are complex and not easily related either to direct inheritance from one or both parents or to vocal learning from parents (Geissmann 1984; Tenaza 1985).

However, with the exception of the gibbons, none of these species reviewed so far are family living. Would developmental processes be different in family-living species? There are two types of examples: the spontaneous babbling-like behavior of pygmy marmosets (*Cebuella pygmaea*) and the naturalistic study of vocal development combined with some experimental manipulations in pygmy marmosets, common marmosets, and cotton-top tamarins. Little is known about other family-living species, and this material is reviewed in the final section.

6.6.2 Babbling-Like Behavior

From the first two weeks of life, young pygmy marmosets engage in long vocal bouts that contain a variety of call types (Elowson et al. 1998). These bouts share many characteristics with the babbling behavior of human infants. The majority of the calls produced was similar to adult calls and, indeed, represent a subset of adult calls. The calls (e.g., alarm calls, food calls, contact calls, etc.) are given out of context, given in a haphazard order, and often repeated several times with no relationship to the normal adult context for calls. Finally, adults respond to calling infants by approaching them and making physical contact. The main difference in comparison to human babbling is that the pygmy marmosets do not have a phonetic structure; thus babbling consists of calls rather than phonemes. Often the subsong and plastic song of songbirds is treated as a parallel to human babbling behavior (Marler 1970), but there are some fundamental differences. Song is typically produced only by male birds and subsong and plastic song appear only as birds undergo puberty. In contrast, pygmy marmoset babbling begins in infancy and is seen equally in both sexes.

What are the consequences of babbling? Snowdon and Elowson (2001) reported that greater babbling early in infancy led to improved vocal production and a greater number of adult-like vocalizations after weaning. However, vocal development was not completed at weaning. The most commonly used adult call is the *trill*, and marmosets continued to improve on the production of adult *trills* throughout puberty and adolescence, reaching adult-like trill structure only as breeding adults, much like the food-associated calls of cotton-top tamarins (see below). Interestingly, submissive adult marmosets regress to babbling behavior during aggressive encounters, implying a plasticity of usage of infant vocalizations.

6.6.3 Naturalistic and Noninvasive Experimental Approaches

Studies of cotton-top tamarins found some plasticity in vocal development. In a feeding context, when adults gave specific food-associated calls as approaching and leaving food (Elowson et al. 1991), infant and juvenile tamarins produced calls that did not match adult structure and were considerably more variable. These young animals also produced other vocalizations (not heard from adults) in feeding contexts (Roush and Snowdon 1994). Curiously, there was no developmental progression toward the production of adult-like vocalizations in this context, even in animals that were past puberty. In an experimental study, Roush and Snowdon (1999) recorded feeding vocalizations in cotton-top tamarins while living in family groups and after they were paired with a mate and separated from their natal families. There was a rapid (within 2–3 weeks) change in feeding vocalizations, including the elimination of the other calls and development of a clear adult structure for the food calls. This suggests that social context may serve as a constraint on adult vocal production. As tamarins are cooperative breeders, in which only the adult pair reproduce and other group members act as nonreproductive helpers, it may be that young animals inhibit their adult usage of calls until they become reproductively active themselves.

Cotton-top tamarins produce eight chirp-like vocalizations (short, high-pitched, frequency-modulated calls, see Fig. 6.1) with each chirp type being used in a discrete context (e.g. feeding, mobbing, alarming, responding to a stranger's call, and responding to a group member) (Cleveland and Snowdon 1982). Castro and Snowdon (2000) carried out an experimental study of how infant tamarins used these calls. Adult tamarins used the appropriate chirp type in each of the different contexts. Infants, unlike adults, typically did not produce discrete chirps but instead produced a sequence of chirps with descending frequency. Over the period of infant dependence through weaning, each of the infants tested produced some of the chirp types in an appropriate context, but no one individual produced all of the chirp types and no experimental context elicited an appropriate chirp type from each infant. These results suggest a relatively slow process of development and show that young tamarins are not able to produce adult calls at birth, in marked contrast to non-family-living squirrel monkeys. Although cotton-top tamarins did not show the babbling-like behavior seen in pygmy marmosets, they did show great variation in chirp structure and only rarely produced adult-like calls. If there are innate templates for vocal structures, they need to be shaped and sharpened through experience.

Elowson et al. (1992) recorded pygmy marmoset *trills* throughout ontogeny and found that *trills* changed during the course of development, suggesting they are not produced in adult-like ways at birth. Given that maturational processes are involved in development, all animals should show a similar pattern of vocal development. However, young marmosets, even twins within a litter, showed different patterns of *trill* development that were not consistent with a simple maturational model. Evidence of adult plasticity in vocal production and usage (presented in the next section) suggests that marmosets and tamarins can adjust vocal production throughout their lives.

A study of common marmosets shows quite elegantly that adult caregivers play an important role in shaping the vocal development of their offspring. Takahashi et al. (2015) studied the development of the *phee call*, a frequent call given when marmosets are separated from one another. They found that the calls became more stereotyped over the first two months with increased duration, decreased central frequency, and decreased entropy. Four discrete clusters of calls were seen in neonates, but these were reduced to one or two clusters by two months of age. At first glance this may seem to support a simple maturational model of vocal development. However, changes in *phee* quality were not correlated with age, body weight, or physiological development of the respiratory system. Takahashi et al. (2015) recorded infants both when alone and when in vocal contact with one of their parents. Parents generally respond to infant calls with well-formed adult *phees*. Rates of parental responsiveness to infants correlated directly with the age at which infants began producing well-formed *phees* of their own, suggesting that parental responsiveness to infant cries directly influences an infant's trajectory toward an adult call. Although studies of babbling in pygmy marmosets showed a higher rate of adult social interaction with babbling versus nonbabbling infants (Elowson et al. 1998), this is the first experimental demonstration of parental influence on vocal development in any nonhuman primate. However, there are clear parallels to vocal development in other taxa, including birds and humans (West and King 1988; Goldstein et al. 2003).

6.6.4 Vocal Development in Other Family-living Species

In hybrid gibbons, the song structure was complicated with few direct structural features inherited or learned from parents (Geissmann 1984; Geissmann and Orgeldinger 2000). However, Merker and Cox (1999), studying a single female gibbon, reported that vocal development was a slow process with different components of female *great call* structure appearing at different ages, much like the relatively slow development reported for marmosets and tamarins. There was also increased coordination of the infant's calling with that of its mother as the infant grew older, suggesting that the mother may serve as a model. Further support for mothers serving as models for gibbon vocal development comes from Koda et al. (2013) who found acoustic matching of songs between mothers and daughters. Mothers adjusted their songs to be more stereotyped when co-singing with daughters, especially with daughters who co-sang less. Thus, for female gibbons at least, there appears to be a form of coaching behavior that may serve like the contingent responding in marmosets to shape vocal development in the young.

6.6.5 Summary: Vocal Development

In contrast to the general view that primate vocal structures are innate and not modified through learning processes, the data from family-living primates clearly show that development of adult vocal structures is a gradual process that cannot be attributed solely to maturation. Social variables, such as contingent responding by adults to infant babbling in pygmy marmosets, in response to infant cries by common marmosets and coaching songs by gibbons, can influence the rapidity of acquisition of adult-like calling. At the same time the suppression of breeding in adult helpers, inherent in the structure of cooperative breeding, may also inhibit the expression of some adult-like vocalizations until animals achieve breeding status. There are several parallels between development in family-living primates and that of humans that have not yet been reported in species with other breeding systems. Does this plasticity seen in young animals carry over into adult vocal production?

6.7 Flexible Adult Vocal Structure and Usage

Another characteristic of family-living primates is that vocal communication can be used flexibly by adults, with evidence for change in structure and usage in different social and environmental contexts. This is especially evident in four areas: (1) adjustment and convergence of vocal structures with pair or group formation (Sect. 6.7.1); (2) population specific dialects (Sect. 6.7.2); (3) structural change in response to environmental noise (Sect. 6.7.3); and (4) novel responses to captive environments (Sect. 6.7.4).

6.7.1 *Modification and Convergence of Calls with Pair Formation*

In a wide array of species, ranging from birds through dolphins to humans, there is evidence of vocal convergence with preferred social partners (Snowdon and Hausberger 1997), but there has been little evidence of vocal change in nonhuman primates. The primary examples again occur among family-living primates. Elowson and Snowdon (1994) recorded calls from two different colonies of pygmy marmosets and subsequently combined the colonies. Within 10 weeks of housing the colonies together, adult and subadult members of both colonies showed an increase in bandwidth of *trill* calls as well as an increase in pitch. There is no obvious reason for calls to change in this way, but the results demonstrate vocal flexibility in response to a changed social environment. In a parallel study on Weid's black-tufted-eared marmosets, Ruckstalis et al. (2003) recorded *phee calls* in marmosets under stable social conditions and reported strong individual differences in call structure. Subsequently, some of the animals remained in the same colony room, but others were moved to a different colony room with unfamiliar conspecifics. When *phees* were recorded eight weeks later, *phee calls* of the marmosets in the stable social condition could still be identified, whereas those with changed social environments also exhibited changes in their individual call structure.

Another study by Snowdon and Elowson (1999) examined trill structure of pygmy marmosets while animals were living in their natal family groups and then paired each individual with an unfamiliar mate and tracked trill structure for the first six weeks after pairing. Some pairs were followed for up to three years. In every case where the calls of individual monkeys were distinct before pairing, there was a convergence in call structure to a common "pair trill" within the first six weeks of pairing. Although there were changes in call structure over the course of three years, the similarity in call structure between pair mates remained. Jorgenson and French (1998) also noted that there were clear individual differences in marmoset call structure within a year of pairing, but over the course of three years, the individually distinct signatures changed. Although they could not identify any specific cause of the vocal change, the implications of these studies are clear: marmosets are able to change their individual signatures in response to changes in their social environments, and as a consequence of this, listeners must also be able to track these changes perceptually in order to maintain individual relationships.

There is less evidence concerning vocal convergence in other family-living species. In both coppery titi monkeys (*Callicebus cupreus*) and siamangs (*Hylobates syndactylus*), adult mates alternate in producing duets. Although newly formed pairs appear to engage in duets with their partners soon after pairing, they do not match the duetting ability of long-term pairs. In the case of coppery titi monkeys, the phrases are much more variable in nearly all acoustic features of the duets (Müller and Anzenberger 2002), whereas in siamangs the pair may take several months to reach the level of coordination and pair specificity seen in long-term pairs

(Geissmann and Orgeldinger 2000). In gibbons, several variables appear to influence singing. Although general rewards such as food and water have no effect on vocalizations, social influences, such as a new mate, the maturation of offspring attempting to sing themselves, the ability to adjust to "mistakes" in calling by others, and the presence or absence of familiar or unfamiliar members of other groups, can influence singing patterns in gibbons (Haraway and Maples 1998). As with marmosets and tamarins, we see evidence of adult vocal flexibility in response to social change in other family-living species.

6.7.2 Population-Specific Dialects

Among the best evidence for song learning in birds is when different populations exhibit vocal dialects (e.g., Marler and Tamura 1964). The apparent lack of population-specific calls in nonhuman primates has been another factor in arguing against environmental influences on vocal development. An early study hypothesized the existence of dialects in the food calls of provisioned Japanese macaques (Green 1975), although the differences may have resulted from humans rewarding variant calls with food. Dialects have been described in different populations of squirrel monkeys (*Saimiri* sp.) (Newman and Symmes 1982), and members of each population responded preferentially to playback of infant separation calls from those in the same population but were indifferent to infant calls from the other population (Snowdon et al. 1985). However, genetic analyses have revealed that these phenotypically distinct populations are actually separate species (rather than subspecies). Subspecies-typical calls have been reported for wild saddleback tamarins (*Saguinus fuscicollis*) in Peru (Hodun et al. 1981), although the pelage differences are quite pronounced, again raising the question of whether these should be considered as subspecies or separate species.

The clearest data on population differences in vocalizations come from different populations of pygmy marmosets in Ecuador. De la Torre and Snowdon (2009) analyzed the structure of *trill* and *J-call* vocalizations from five populations. After accounting for individual and pair-specific differences, they showed that there were acoustic differences that differentiated each population from the other for both call types (see Fig. 6.3). Measurements of the spectrum of ambient noise and call playbacks with re-recording at different distances showed different patterns of ambient noise and reverberation in the local habitat of each population. However, the differences in habitat acoustics did not predict the call structures found in each habitat (de la Torre and Snowdon 2009). Preliminary evidence of genetic variability (de la Torre, personal communication) provides no evidence for any genetic diversity in parallel with the vocal diversity. Given the results on vocal flexibility in captive marmosets (described above), the most parsimonious interpretation of the results is that social learning or socially induced plasticity is responsible for the dialects.

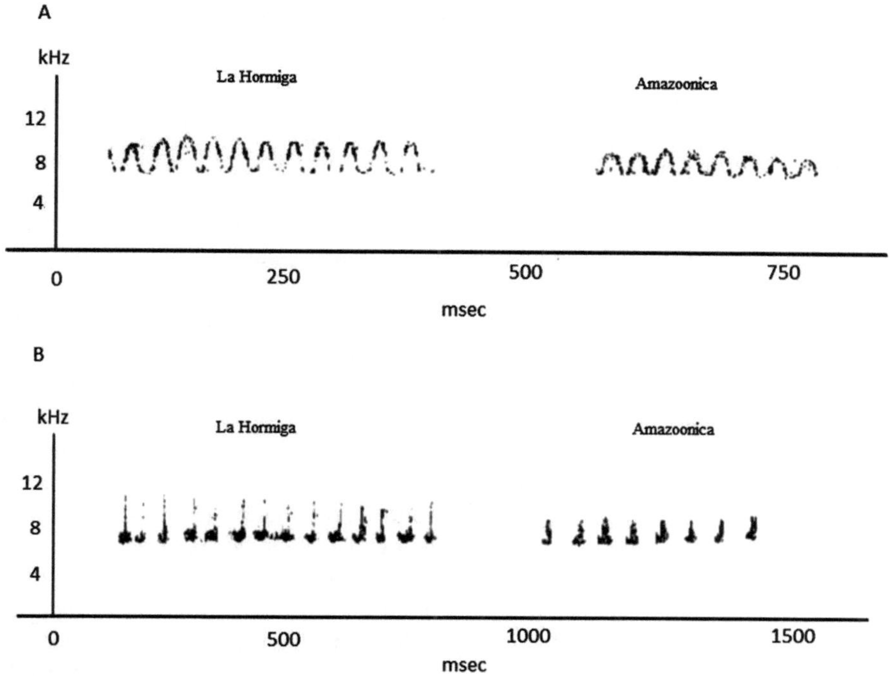

Fig. 6.3 Dialect differences in wild pygmy marmoset *trill* (**A**) and *J-calls* (**B**). Calls are from the La Hormiga population (*left*) and the Amazoonica population (*right*). (Modified from de la Torre and Snowdon 2009)

Maeda and Masataka (1987) also reported the presence of dialects in the *long calls* of two populations of the red-bellied tamarin (*Saguinus labiatus*) in Bolivia with peak frequency and the range of frequency modulation differing between the populations. Although they did not evaluate possible effects of habitat acoustics on call structure, it seems unlikely that forest vegetation differed significantly between the two populations. In a follow-up study, Masataka (1988) played back calls to captive monkeys of each population and reported that females responded selectively to the *long calls* of males from their own population. However, females showed no difference in response to female *long calls*, and males failed to differentiate between male and female calls of their own population versus the other population. This suggests that either differentiation between populations was not yet well-developed or that there was no functional significance to the dialect differences, at least for males.

An interesting experimental study played back the affiliative calls of common marmosets to simulate amicable neighbors (Watson et al. 2014). Over the course of the playback, the listeners demonstrated increased affiliation, decreased aggression, and decreased anxious behavior. Although the behavioral changes did not continue after the playbacks ended, the study does suggest how vocalizations might vary between groups and lead to distinct cultural styles.

6.7.3 Structural Change in Response to Ambient Noise

Environmental noise can have an important influence on vocal signals (Brown and Waser, Chap. 4). In the natural environments of pygmy marmosets in Peru and Ecuador, the principal frequency of marmoset calls was above the spectral range of the majority of ambient noise (mainly from birds, frogs, and insects) (Snowdon and Hodun 1981; de la Torre and Snowdon 2002). These calls were also above the hearing range of many birds that might be predators, and natural selection appears to have influenced the frequency range of calls in this species.

But are nonhuman primates capable of responding to short-term changes in the acoustic environment? One common response seen in humans and some Old World primates is the Lombard effect, an increase in vocal amplitude with increased ambient noise. Common marmosets showed evidence of the Lombard effect, increasing the amplitude of their *twitter calls* with increasing amplitude of ambient noise and increasing the duration of individual units within their *twitter calls* (Brumm et al. 2004). Similar results were found with cotton-top tamarins (Egnor and Hauser 2006), which also adjusted the timing of their calls amidst bursts of white noise to call during the silent periods (Egnor et al. 2007). Using a different methodology involving presentation of a burst of white noise in the middle of an on-going *long call* in cotton-top tamarins, Miller et al. (2003) found that the noise would interrupt the production of *long calls*, with the call terminating after completion of the syllable that was interrupted. This led the authors to conclude that the *long call* was not organized as a complete call, either cognitively or with respect to motor pattern, but rather the syllables of the call were formed discretely, suggestive of grammar.

In an extension of this paradigm, Egnor et al. (2006) found that white noise bursts during *long call* production lead to shorter notes and calls with higher amplitude and longer interpulse intervals, consistent with both the Lombard effect and the idea that tamarins can adjust their calling in a flexible way to environmental noise. Roy et al. (2011) played noise bursts that varied in duration and predictability to common marmosets. They found that the monkeys initiated calling in silent intervals under all conditions, suggesting vocal control with respect to noise. The Lombard effect, the truncation of a call in response to a burst of white noise, and the ability to initiate calls during quiet periods imply that marmosets and tamarins must have some degree of control over the structure of their vocalizations. Vocalizations are not simply due to fixed motor patterns.

6.7.4 Responses in Novel Environments

Although the structure, usage, and understanding of vocal signals by conspecifics have been shaped by natural selection in wild populations, many nonhuman primates are faced with novel environments either through captivity in zoological parks and research institutions or through increasing anthropogenic influences on natural environments. How do nonhuman primates adjust to these novel environments?

A study of pygmy marmosets in Ecuador compared groups living with high levels of anthropogenic noise before and after the capture of one or more animals for the pet trade with another population that experienced little anthropogenic influence (de la Torre et al. 2000). In groups with extensive human exposure, social play was greatly reduced, and the monkeys used higher strata within the forest compared with more isolated groups. After the capture of animals from a group, calling rates were greatly reduced. Duarte et al. (2011) reported that black-tufted marmosets (*Callithrix penicillata*) living in a park in the middle of the city of Belo Horizonte, Brazil, actively avoided the periphery of the park and more often frequented the central areas away from traffic noise. Although the authors did not record vocal activity, the monkeys may have been minimizing potential masking noise from human activities.

Captivity can be viewed as a novel environment and one can ask whether features of the captive environment can affect vocal communication. Although comparative field studies of captive versus wild populations of pygmy marmosets and cotton-top tamarins have failed to reveal any significant differences in vocal structure or in vocal repertoire, the use of calls in captivity does vary from usage in the wild. This is best illustrated in two examples. In the first example (also see Sect. 6.5), cotton-top tamarins that sampled a familiar, highly preferred food that had been adulterated by invisible white pepper produced alarm calls in this completely novel context (Snowdon and Boe 2003). The second example focused on how captive-born tamarins would react when exposed to cues of natural predators, either a live snake or audio recordings of natural predators. Captive-born tamarins did not give alarm or mobbing calls when exposed to live boa constrictors (Hayes and Snowdon 1990; Campbell and Snowdon 2007), but they did give mobbing calls to a human dressed as a veterinarian and also to a brush used to clean the light fixtures (Campbell and Snowdon 2009). Thus, captive-born monkeys failed to give alarms to a natural predator but did alarm to features of the captive environment. Despite attempting to use several different conditioning paradigms, Campbell and Snowdon (2009) were unable to train captive-born tamarins to fear snakes by associating snakes with conspecific alarm calls. When captive tamarins were played calls of natural predators and harmless sympatric herbivores, they responded to vocalizations that had low-frequency and broadband components whether or not the calls were from a predator or herbivore (Friant et al. 2008). The lack of response to natural predators or calls of natural predators suggests strongly that these captive-born monkeys do not have an innate response to predators but learn to use alarm calls in contexts that are appropriate to their captive environments.

6.7.5 Summary: Vocal Flexibility in Adults

This section has shown that adult family-living monkeys have a great deal of flexibility in vocal production and usage. When new pairs are formed or previously separated colonies are merged, there is evidence of vocal convergence toward a common pair or group structure. In duetting species, the development of a well-coordinated pair song may take several weeks or months and might be an indicator of the state of pair bonding. Whereas evidence of vocal dialects in squirrel monkeys and Japanese macaques is questionable, the presence of dialects in pygmy marmosets is clear and at present cannot be explained by adaptation to habitat acoustics or genetic divergence. The results are most parsimoniously explained as reflecting social learning processes, given the evidence of social influences from captive studies. Marmosets and tamarins are sensitive to their auditory environments and either avoid areas of possible masking, reduce calling when human activities have been disruptive, or adjust call structure by making calls louder or longer, or they interrupt calling. Finally, marmosets and tamarins adjust to captivity as an ecological niche and fail to respond to stimuli from nature, but they can direct alarm calling to novel situations found only in captivity.

6.8 Primates as Psychophysicists

Are monkeys able to adjust their calling according to principles of psychophysics? One problem faced by all species that depend on vocal communication is the accurate localization of sound sources. This is important not only for localizing predators but also for locating conspecifics. Most research on sound localization has involved the two-dimensional space in which humans and other terrestrial animals live, but localization in three dimensions creates additional problems.

An early study of vocalizations in captive pygmy marmosets described several trill-like vocalizations that were sinusoidal frequency-modulated calls that varied in bandwidth, duration, and whether the modulation was continuous or interrupted (Fig. 6.2) (Pola and Snowdon 1975). Three of these trill variants appeared to be used in similar contexts of vocal contact with other group members, but the structural differences between the calls suggested that they contained different cues for sound localization. The softest call, the *quiet trill*, was short and continuous and had a narrow bandwidth. The *closed mouth trill* was also continuous but had a larger bandwidth, and the *J-call* was a series of separate sinusoidal frequency-modulated notes with an even greater bandwidth. Based on principles of sound localization (Thurlow 1971), these three calls represent a continuum that is increasingly locatable.

Since vocal communication is risky in natural environments, one might predict that, ideally, callers would use the most cryptic calls when close to other group members and reserve the calls most easily localized to contexts when group members are widely separated. To test this prediction, two field studies on pygmy marmosets in the Amazon have been completed. When one animal called, the location between the caller and the closest visible conspecific was measured and the distances between animals plotted as a function of the call type recorded (Snowdon and Hodun 1981). With the most cryptic call, the *quiet trill*, the vast majority of nearest conspecifics was located within 5 m, whereas with the more locatable *J-call*, the majority of nearest neighbor distances was 10–15 m. In a second study, an additional call, the *long call*, was added and was primarily used when caller and recipient were more than 16 m apart (de la Torre and Snowdon 2002).

Broadcasts of pygmy marmoset calls were made in the habitat with re-recording done between 1 and 40 m from the speaker (de la Torre and Snowdon 2002). *Trills* and *J-calls* were highly distorted at 20 m, and only *long calls* could be re-recorded with minimal distortion at 40 m. The upper frequency range of each call type was degraded, as predicted by the inverse square law and the excess attenuation found in an arboreal habitat. The reduction of the upper frequency range with increasing distance provides a potential mechanism for ranging distance (Owings and Morton 1998). By using the amount of attenuation in higher frequency components, the listener could compute the distance to the caller. This distance estimating ability may allow those who respond to vocalizations to select from the repertoire of *trills* and *J-calls* the one that is most likely to be heard by others while minimizing risk of detection by potential predators.

Thus, pygmy marmosets appear to be good psychophysicists, adjusting the structure of the calls they use to maintain contact with other group members based on how far away the recipient is and how far the call is likely to travel. These results imply another type of vocal flexibility not described in many other primates, the selective use of contact call types depending on the distance from others.

6.9 Chapter Summary and Future Directions

This chapter emphasized vocal communication in family-living primates and illustrated several unique features in these species. Compared with primates with other forms of social organization, family-living primates display less sexual dimorphism in vocal communication and often use vocal signals to maintain spacing and to defend territories and mates. Vocalizations from partners reduce the stress of separation. Nonetheless, these species show similar cognitive aspects of vocal communication as seen in other species with referential signals, syntax, turn taking, and long-term vocal memory. Among the consequences of family living is an increased role for vocal signals in social learning and teaching, and to date, some of the strongest evidence for social influences on learned vocal development arises from research on family-living primates. There is good evidence for vocal flexibility and plasticity throughout adulthood as these primates are able to adjust to anthropogenic noise and the ecological niche of captivity, and they are able to apply calls in novel settings. There is good evidence of population-specific dialects, at least in pygmy marmosets. All of these findings suggest that if one is to understand the evolution of human communication, one needs to look not only at the vocal signaling of our closest relatives but also to consider the role of evolutionary convergence as illustrated by family-living primates.

However, there remain several gaps in the literature that require future research. Most of the work on social influences on vocal development and on the role of vocalization in social learning and teaching has been carried out on marmosets and tamarins, which are cooperative breeders. Will these same findings also be seen in family-living species without cooperative infant care or are these adaptations unique to cooperative breeders? Although duetting and coordinated singing in titi monkeys and gibbons appear to be important in pair bonding as well as territory defense, the current evidence provides stronger support for these calls being used by each sex to defend against same-sex intruders. More studies are needed of these species to see if duetting calls serve to strengthen or form a pair bond and whether calls from a mate can have a stress-reducing effect as seen in marmosets. Research on titi monkeys has shown the physiological and behavioral effects of mate separation, but there appears to be no work on vocal communication. The major work has been carried out in field studies on gibbons and titi monkeys and in captive studies with marmosets and tamarins. Increasing the breadth of research to include captive titi monkeys and gibbons and determining whether results from captive studies are seen in wild populations of marmosets and tamarins would be most welcome.

Acknowledgments I thank the many students and colleagues who have served as collaborators on my own research, especially Yvonne Pola, Jayne Cleveland, A. Margaret Elowson, Rebecca Roush, Cristina Lazaro-Perea, Stella de la Torre, Stella Joyce, and Tatyana Humle. Much of my research reported here was supported by USPHS Grant MH029775. I thank the editors for their critical feedback on an earlier version of this chapter.

Compliance with Ethics Requirements
Charles T. Snowdon states that he has no conflicts of interest.

References

Barelli, C., Mundry, R., Heistermann, M., & Hammerschmidt, K. (2013). Cues to androgen and quality in male gibbon songs. *PLoS One*, 8, e82748.

Bauers, K., & Snowdon, C. T. (1990). Discrimination of chirp vocalizations in the cotton-top tamarin. *American Journal of Primatology*, 21, 53–60.

Benz, J. (1993). Food-elicited vocalizations in golden lion tamarins: Design features for representational communication. *Animal Behaviour*, 45, 443–455.

Brown, G. R., Almond, R. E. A., & Bates, N. J. (2005). Adult-infant food transfer in common marmosets: An experimental study. *American Journal of Primatology*, 65, 301–312.

Brown, K., & Mack, D. S. (1978). Food sharing among captive *Leontopithecus rosalia*. *Folia Primatologica*, 29, 268–290.

Brumm, H., Voss, K., Köllmer, I., & Todt, D. (2004). Acoustic communication in noise: Regulation of call characteristics in a New World monkey. *Journal of Experimental Biology*, 207, 443–448.

Burkart, J. M., Fehr, E., Efferson, C., & van Schaik, C. P. (2007). Other-regarding preferences in a non-human primate: Common marmosets provision food altruistically. *Proceedings of the National Academy of Sciences of the United States of America*, 104, 19762–19766.

Caine, N. G., Addington, R. L., & Windfelter, T. L. (1995). Factors affecting rates of food calls given by red-bellied tamarins. *Animal Behaviour*, 50, 53–60.

Campbell, M. W., & Snowdon, C. T. (2007). Vocal response of captive-reared *Saguinus oedipus* during mobbing. *International Journal of Primatology*, 28, 257–270.

Campbell, M. W., & Snowdon, C. T. (2009). Can auditory playback condition predator mobbing in captive-reared cotton-top tamarins? *International Journal of Primatology*, 30, 93–102.

Caro, T. M., & Hauser, M. D. (1992). Is there teaching in nonhuman animals? *Quarterly Review of Biology*, 67, 151–174.

Cäsar, C., Byrne, R., Young, R. J., & Zuberbühler, K. (2012a). The alarm call system of wild black-fronted titi monkeys, *Callicebus nigrifrons*. *Behavioral Ecology and Sociobiology*, 66, 653–667.

Cäsar, C., Byrne, R. W., Hoppitt, W., Young, R. J., & Zuberbühler, K. (2012b). Evidence for semantic communication in titi monkey alarm calls. *Animal Behaviour, **84***, 405–411.

Caselli, C. B., Mennill, D. J., Bicca-Marques, J., & Setz, E. Z. F. (2014). Vocal behavior of black-fronted titi monkeys (*Callicebus nigrifrons*): Acoustic properties and behavioral contexts of loud calls. *American Journal of Primatology, **76***, 788–800.

Castro, N. A., & Snowdon, C. T. (2000). Development of vocal responses in infant cotton-top tamarins. *Behaviour*, 137, 629–646.

Chow, C. P., Mitchell, J. F., & Miller, C. T. (2015). Vocal turn-taking in a non-human primate is learned during ontogeny. *Proceedings of the Royal Society of London B: Biological Sciences*, 282, 20150069.

Clarke, E., Reichard, U., & Zuberbühler, K. (2006). The syntax and meaning of gibbon song. *PLoS One*, 1(1):e73. https://doi.org/10.1371/journal.pone.0000073.

Clarke, E., Reichard, U., & Zuberbühler, K. (2012). The anti-predator behaviour of wild white-handed gibbons (*Hylobates lar*). *Behavioral Ecology and Sociobiology*, 66, 85–96.

Clarke, E., Reichard, U., & Zuberbühler, K. (2015). Context-specific close-range "hoo" calls in wild gibbons (*Hylobates lar*). *BMC Evolutionary Biology*, 15, 56.

Cleveland, J., & Snowdon, C. T. (1982). The complex vocal repertoire of the adult cotton-top tamarin (*Saguinus oedipus oedipus*). *Zeitschrift für Tierpsychologie*, 58, 231–270.

de la Torre, S., & Snowdon, C. T. (2002). Environmental correlates of vocal communication of wild pygmy marmosets, *Cebuella pygmaea*. *Animal Behaviour*, 63, 847–856.

de la Torre, S., & Snowdon, C. T. (2009). Dialects in pygmy marmosets? Population variation in call structure. *American Journal of Primatology*, 71, 333–342.

de la Torre, S., Snowdon, C. T., & Bejarano, M. (2000). Effects of human activities on wild pygmy marmosets in Ecuadoran Amazon. *Biological Conservation*, 94, 153–163.

Deputte, B. L. (1982). Duetting in male and female songs of the white-cheeked gibbon (*Hylobates concolor leucogenys*). In C. T. Snowdon, C. H. Brown, & M. R. Petersen (Eds.), *Primate communication* (pp. 67–93). New York: Cambridge University Press.

Dooley, H. M., Judge, D. S., & Schmitt, L. H. (2013). Singing by male and female Kloss gibbons (*Hylobates klossii*) in the Pelonian Forest, Siberut Island, Indonesia. *Primates*, 54, 39–48.

Duarte, M. H. L., Vecci, M. A., Hirsch, A., & Young, R. J. (2011). Noisy human neighbours affect where urban monkeys live. *Biology Letters*, 7, 840–842.

Egnor, S. E. R., & Hauser, M. D. (2006). Noise-induced vocal modulation in cotton-top tamarins (*Saguinus oedipus*). *American Journal of Primatology*, 68, 1183–1190.

Egnor, S. E. R., Iguina, C. G., & Hauser, M. D. (2006). Perturbation of auditory feedback causes systematic perturbation in vocal structure in adult cotton-top tamarins. *Journal of Experimental Biology*, 209, 3652–3663.

Egnor, S. E. R., Wickelgren, J. G., & Hauser, M. D. (2007). Tracking silence: Adjusting vocal production to avoid acoustic interference. *Journal of Comparative Physiology A: Neuroethology, Sensory, Neural, and Behavioral Physiology*, 193, 477–483.

Elowson, A. M., & Snowdon, C. T. (1994). Pygmy marmosets, *Cebuella pygmaea*, modify vocal structure in response to changed social environment. *Animal Behaviour*, 47, 1267–1277.

Elowson, A. M., Tannenbaum, P. L., & Snowdon, C. T. (1991). Food-associated calls correlate with food preferences in cotton-top tamarins. *Animal Behaviour*, 42, 931–937.

Elowson, A. M., Snowdon, C. T., & Sweet, C. S. (1992). Ontogeny of trill and J-call vocalizations in pygmy marmosets, *Cebuella pygmaea*. *Animal Behaviour*, 43, 703–715.

Elowson, A. M., Snowdon, C. T., & Lazaro-Perea, C. (1998). Infant "babbling" in a nonhuman primate: Complex vocal sequences with repeated call types. *Behaviour*, 135, 643–664.

Friant, S. C., Campbell, M. A., & Snowdon, C. T. (2008). Captive-born cotton-top tamarins (*Saguinus oedipus*) respond similarly to vocalizations of predators and non-predators. *American Journal of Primatology*, 70, 707–710.

Galef, B. G., & Giraldeau, L. A. (2001). Social influences on foraging in vertebrates: Causal mechanisms and adaptive functions. *Animal Behaviour*, 61, 3–15.

Gauther, J. P., & Gauthier, A. (1977). Communication in Old World monkeys. In T. E. Sebeok (Ed.), *How animals communicate* (pp. 894–964). Bloomington, IN: Indiana University Press.
Gauthier, J. P., & Gauthier-Hion, A. (1982). Vocal communication within a group of monkeys: An analysis by biotelemetry. In C. T. Snowdon, C. H. Brown, & M. R. Petersen (Eds.), *Primate communication* (pp. 5–29). New York: Cambridge University Press.
Geissmann, T. (1984). Inheritance of song parameters in the gibbon song analysed in two hybrid gibbons (*Hylobates pileatus* X *H. lar*). *Folia Primatalogica*, 42, 216–235.
Geissmann, T., & Orgeldinger, M. (2000). The relationship between duet songs and pair bonds in siamangs, *Hylobates syndactylus*. *Animal Behaviour*, 60, 805–809.
Ghazanfar, A. A., Flonbaum, J. I., Miller, C. T., & Hauser, M. D. (2001). The units of perception in the antiphonal calling behavior of cotton-top tamarins (*Saguinus oedipus*): Playback experiments with long calls. *Journal of Comparative Physiology A: Neuroethology, Sensory, Neural, and Behavioral Physiology*, 187, 27–35.
Goldstein, M. H., King, A. P., & West, M. J. (2003). Social interaction shapes babbling: Testing parallels between birdsong and speech. *Proceedings of the National Academy of Sciences of the United States of America*, ***100***, 8030–8035.
Green, S. (1975). Dialects in Japanese monkeys: Vocal learning and cultural transmission of locale-specific behavior? *Zeitschrift für Tierpsychologie*, ***38***, 304–314.
Haimoff, E. H. (1986). Convergence in the duetting of monogamous Old World primates. *Journal of Human Evolution*, 15, 51–59.
Halloy, M., & Kleiman, D. G. (1994). Acoustic structure of long calls in free-ranging groups of golden lion tamarins, *Leontopithecus rosalia*. *American Journal of Primatology*, 32, 303–310.
Hammerschmidt, K., & Fischer, J. (2008). Constraints on primate vocal production. In D. K. Oller & U. Griebel (Eds.), *Evolution of communicative flexibility* (pp. 93–119). Cambridge, MA: MIT Press.
Haraway, M. M., & Maples, E. G. (1998). Flexibility in the species-typical songs of gibbons. *Primates*, 39, 1–12.
Hayes, S. L., & Snowdon, C. T. (1990). Predator recognition in cotton-top tamarins (*Saguinus oedipus*). *American Journal of Primatology*, 20, 283–291.
Herzog, M., & Hopf, S. (1983). Effects of species-specific vocalizations on the behavior of surrogate reared squirrel monkeys. *Behaviour*, 86, 197–214.
Herzog, M., & Hopf, S. (1984). Behavioral responses to species-specific warning calls in infant squirrel monkeys reared in social isolation. *American Journal of Primatology*, 7, 99–106.
Hodun, A., Snowdon, C. T., & Soini, P. (1981). Subspecific variation in the long call of the tamarin, *Saguinus fuscicollis*. *Zeitschrift für Tierpsychologie*, 57, 97–110.
Humle, T., & Snowdon, C. T. (2008). Socially biased learning in the acquisition of a complex foraging task in juvenile cottontop tamarins (*Saguinus oedipus*). *Animal Behaviour*, 75, 267–277.
Humle, T., Snowdon, C. T., & Matsuzawa, T. (2009). Social influences on the acquisition of ant dipping among the wild chimpanzees (*Pan troglodytes verus*) of Bossou, Guinea, West Africa. *Animal Cognition*, 12(1), 37–48.
Janik, V. M., & Slater, P. J. B. (1997). Vocal learning in mammals. *Advances in the Study of Behavior*, 26, 59–99.
Janik, V. M., & Slater, P. J. B. (2000). The different roles of social learning in vocal communication. *Animal Behaviour*, 60, 1–11.
Jorgenson, D. D., & French, J. A. (1998). Individuality but not stability in marmoset long calls. *Ethology*, 104, 729–742.
Joyce, S. M., & Snowdon, C. T. (2007). Developmental changes in food transfers in cotton-top tamarins (*Saguinus oedipus*). *American Journal of Primatology*, 28, 257–270.
Kinzey, W. G., & Robinson, J. G. (1983). Intergroup loud calls, range size and spacing in *Callicebus torquatus*. *American Journal of Physical Anthropology*, 60, 539–544.
Kirchhof, J., & Hammerschmidt, K. (2006). Functionally referential alarm calls in tamarins (*Saguinus fuscicollis* and *Saguinus mystax*)—evidence from playback experiments. *Ethology*, 112, 346–354.

Koda, H., Lemason, A., Oyakawa, C., Rizaldi, Pamungkas, J., & Masataka, N. (2013). Possible role of mother-daughter vocal interactions on the development of species-specific song in gibbons. *PLoS One*, 8, e71432.
Lan, D. (1993). Feeding and vocal behaviours of back gibbons (*Hylobates, concolor*) in Yunnan A preliminary study. *Folia Primatologica*, 60, 94–105.
Lazaro-Perea, C., Arruda, M. F., & Snowdon, C. T. (2004). Grooming as reward? Social functions of grooming in cooperatively breeding marmosets. *Animal Behaviour*, 67, 627–636.
Lukas, D., & Clutton-Brock, T. H. (2013). The evolution of social monogamy in mammals. *Science*, 341, 526–529.
Maeda, T., & Masataka, N. (1987). Locale-specific behaviour of the tamarin (*Saguinus l. labiatus*). *Ethology*, 75, 25–30.
Marler, P. (1970). Bird song and speech development: Could there be parallels? *American Scientist*, 58, 669–673.
Marler, P., & Tamura, M. (1964). Song dialects in three populations of white-crowned sparrows. *Science*, 146, 1483–1486.
Marshall, J. T., & Marshall, E. R. (1976). Gibbons and their territorial songs. *Science*, 193, 235–237.
Masataka, N. (1983). Categorical responses to natural and synthesized alarm calls in Goeldi's monkeys (*Callimico goeldii*). *Primates*, 24, 40–51.
Masataka, N. (1988). The response of red-chested moustached tamarins to long calls from their natal and alien populations. *Animal Behaviour*, 36, 55–61.
Matthews, S. L., & Snowdon, C. T. (2011). Long-term memory for the calls of relatives in cotton top tamarins (*Saguinus oedipus*). *Journal of Comparative Psychology*, 125, 366–369.
McConnell, P. B., & Snowdon, C. T. (1986). Vocal interactions between unfamiliar groups of captive cotton top tamarins. *Behaviour*, 97, 273–296.
Mendoza, S. P., & Mason, W. A. (1986). Parental division of labour and differentiation of attachments in a monogamous primate (*Callicebus moloch*). *Animal Behaviour*, 34, 1336–1347.
Merker, B., & Cox, C. (1999). Development of the female great call in *Hylobates gabrielliae* A case study. *Folia Primatologica*, 70, 97–106.
Miller, C. T., & Thomas, A. W. (2012). Individual recognition during bouts of antiphonal calling in common marmosets. *Journal of Comparative Physiology A: Neuroethology, Sensory, Neural, and Behavioral Physiology*, 198, 337–346.
Miller, C. T., Flusberg, S., & Hauser, M. D. (2003). Interruptibility of long call production in tamarins: Implications for vocal control. *Journal of Experimental Biology*, 206, 2629–2639.
Miller, C. T., Scarl, J., & Hauser, M. D. (2004). Sensory biases underlie sex differences in tamarin long call structure. *Animal Behaviour*, 68, 713–720.
Miller, C. T., Iguina, C. G., & Hauser, M. D. (2005). Processing vocal signals for recognition during antiphonal calling in tamarins. *Animal Behaviour*, 69, 1387–1398.
Miller, C. T., Mandel, K., & Wang, X. (2010). The communicative content of the common marmoset phee call during antiphonal calling. *American Journal of Primatology*, 72, 974–980.
Miller, R. E. (1967). Experimental approaches to the physiological and behavioral concomitants of affective communication in rhesus monkeys. In S. A. Altmann (Ed.), *Social communication among primates* (pp. 125–134). Chicago: University of Chicago Press.
Mitani, J. C. (1984). The behavioral regulation of monogamy in gibbons (*Hylobates muelleri*). *Behavioral Ecology and Sociobiology*, 15, 225–229.
Mitani, J. C. (1985). Location-specific responses of gibbons (*Hylobates muelleri*) to male songs. *Zeitschrift für Tierpsychologie*, 70, 219–224.
Müller, A. E., & Anzenberger, G. (2002). Duetting in the titi monkey *Callicebus cupreus*: Structure, pair specificity and development of duets. *Folia Primatologica*, 73, 104–115.
Newman, J. D., & Symmes, D. (1974). Vocal pathology in socially-deprived monkeys. *Developmental Psychobiology*, 7, 351–358.
Newman, J. D., & Symmes, D. (1982). Inheritance and experience in the acquisition of primate acoustic behavior. In C. T. Snowdon, C. H. Brown, & M. R. Petersen (Eds.), *Primate communication* (pp. 259–278). Cambridge: Cambridge University Press.

Norcross, J. L., & Newman, J. D. (1993). Context and gender-specific differences in the acoustic structure of common marmoset (*Callithrix jacchus*) phee call. *American Journal of Primatology*, 30, 37–54.
Norcross, J. L., & Newman, J. D. (1997). Social context affects phee call production by nonreproductive common marmosets (*Callithrix jacchus*). *American Journal of Primatology*, 43, 135–146.
Norcross, J. L., & Newman, J. D. (1999). Effects of separation and novelty on distress vocalizations and cortisol in the common marmoset (*Callithrix jacchus*). *American Journal of Primatology*, 47, 209–223.
Owings, D. H., & Morton, E. (1998). *Animal vocal communication: A new approach*. Cambridge: Cambridge University Press.
Owren, M. J., Dieter, J. A., Seyfarth, R. M., & Cheney, D. L. (1993). Vocalizations of rhesus *(Macaca mulatta)* and Japanese *(M. fuscata)* macaques cross-fostered between species show evidence of only limited modification. *Developmental Psychobiology*, 26, 389–406.
Pola, Y. V., & Snowdon, C. T. (1975). The vocalizations of the pygmy marmoset (*Cebuella pygmaea*). *Animal Behaviour*, 23, 826–842.
Porter, T. A. (1994). *The development and maintenance of heterosexual pair associations in cottontop tamarins (Saguinus oedipus oedipus)*. Ph.D. dissertation, University of Wisconsin, Madison. Proquest Dissertation Publishing 9503902.
Price, E. C., & Piedade, H. M. (2001). Ranging behavior and intraspecific relationships of masked titi monkeys (*Callicebus personatus personatus*). *American Journal of Primatology*, 53, 87–92.
Puts, D. A., Jones, B. C., & de Bruine, L. M. (2012). Sexual selection on human faces and voices. *Journal of Sex Research*, 49, 227–243.
Raemakers, J. J., & Raemakers, P. M. (1985). Field playbacks of loud calls to gibbons (*Hylobates lar*): Territorial, sex-specific and species-specific responses. *Animal Behaviour*, 33, 481–493.
Rapaport, L. G. (1999). Provisioning of young in golden lion tamarins (Callitrichidae, *Leontopithecus rosalia*): A test of the information hypothesis. *Ethology*, 105, 619–636.
Rapaport, L. G. (2006). Provisioning in wild golden lion tamarins (*Leontopithecus rosalia*): Benefits to omnivorous young. *Behavioral Ecology*, 17, 212–221.
Rapaport, L. G., & Ruiz-Miranda, C. R. (2006). Ontogeny of provisioning in two populations of wild golden lion tamarins (*Leontopithecus rosalia*). *Behavioral Ecology and Sociobiology*, 60, 724–735.
Robinson, J. G. (1979a). An analysis of the organization of vocal communication in the titi monkey, *Callicebus moloch*. *Zeitschrift für Tierpsychologie*, 49, 381–405.
Robinson, J. G. (1979b). Vocal regulation of the use of space by groups of titi monkeys, *Callicebus moloch*. *Behavioral Ecology and Sociobiology*, 6, 1–15.
Robinson, J. G. (1981). Vocal regulation of inter- and intra-group spacing during boundary encounters in the titi monkeys, *Callicebus moloch*. *Primates*, 22, 161–172.
Roush, R. S., & Snowdon, C. T. (1994). Ontogeny of food associated calls in cotton-top tamarins. *Animal Behaviour, 47*, 263–273.
Roush, R. S., & Snowdon, C. T. (1999). The effects of social status on food associated calls in captive cotton-top tamarins. *Animal Behaviour*, 58, 1299–1305.
Roush, R. S., & Snowdon, C. T. (2001). Food transfer and development of feeding behavior and food-associated vocalizations in cotton-top tamarins. *Ethology*, 107, 415–429.
Roy, S., Miller, C. T., Gottsch, D., & Wang, X. (2011). Vocal control by the common marmoset in the presence of interfering noise. *Journal of Experimental Biology*, 214, 3619–3629.
Ruckstalis, M., & French, J. A. (2005). Vocal buffering of the stress response: Exposure to conspecific vocalizations moderates urinary cortisol excretion in isolated marmoset. *Hormones and Behavior*, 47, 1–7.
Ruckstalis, M., Fite, J. E., & French, J. A. (2003). Social change affects vocal structure in a Callitrichid primate (*Callithrix kuhlii*). *Ethology*, 109, 327–340.
Savage, A., Snowdon, C. T., Giraldo, H., & Soto, H. (1996). Parental care patterns and vigilance in wild cotton-top tamarins (*Saguinus oedipus*). In M. Norconk, A. Rosenberger, & P. A. Garber (Eds.), *Adaptive radiations of neotropical primates* (pp. 187–199). New York: Plenum Press.

Savage, A., Thomas, L., Leighty, K. A., Soto, L. H., & Medina, F. S. (2010). Novel survey method finds dramatic decline of wild cotton-top tamarin population. *Nature Communications*, 1(3), 30. doi:10.1038/ncomms1030.

Scott, J. J., Carlson, K. L., & Snowdon, C. T. (2006). Lability of sex differences in long calling in cotton-top tamarins. *American Journal of Primatology*, 68, 153–160.

Seyfarth, R. M., & Cheney, D. L. (1997). Vocal development in nonhuman primates. In C. T. Snowdon & M. Hausberger (Eds.), *Social influences on vocal development* (pp. 249–273). Cambridge: Cambridge University Press.

Seyfarth, R. M., Cheney, D. L., & Marler, P. (1980). Monkey responses to three different alarm calls: Evidence for predator classification and semantic communication. *Science*, 210, 801–803.

Shepherd, R. E., & French, J. A. (1999). Comparative analysis of sociality in lion tamarins (*Leontopithecus rosalia*) and marmosets (*Callithrix kuhlii*): Responses to separation from long term pair mates. *Journal of Comparative Psychology*, 113, 24–32.

Smith, A. S., Birnie, A. K., Lane, K. R., & French, J. A. (2009). Production and perception of sex differences on vocalizations of Weid's black-tufted-ear marmosets (*Callithrix kuhlii*). *American Journal of Primatology*, 71, 324–332.

Smith, T. E., McGreer-Whitworth, B., & French, J. A. (1998). Close proximity of the heterosexual partner reduces the physiological and behavioral effects of novel-cage housing in Weid's black-tufted-ear marmoset (*Callithrix kuhlii*). *Hormones and Behavior*, 34, 211–222.

Snowdon, C. T. (1982). Linguistic and psycholinguistic approaches to primate communication. In C. T. Snowdon, C. H. Brown, & M. R. Petersen (Eds.), *Primate communication* (pp. 212–238). New York: Cambridge University Press.

Snowdon, C. T. (2009). Plasticity of communication in nonhuman primates. In M. Naguib & V. M. Janik (Eds.), *Advances in the study of behavior* (Vol. 40, pp. 239–276). Burlington, VT: Academic Press.

Snowdon, C. T., & Boe, C. Y. (2003). Social communication about unpalatable foods in tamarins (*Saguinus oedipus*). *Journal of Comparative Psychology*, 117, 142–148.

Snowdon, C. T., & Cleveland, J. (1980). Individual recognition of contact calls by pygmy marmosets. *Animal Behaviour*, 28, 717–727.

Snowdon, C. T., & Cleveland, J. (1984). "Conversations" among pygmy marmosets. *American Journal of Primatology*, 7, 15–20.

Snowdon, C. T., & Elowson, A. M. (1999). Pygmy marmosets modify vocal structure when paired. *Ethology*, 105, 893–908.

Snowdon, C. T., & Elowson, A. M. (2001). 'Babbling' in pygmy marmosets: Development after infancy. *Behaviour*, 138, 1235–1248.

Snowdon, C. T., & Hausberger, M. (1997). *Social influences on vocal development*. Cambridge: Cambridge University Press.

Snowdon, C. T., & Hodun, A. (1981). Acoustic adaptations in pygmy marmoset contact calls: Locational cues vary with distances between conspecifics. *Behavioral Ecology and Sociobiology*, 9, 295–300.

Snowdon, C. T., & Pola, Y. V. (1978). Intra- and inter-specific responses to synthesized pygmy marmoset vocalizations. *Animal Behaviour*, 26, 192–206.

Snowdon, C. T., Cleveland, J., & French, J. A. (1983). Responses to context- and individual-specific cues in cotton-top tamarin long calls. *Animal Behaviour*, 31, 92–101.

Snowdon, C. T., Coe, C. L., & Hodun, A. (1985). Population recognition of infant isolation peeps in the squirrel monkey. *Animal Behaviour*, 33, 1145–1151.

Takahashi, D. Y., Narayanan, D. Z., & Ghazanfar, A. A. (2013). Coupled oscillator dynamics of vocal turn-taking in monkeys. *Current Biology*, 23, 2162–2168.

Takahashi, D. Y., Fenley, A. R., Teramoto, Y., Narayanan, D. Z., et al. (2015). The developmental dynamics of marmoset monkey vocal production. *Science*, 349, 734–738.

Tenaza, R. (1985). Songs of hybrid gibbons (*Hylobates lar* X *H. muelleri*). *American Journal of Primatology*, 8, 249–253.

Thurlow, W. R. (1971). Audition. In J. W. Kling & L. A. Riggs (Eds.), *Experimental psychology* (3rd ed., pp. 267–268). New York: Holt Rinehart & Winston.

Visalberghi, E., & Addessi, E. (2000). Response to changes in food palatability in tufted capuchin monkeys, *Cebus apella*. *Animal Behaviour*, 59, 231–238.
Watson, C. F. I., Buchanan-Smith, H. M., & Caldwell, C. A. (2014). Call playback generates a temporary cultural style of high affiliation in marmosets. *Animal Behaviour*, 93, 163–171.
West, M. A., & King, A. J. (1988). Female visual displays affect the development of male song in the cowbird. *Nature*, 334, 244–246.
Winter, P., Hanley, P., Ploog, D., & Schott, D. (1973). Ontogeny of squirrel monkey calls under normal conditions and under acoustic isolation. *Behaviour*, 47, 230–239.
Zuberbühler, K. (2000). Interspecies semantic communication in two forest primates. *Proceedings of the Royal Society of London B: Biological Sciences*, 267, 713–718.

Chapter 7
The Primate Roots of Human Language

Klaus Zuberbühler

Abstract Human language is largely a vocal behavior, but its evolutionary origins remain elusive. Although vocalizations are also the main way by which nonhuman primates communicate and interact socially, it has been difficult to demonstrate direct transitions from nonlinguistic primate vocal communication to human language. Nonhuman primates produce and perceive sounds by specialized anatomical and neural structures also present in humans. Compared to humans, however, nonhuman primates are severely limited in the control they have over vocal production, which restricts their ability to produce rapid sound combinations and limits vocal learning. But language is also a cognitive capacity, and there is good evidence that nonhuman primates understand others' calls as given by specific individuals to specific events or as part of specific social interactions. In great apes, callers can take the past history with their audience into account by suppressing, exaggerating, and socially directing their calls in seemingly strategic ways. But there is no clear evidence that primates, apart from humans, perceive others as governed by complex mental states, especially knowledge, during acts of communication, nor is there evidence that they are motivated to seek common ground and actively inform their audience accordingly. There is also no clear indication that nonhuman primates use vocalizations for the sole purpose of social bonding. One hypothesis is that these differences in cognitive ability and social motivation may have prevented the evolution of flexible and combinatorial vocal communication in nonhuman primates.

Keywords Audience effect • Communication • Intention • Pragmatics • Primate cognition • Semanticity • Speech evolution • Syntax • Vocal tract • Vocalization

---

K. Zuberbühler (✉)
Department of Comparative Cognition, Institute of Biology, University of Neuchatel, Rue Emile Argand 11, Neuchatel CH-2000, Switzerland

School of Psychology and Neuroscience, University of St Andrews, Fife, Scotland, UK
e-mail: klaus.zuberbuehler@unine.ch

R.M. Quam et al. (eds.), *Primate Hearing and Communication*, Springer Handbook of Auditory Research 63, DOI 10.1007/978-3-319-59478-1_7

7.1 Introduction

7.1.1 *Primate Vocal Behavior*

There is something bewildering about primate vocal behavior. Monkeys, apes, and humans resemble each other in so many ways, but in terms of their vocal communication, they seem to be fundamentally different. From an early age, humans play, learn, combine, and communicate with sounds at will, while nonhuman primates are limited to finite sets of calls that develop under relatively strong genetic control and serve distinct biological functions (Fitch and Zuberbühler 2013). Call repertoires are usually very species specific with closely related species resembling each other and with few acoustic modifications within the different call types (Gautier and Gautier 1977; Snowdon et al. 1982). Even for the closest relatives of humans, chimpanzees (*Pan troglodytes*) and bonobos (*Pan paniscus*), no major differences have been reported relative to monkey vocal behavior, suggesting that humans are truly unique in terms of vocal communication. Of course, humans also possess a repertoire of species-specific vocalizations—cries, laughter, grunts—the possible remnants of a more ancestral communication system not unlike what is seen in modern nonhuman primates. For most social interactions, however, these sounds only play a minor role, at least in adult humans. Thus, the main question is how and why humans evolved an additional layer of vocal control enabling them to produce speech, a vocal behavior characterized by highly coordinated and socially learned movements of the jaws, lips, and tongue in conjunction with highly controlled laryngeal phonation, to express vast numbers of mental concepts.

7.1.2 *Transitions to Language*

At least three major evolutionary transitions toward speech-based vocal communication appear to have occurred in the human lineage (Fitch and Zuberbühler 2013). First, humans have evolved unusually high control over their sound production apparatus while keeping some of their more ancestral primate-like vocal behavior. Second, human communication is based on advanced cognitive abilities, which involve mental state attributions that allow a signaler to take other beliefs and knowledge into account. And third, human communication is based on a highly cooperative motivation by which signalers experience an urge to inform others and to use vocal behavior as a means to interact socially.

This chapter addresses each of these three evolutionary transitions in detail. The focus will be on comparative research based on the assumption that data on closely related species, especially the nonhuman primates, can reveal the evolutionary origins of the anatomical structures required for communication, the brain's cognitive capacities, and the social motivations required for spoken language.

7.2 The Comparative Approach to Studying Language Evolution

7.2.1 Human Uniqueness

Can language evolution be understood by studying nonhuman primate communication? One group of researchers has argued that "…animal communication systems have thus far failed to demonstrate anything remotely like our systems of phonology, semantics, and syntax" (see p. 8 in Hauser et al. 2014). In their view, this is not only true for animal research but also for other research disciplines as none of the disciplines concerned with language evolution, including genetics, archaeology, and evolutionary modeling, has provided relevant evidence for how and why humans have evolved language. Obviously, not everyone shares this pessimistic view, so what is the main argument?

Hauser et al. (2014) argue that the only relevant competence in language is the capacity to combine linguistic elements (phonemes, morphemes, words) into more complex structures. This is carried out by a recursive operation known as *merge*, which in its simplest form takes two items (e.g., a syntactic unit, a mental concept) to construct a set (another syntactic unit, a more complex mental concept), leading to an infinite variety of hierarchically structured expressions and mental concepts (Fig. 7.1) (Chomsky 1995). The claim is that the ability to *merge* is uniquely human and only found in linguistic processing; by default, comparative animal research cannot contribute in any meaningful way to understanding how human language evolved.

However, there is considerable disagreement in the linguistic community regarding the relevance of *merge* on both theoretical and empirical grounds. For example, linguists studying the diversity of the world's roughly 6,000 languages continue to emphasize the sheer diversity of structures at every level of linguistic organization with no evidence for any universals in language (Evans and Levinson 2009). Consequently, at this time it appears worthwhile to continue exploring the nonhuman roots of human language despite the proposals of human uniqueness by Hauser et al. (2014).

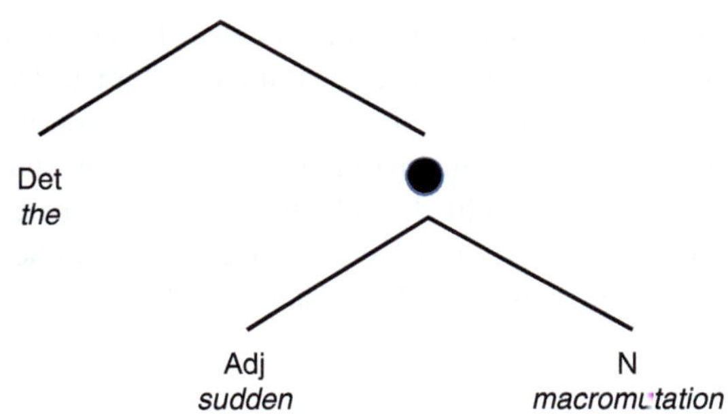

Fig. 7.1 An example of two *merge*: A lexical item ("*the*") is merged with a set previously formed by merge (*black circle*: *sudden macromutation*). Abbreviations: *N*, noun; *Adj*, adjective; *Det*, determiner

7.2.2 *Primate Models*

Perhaps more relevant is the evaluation of the empirical animal literature by Hauser et al. (2014), which led them to conclude that "… the gap between us and them is simply too great to provide any understanding of evolutionary precursors or the evolutionary processes (e.g., selection) that led to change over time." The gap between nonhuman animal and human communication is clearly great, but the purpose of this chapter is to demonstrate that this cannot be a serious argument against comparative research. The following sections are devoted to exploring the chasms between human and nonhuman primate communication, including differences in vocal control and learning, sound-meaning linkages, combinations of signal units, and the social cognition underlying human and nonhuman primate communication.

7.3 Evolution of Vocal Control

7.3.1 *Primate Vocal Tracts*

Human and nonhuman primate sounds are produced by a specialized vocal tract consisting of a sound-producing source and an acoustic filter apparatus (Fant 1960). During sound production, the larynx oscillates in response to airflow from the lungs, and this creates a basic acoustic signal, which then travels through the supralaryngeal vocal tract. The acoustic properties of the signal emitted into the environment thus are determined not only by the activity of the larynx but also by the spatial configurations of the vocal tract (the shape of the nasal and oral cavities), which determines the resonance properties and acoustic quality of the emitted sounds (Fitch and Hauser 1995).

Although only limited comparative data are available, the evidence suggests that there is a fundamental similarity in the morphological structures of the sound-producing apparatus across primates, including humans, and many other mammals (Fig. 7.2) (Riede et al. 2005; Taylor and Reby 2010; Fitch et al. 2016). One main difference is that, in adult humans, the larynx is in a permanently low position, which gives the human vocal tract a characteristic, perpendicular, two-tube shape. Whether or not this anatomical specialization is crucial for speech production has been the topic of much ongoing debate (Fitch and Reby 2001; Lieberman 2012), but arguably it is unlikely to be the key prerequisite for the evolution of vocal control and, by extension, the production of intelligible speech (Fitch et al. 2016; Quam, Martínez, Rosa, and Arsuaga, Chap. 8).

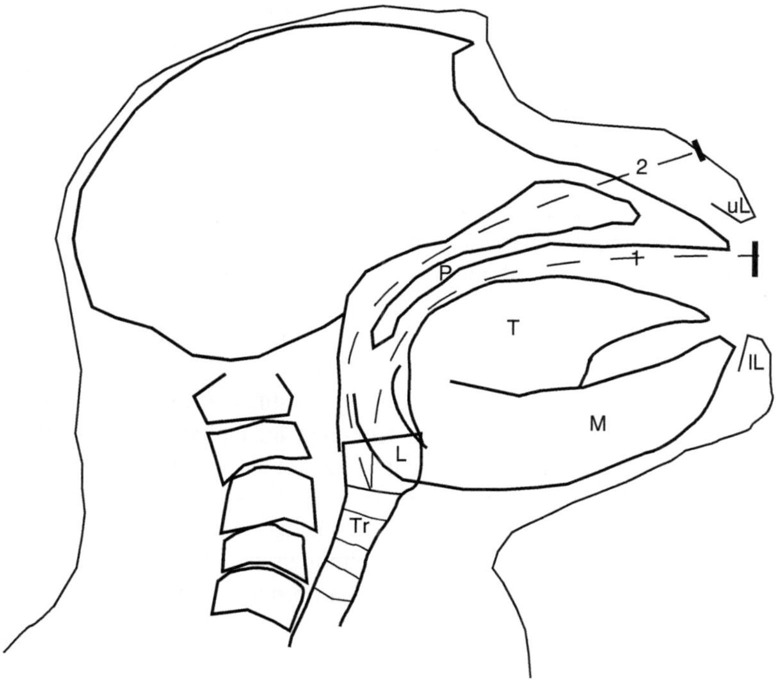

Fig. 7.2 Schematic drawing of the head-neck region of a Diana monkey with details from dissection and lateral x-ray. Abbreviations: *L*, larynx; *lL*, lower lip; *P*, palate; *T*, tongue; *Tr*, trachea; *uL*, upper lip; *dashed line 1*, oral vocal tract length; *dashed line 2*, nasal vocal tract length; *arrows* indicate the dorsoventral distances of the oral vocal tract. (Reprinted with permission from Riede et al. 2005)

7.3.2 Vocal Flexibility

Humans are undoubtedly unusual in their extraordinarily high degree of motor control of both larynx and vocal tract (Ackermann et al. 2014). Nonhuman primates are different, which impedes them from acquiring new sound patterns through vocal learning (but see Snowdon, Chap. 6). One manifestation of this is that chimpanzees all over Africa possess the same basic vocal repertoire regardless of habitat and social upbringing (Goodall 1986; Slocombe and Zuberbühler 2010). This finding is often contrasted with evidence for dialects in some marine mammals, such as killer whales (*Orcinus orca*) and sperm whales (*Physeter macrocephalus*), which serve as acoustical "badges" to secure group cohesion (Ford 2009).

Another manifestation of low vocal control in primates is that it has been nearly impossible to get chimpanzees and other primates to mimic human speech sounds even after extensive training, and learning to produce sounds on command has turned out to be a very difficult task for them. In a classic study, Hayes and Hayes

(1951) describe the vocal abilities of their home-raised and speech-trained chimpanzee "Viki" as follows: "…we began a speech training program when she was five months old. The first step was aimed at teaching her merely to vocalize on command, in order to obtain a reward. … Although she seemed to learn what was required quickly, she had serious trouble with the motor skill of voluntary vocalization. It took her five months to learn to produce a hoarse, staccato grunt, quite unlike her normal spontaneous sounds. She could do this quickly and dependably, when told to "speak," but only with much grimacing and straining. This phase of the training was also given to several laboratory chimpanzees, with similar results" (p. 107 in Hayes and Hayes 1951).

Nonhuman primate natural vocal communication is characterized by species-specific repertoires, which consist of a limited number of basic call types that are produced in relatively specific situations to serve distinct biological and social functions. In our closest relatives, the chimpanzees, the vocal repertoire consists of a few basic call types, many of them blending into each other, which makes classification difficult (Slocombe and Zuberbühler 2010). This limited flexibility in nonhuman primate vocal behavior is also striking when considering the fact that vocal learning is not uncommon in the animal kingdom, although usually restricted to courtship behavior or contact, which often involves sound-producing mechanisms other than the larynx (Janik and Slater 1997; Janik 2014).

Why are humans the only primates that have evolved such a high degree of vocal control? Although the differences are vast, there is evidence for limited vocal flexibility in some primate species (Snowdon, Chap. 6). In adult Campbell's monkeys (*Cercopithecus campbelli*), for example, contact calls of closely affiliated pairs of females are more similar than calls of socially less close individuals (Lemasson and Hausberger 2004). In chimpanzees, *pant hoot* vocalizations, a long-distance contact and display signal, are affected in similar ways, with several studies showing acoustic convergence of calls between closely affiliated males (Marshall et al. 1999; Crockford et al. 2004).

External events can further influence the acoustic variation of primate calls. For example, chimpanzee *rough grunts*, given when discovering food, vary in their acoustic structure depending on the caller's perception of the quality of the food, which is something that other group members can discriminate (Slocombe and Zuberbühler 2005) and some of which may be subject to social learning. According to one study, a group of chimpanzees brought in from a Dutch facility to Edinburgh Zoo gradually adjusted the acoustic structure of *rough grunts* to match the calls given by resident group members, as if adapting to the local communicative convention (Watson et al. 2015).

Another way by which primates can create acoustic variation is by combining acoustic units within calls. One example is the alarm call of Campbell's monkeys (*Cercopithecus campbelli*). Males produce three basic alarm calls, *krak*, *hok*, and *wak* calls, all of which can be combined with an acoustically invariable vocal suffix (*oo*) to generate a combined call (*krak-oo*, *hok-oo*, *wak-oo*) (Fig. 7.3) (Ouattara et al. 2009a, b). Unsuffixed calls are typically given in response to dangerous preda-

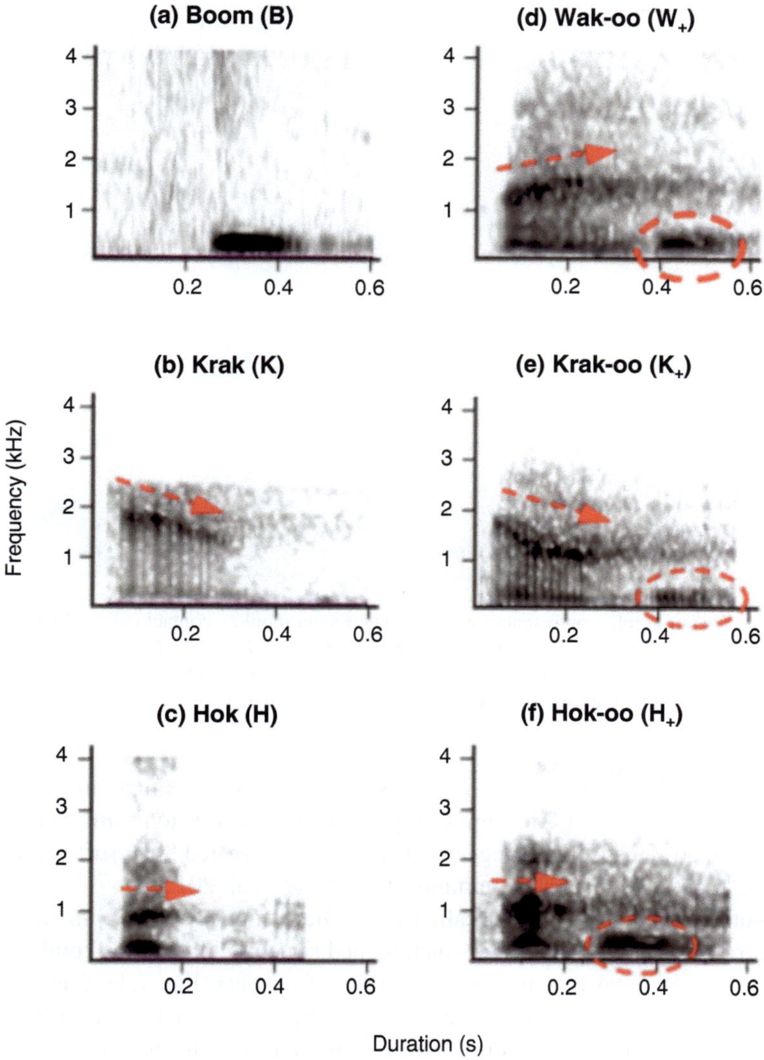

Fig 7.3 Spectrographic illustrations of the different loud call types produced by male Campbell's monkeys in different contexts. (**a**) *Boom call [B]*: a low-pitched loud call produced by the vocal sac with no frequency modulation; (**b**) *Krak call [K]*: a single loud tonal utterance of ø = 0.176 s duration with a decreasing main frequency band, starting at about 2.2 kHz; (**c**) *Hok call [H]*: a single loud tonal utterance of ø = 0.070 s with no frequency modulation at about 1.0 kHz; (**d**) *Wak-oo call [W_+]*: a suffixed loud tonal utterance of 0.330 s consisting of a call stem with an increasing main frequency band, rising from about 1.0 to 1.3 kHz, followed by a compulsory *oo suffix*; (**e**) *Krak-oo call [K_+]*: a *krak* call followed by the *oo suffix*; (**f**) *Hok-oo [H_+]*: a *hok* call followed by the *oo suffix. Dashed red arrow* indicates direction of frequency transition; *dashed red oval* indicates the *oo* suffix. (Reproduced with permission from Ouattara et al. 2009b)

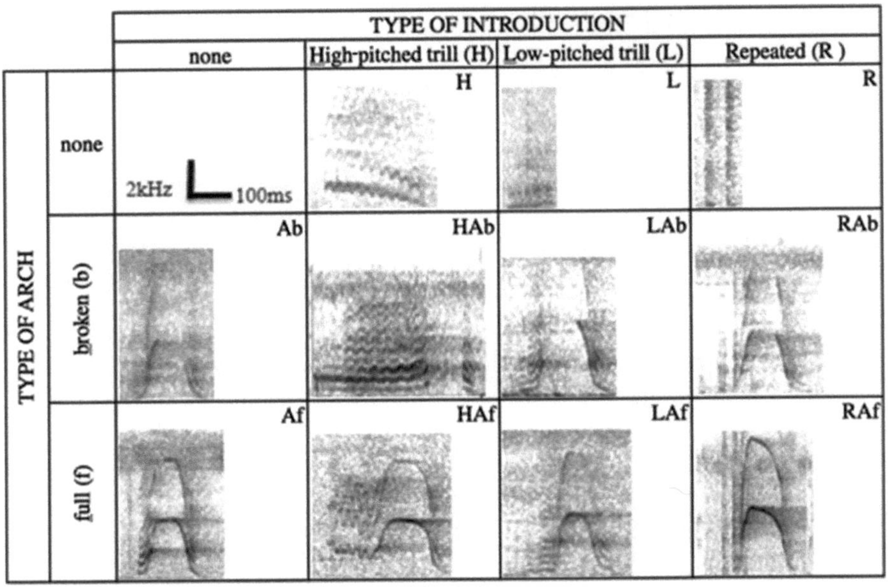

Fig. 7.4 Spectrographic representations of female Diana monkey contact calls, which consist of an optional introductory unit (*High-pitched trill, H*; *Low-pitched trill, L*; *Repeated unit, R*) followed by a broken (*b*) or full (*f*) arch (*A*). Introductory units and arches can also be produced on their own. (Reprinted with permission from Candiotti et al. 2012)

tors, while suffixed calls are associated with less dangerous situations. In playback experiments, monkeys gave significantly stronger responses to unsuffixed (leopard) than suffixed (unspecific danger) calls, which suggested that suffixation is an evolved function in primate communication (Coye et al. 2015).

Another example is contact calls by the Diana monkey (*Cercopithecus diana*) that consist of an individually distinct, arched structure that can be combined with three other call types that are linked with specific events (Fig. 7.4) (Candiotti et al. 2012). Importantly, R, L, and A call units can be given either singly or merged as RA or LA combinations. While R and L units refer to information about external events, the A units convey information about caller identity. In playback experiments, subjects responded in ways that suggested that both event type and identity information were perceived by listeners, which was an empirical demonstration of morphosemantic properties in primate social calls (Coye et al. 2016).

Despite these findings, human speech goes much beyond such phenomena, so what structures enable it? As mentioned earlier, initial explanations have highlighted differences in vocal tract anatomy, in particular the fact that humans have a permanently lowered larynx (Lieberman 2012). However, it now seems unlikely that this is sufficient to explain differences in vocal behavior between human and nonhuman primates (Quam, Martínez, Rosa, and Arsuaga, Chap. 8).

A more plausible hypothesis is that humans possess a direct cortical innervation of the nucleus ambiguous in the brain, the site of laryngeal motor control, which yields a high degree of laryngeal control during phonation (Jürgens 2002). Motor control of the filter, the supralaryngeal vocal tract, is evolutionarily more ancient since it is shared with at least the great apes. Various lines of evidence suggest that great apes have good motor control over the facial musculature, including those muscles involved in producing speech (Lameira et al. 2014). For example, captive orangutans (*Pongo pygmaeus*) can learn to mimic a caretaker's whistles by controlling the airflow passing through their lips (Lameira et al. 2013), although it is less clear whether the control of the tongue is equally advanced. However, the main point here is that parts of the speech apparatus appear to have been in place prior to the evolution of speech in humans.

Comparative ontogenetic research has also contributed to this discussion. In humans, the larynx descends during early infant development, and this process is related to the onset of speech production. However, in infant chimpanzees, the larynx also descends during early development, suggesting that relevant anatomical changes of the vocal tract during development are phylogenetically ancient (Nishimura et al. 2003). Of course, the adult vocal tract anatomy of humans and chimpanzees still differs considerably. In chimpanzees, the horizontal part of the vocal tract grows relatively more than the vertical part, while the pattern is the opposite for humans with the larynx descending more rapidly in human infants. The human-specific laryngeal descent thus may simply be a by-product of more general differences in facial developments of humans and chimpanzees (Nishimura 2005).

Another line of argument has been that the human FOXP2 gene, which plays a role in speech production in humans, is structurally different from the gene in all other primates. This is due to two relatively recent mutations during human evolution that became stabilized around 200 Ka, approximately coinciding with when modern humans evolved in Africa (Enard et al. 2002). In modern humans, deleterious mutations in the FOXP2 gene lead to severe speech disorders, apparently by affecting orofacial control during speech production (Fisher and Scharff 2009). Control of the larynx, however, does not seem to be impaired in affected patients, suggesting that FOXP2 evolution has little to add to the basic problem of what, how, and why humans evolved the capacity to control sound production. The human-specific FOXP2 gene also has been found in two Neandertal specimens (Krause et al. 2007), suggesting that the key mutations occurred before the advent of modern humans.

In sum, like all other primates, humans possess a repertoire of species-specific vocalizations – the possible remnants of an ancestral, nonhuman primate-like communication system. But humans also have evolved an additional layer of vocal control that is characterized by highly coordinated movements of the jaws, lips, and tongue in union with highly controlled sound production. While motor control of parts of the supralaryngeal vocal tract appears to be phylogenetically older and shared at least with the great apes, motor control of the larynx appears to be a recent human invention. How brain evolution and the associated laryngeal innervation changed to foster the transition from nonhuman primate to human vocal behavior is unclear. A potentially relevant point is the loss of laryngeal air sacs, present in nonhuman primates but absent in humans, which may have further facilitated the production of fine-grained vocalizations in humans (Quam, Martínez, Rosa, and Arsuaga, Chap. 8).

7.4 Reference, Inference, and Meaning in Communication

7.4.1 Information About External Entities

Much research has been devoted to the question of whether primate calls are meaningful (i.e., have an informational content), similar to how human words are meaningful (Fedurek and Slocombe 2011). This line of work has been inspired by early results from East African vervet monkeys (*Chlorocebus aethiops*), which produce acoustically distinct alarm calls to their main predators: pythons, leopards, and predatory eagles (Seyfarth et al. 1980). With playback experiments it was possible to demonstrate that vervet monkeys responded to the different calls as if the corresponding predators were present (e.g., standing bipedally to visually search the ground in response to a snake alarm). Comparable findings have been reported from other primate species, including Campbell's monkeys (Zuberbühler 2001), black-and-white Colobus monkeys (*Colobus guereza*) (Schel et al. 2010), and several lemur species (Pereira and Macedonia 1991; Fichtel and Kappeler 2002), suggesting that predator-specific alarm calls are a general feature of primate communication (Zuberbühler 2001). It is also relevant that primates (and other groups of animals) recognize alarm calls of other species, a demonstration that call recognition and comprehension is not based on some innate capacity but is acquired by observing behavioral interactions of other individuals (Zuberbühler 2000; Rainey et al. 2004) (Snowdon, Chap. 6).

Alarm calls are not the only class of signals that refer to external entities. Some animals also produce acoustically distinct calls when finding food, with acoustic variations that sometimes convey something about the perceived value of the food (Fig. 7.5) (Scarantino and Clay 2015). Similar to alarm calling, food calls thus refer to distinct external events, probably mediated by specific internal emotional/psychological states, which, on the surface, appear to have negative consequences for the caller, since these calls are likely to increase feeding competition for the caller or attract a predator's attention, respectively. However, observations and field experiments with chimpanzees have shown that callers are very selective in when they produce alarm or food calls, ensuring that social allies and other important group members are the main beneficiaries (Crockford et al. 2012; Fedurek and Slocombe 2013; Schel et al. 2013a).

These examples go to the heart of the difficulties in deciding whether primate calls reflect the emotional/psychological state of a caller or whether they have an informational content. In this and many other cases both aspects seem to matter, suggesting that primate calls have a dual nature.

Primate vocal responses to external events, such as to foods or predators, are part of a more general pattern seen across nonhuman primate signaling systems. Most nonhuman primate calls serve relatively specific biological functions: they are given in very specific social situations or given to specific external events to the effect that recipients can draw inferences about the event experienced by the caller almost by default. For example, primates, including humans, produce specific vocalizations during aggressive interactions with aversive effects on opponents, probably to facilitate rapid learning by operant conditioning (Gouzoules et al. 1984; Owren and

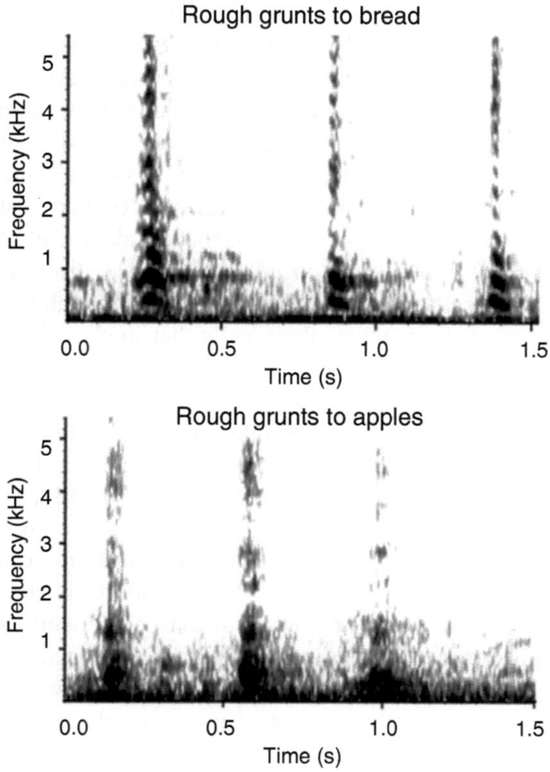

Fig. 7.5 Time-frequency spectrograms of chimpanzee food calls (*rough grunts*) given by an adult male at Edinburgh Zoo to bread and apples. Bread is the more preferred food, and the corresponding grunts have more energy (*depicted by the darkness of the image*) at higher frequencies and a clearer harmonic structure in comparison to the lower-pitched, noisier grunts to apples. (Reproduced with permission from Slocombe and Zuberbühler 2005)

Rendall 2001). At the same time, any such tight signal-event link allows nearby listeners to draw inferences about the nature of the ongoing event. Calls come to convey information about an external entity or social event (Slocombe et al. 2010a).

Interestingly, during fights, chimpanzees sometimes produce sequences consisting of two different types of calls: *barks* directed at the aggressor to signal readiness to retaliate and *screams* directed at allies to solicit their help (Fedurek et al. 2015). *Screams* also show event-related acoustic variation that roughly encodes the severity of the attack, and field experiments have shown that listeners can discriminate this information readily (Slocombe and Zuberbühler 2007). Chimpanzees that are victims of aggression, in other words, appear to address two different audiences with their calls with two different intentions.

7.4.2 *Symbolic Information*

Are primate calls symbolic? Most definitions of "symbol" are based on notions of signal arbitrariness and reference to something else, either by association or by convention. A symbol thus represents, stands for, or suggests something else, usually an idea or an object. Since it is clear that primate alarm calls can refer to relatively specific predator classes (Marler 1998; Zuberbühler et al. 1999), discussions about the symbolic nature of primate calls usually center around the notion of signal arbitrariness. From a signaler's point, alarm calls are not really arbitrary because nonhuman primates are predisposed from birth to produce alarm calls to some classes of events, such as "flying things," and not others (Seyfarth and Cheney 1986). From a recipient's point, however, alarm calls are entirely arbitrary, as demonstrated by research on interspecies alarm call recognition. Black-casqued hornbills (*Ceratogymma atrata*), for instance, discriminate between eagle and leopard alarm calls given by Diana monkeys, although there is nothing in the signal structure of the monkey alarm calls that implies the predator referred to by the calls (Rainey et al. 2004).

However, the *one call-one meaning model* of nonhuman primate communication is not always accurate. Similar calls are often given to a range of different and sometimes seemingly unrelated events, suggesting that recipients need to interpret the meaning of a call by making pragmatic decisions (Wheeler and Fischer 2012). For example, the most common call type in bonobos, the *peep*, is given by individuals in response to a wide range of social situations, as if to comment on the high significance of an event, rather than its nature (Clay et al. 2015), similar to human pointing. Also, many primates have unspecific alert calls that are given to a range of disturbances, including intraspecies conflicts, and terrestrial alarms are usually given to a range of animals, which can include nonpredators, suggesting that listeners need to rely on context to extract the exact meaning of a call (Arnold and Zuberbühler 2013).

7.4.3 Information about Caller Identity

Primate vocalizations are meaningful at multiple levels. For example, many call types carry individual acoustic signatures that enable receivers to identify the caller (Lemasson et al. 2005; Clay and Zuberbühler 2012). In chimpanzees, individuals recognize each other by their loud *pant hoot* vocalizations (Fig. 7.6) and can discriminate the calls of neighboring males from the calls of unknown stranger males (Herbinger et al. 2009). *Pant hoots* are different from all other calls within the chimpanzee repertoire, including *pant grunts* (given to food), in that they consist of four distinct units, at least one of which (the climax) carries over very large distances. Different units contain different information, including caller identity, age, rank, and behavioral context (e.g., arriving at a food tree versus traveling) (Fedurek et al. 2016).

Although there is widespread evidence for individually distinct calls in almost all primate communication systems that have been analyzed, it is important to point out that there are exceptions. For example, male Gelada baboons (*Theropithecus gelada*) do not react more strongly to experimentally presented grunts of rival males (simulating their approach) compared to nonrival males, suggesting that they do not use these vocalizations to recognize other group members (Bergman 2010).

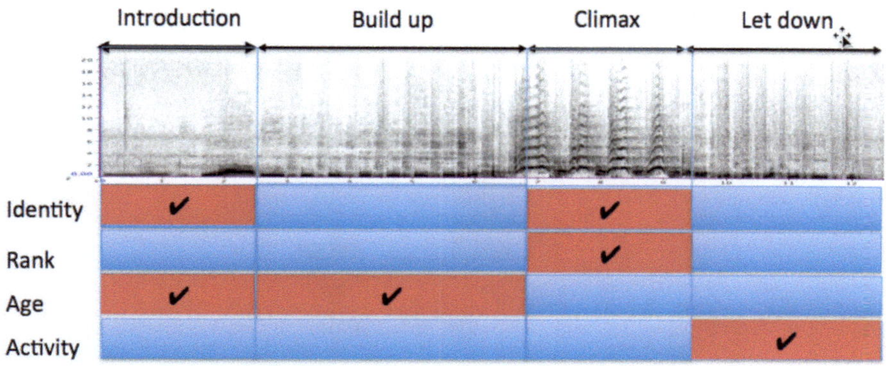

Fig. 7.6 Chimpanzee pant hoots are acoustically complex, long-distance calls, mainly produced by the adult males. They consist of four acoustically distinct units: *Introduction*, *Build up*, *Climax*, and *Let down*. Each unit contains distinct information, including caller *identity*, *social rank*, *age*, and *activity* (travel versus food), as indicated with the checked pink boxes. (Modified from Fedurek et al. 2016)

7.4.4 Call Sequences

Another line of research has found that the relevant information units in animal communication are sometimes not at the level of individual calls but can reside in sequences of calls (Kershenbaum et al. 2016). Primate examples of meaningful call sequences include the alarm call system of black-and-white Colobus monkeys in which sequence length correlates with predator type (Schel et al. 2009); putty-nosed monkey (*Cercopithecus nictitans*) alarm calls (Fig. 7.7) in which different call combinations encode predator class and travel intention (Arnold and Zuberbühler 2006, 2008); Campbell's monkey alarm calls in which call combinations discriminate between predatory and nonpredatory dangers and also predator type (Ouattara et al., 2009a, b); and black-fronted titi monkeys (*Callicebus nigrifrons*) in which different call combinations encode predator class and location (Cäsar et al. 2013).

Apart from the putty-nosed monkeys, it is still largely unclear whether these sequences have evolved specifically to convey meaning or whether they are a by-product of a caller's changing perceptions as an event unfolds—something that needs to be addressed with targeted experiments (Schlenker et al. 2014). For apes, the songs of gibbons are of special interest, representing a vocal behavior with complex sequential structure by which the mated pair advertises social information relevant to neighboring individuals (Geissmann and Orgeldinger 2000; Geissmann 2002). Lar gibbons (*Hylobates lar*) also sing when encountering predators, and acoustic analyses have demonstrated that predator-induced songs and duet songs are assembled from the same song unit repertoire but with different syntactic structures (Clarke et al. 2006). The bonobos are another example of primates who produce acoustically variable calls when finding food. The different call variants are given in combinations, and the value of the food source determines the composition of the

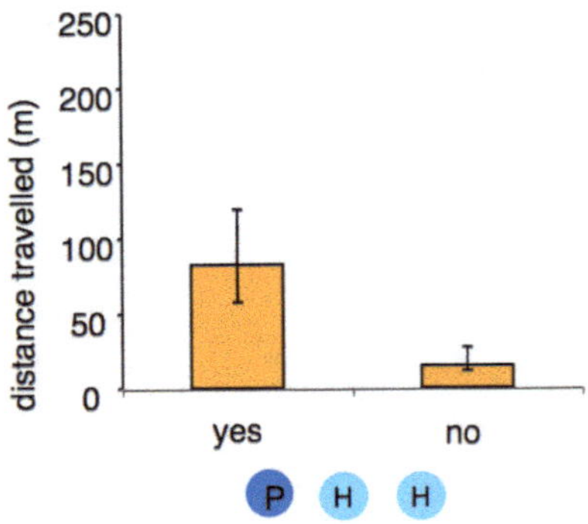

Fig. 7.7 Male putty-nosed monkeys produce two basic types of alarm calls: *pyows* (*P*) and *hacks* (*H*). Males produce different sequences of calls, including series of *hacks* (to eagles), series of *pyows* (to ground predators), and short *pyow-hack combinations*, consisting of one or a few *pyows* followed by a few *hacks*. The distance traveled refers to the group movement once a call sequence has been emitted. Call sequences can (*yes*) or cannot (*no*) contain *pyow-hack combinations*. Sequences with *pyow-hack combinations* consistently lead to more group travel than sequences without *pyow-hack combinations*, suggesting that males produce them to initiate group travel. (Reprinted with permission from Arnold and Zuberbühler 2006)

sequence, which is perceived and discriminated by others (Clay and Zuberbühler 2011). These empirical data have generated the hypothesis that some basic linguistic principles also apply to animal communication (Kershenbaum et al. 2016). Human language follows a number of linguistic laws that may also explain patterns in nonlinguistic animal communication (Schlenker et al. 2016). There is a considerable literature on this problem, already recognized by the pioneers of animal communication research (e.g., Sebeok 1977) and linguists interested in evolutionary questions (e.g., Hockett 1960).

One interesting problem is whether the patterns found in animal sound combinations are more similar to the notions of phonology or of syntax. In language, phonology refers to the process of forming meaningful units from meaningless sounds, an arguably simpler layer of combination than syntax, which refers to the combination of meaningful units. However, Collier and colleagues have reviewed examples of sound combinations in animal communication and concluded, surprisingly, that they are better explained as syntactic rather than phonological systems, suggesting that syntax evolved before phonology (Collier et al. 2014). Another linguistic principle, *Menzerath's Law* (Cramer 2005), states that in linguistic structures there is a negative relationship between the number of syllables per word and the size of individual syllables. In an empirical study on male gelada baboons, the vocal sequence length negatively correlated with the duration of the composite calls, partly because call types were more abbreviated in longer sequences, suggesting that this principle is not restricted to linguistic constituents (Gustison et al. 2016).

In sum, nonhuman primates have managed to overcome some of the constraints of their limited vocal repertoires by combinatorial organization of some call types. There is evidence that receivers benefit from this information (which is probably not provided intentionally) as if the sequences and combinations are meaningful to them, suggesting that some basic principles that govern human language structure are also found in primate communication. Second, although signalers may not produce signals with the sole intention to inform others, receivers are capable of inferring the likely cause of the events associated with vocal signals, sometimes by taking contextual information into account. Primate signals are produced to serve specific biological and social functions, such as to dissuade predators with alarm calls, to recruit help with agonistic calls, or to induce male competition with copulation calls, but these contingencies are readily absorbed and actively interpreted by listeners, representing a sort of by-product semantics.

7.5 Socially Aware Communication

7.5.1 The Question of Meaning

What exactly is "meaning" in primate calls? The evidence reviewed so far provides little doubt that nonhuman primates, and perhaps many other species of animals, can extract useful information from signals (Kaplan 2014). But maybe there is nothing special about these studies because, as stated by Tomasello (2008, p. 19): "… the monkey has simply learned that one thing predicts another, or even causes another, in the same basic way as many other phenomena in their daily lives."

It is true that humans extract meaning from speech signals, not just in terms of the eliciting stimuli that trigger the behavior but as part of ongoing sociocognitive interactions between signalers and receivers. More specifically, humans will make semantic inferences based on the current circumstances shared with the signaler, their prior interaction history and, most importantly, what they perceive as the signaler's intended meaning (Tomasello et al. 2005). A basic maxim of human communication, then, is that interlocutors intend to say what is relevant to their recipients, while recipients assume that their interlocutors intend to say something relevant to them, which requires higher levels of intentionality than simple goal directedness (Grice 1969; Dennett 1983). Underlying this mutual ascription of intention to be meaningful is a cognitive ability to attribute mental states to others. A key question in comparative research, therefore, concerns the degree to which nonhuman primates are able to base acts of communication on the knowledge and mental states of others; that is, whether they can perceive others as independent minds with their own intentions, beliefs, and knowledge (Call and Tomasello 2008).

7.5.2 Audience Effects

Although there is little targeted research on the problem of mental state attribution in communication studies, it has certainly been demonstrated that nonhuman primate vocal behavior can be influenced by the nearby audience (Zuberbühler 2008).

In social interactions, for instance, chimpanzees can time the delivery of calls as a function of the partner's attention, and they can inhibit call production in the presence of unfavorable audiences (Hostetter et al. 2001). Also, chimpanzee *pant grunts*, a greeting call given by low-ranking individuals when encountering higher ranking group members, are suppressed by the presence of the top-ranking alpha male (Laporte and Zuberbühler 2010). If chimpanzees are victims of aggression, they can give acoustically variable victim screams and use them to their own advantage by producing calls that indicate more severe aggression than actually happened if high-ranking bystanders are nearby to interfere on their behalf (Slocombe and Zuberbühler 2007). Similarly, chimpanzee females are more likely to produce copulation calls if high-ranking males are in the vicinity (Townsend et al. 2008), presumably to encourage promiscuity and diffuse paternity, which is an adaptive strategy for females in multimale groups. At the same time, copulation calling decreases with increasing numbers of females in the audience, presumably due to intrasexual competition.

7.5.3 Intentional Communication

Chimpanzee call production also shows signs of basic intentionality in the sense of goal-directed, socially targeted behavior (Crockford et al. 2015). In field experiments and naturalistic observations, callers preferentially call upon the arrival of some group members, mainly allies and high-ranking individuals, but they remain silent to others (Slocombe et al. 2010b; Schel et al. 2013b). Also relevant is that the grunts of young chimpanzees do not show any of the context-specific acoustic variation seen in adults, including *rough grunts* to food (Laporte and Zuberbühler 2011), suggesting that the production of acoustically distinct food grunts only emerges after they become aware of the social importance of signaling food discoveries to others.

Audience-aware communication is also seen during gestural communication, such as during begging for food in orangutans (Cartmill and Byrne 2007) or when bonobo males and females vocally advertise their activities to selected audiences (Clay et al. 2011; Genty et al. 2014). In monkeys, audience-aware communication has also been found in various species and contexts, including studies on alarm calls (Wich and de Vries 2006; Papworth et al. 2008) or social interactions (Semple et al. 2009).

In sum, nonhuman primate signal production is governed by social factors with evidence for partial awareness of the social consequences of signal production. However, the cognitive processes responsible for these effects remain unclear. Possibly, behavioral patterns are based on simple associations of social categories, such as dominance or affiliation, rather than the identity, shared history, and mental states of their recipients. In other words, although nonhuman primates are able to perceive others as intentional beings (Call et al. 2004), they do not necessarily keep track of what others know or believe (Tomasello et al. 2003).

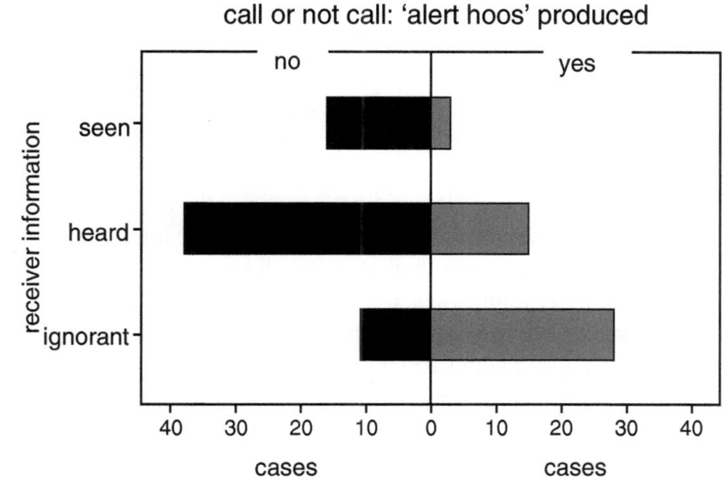

Fig. 7.8 Wild chimpanzees take the knowledge state of their audience into account when encountering snakes and are more likely to produce alarm calls to ignorant than to knowledgeable individuals. *Black* indicates no alarm calls produced; *gray* indicates at least one alarm call produced. *Receiver information* indicates receiver ignorance or knowledge from the perspective of the subject who produced the call. *Seen* indicates knowledgeable receivers: the subject had seen all receivers see the snake model. *Heard* indicates partially knowledgeable receivers: the subject had heard an alarm call when all receivers were within 50 m of the snake model but could not have seen all receivers see the snake model. *Ignorant* indicates that the subject could not have seen all receivers see the snake and had not heard an *alert hoo* when all current receivers were within earshot (50 m) of the *alert hoo*. (Reprinted with permission from Crockford et al. 2012)

A noteworthy exception is a series of studies on snake alarms in wild chimpanzees (Fig. 7.8), with evidence that callers are more likely to call to benefit ignorant than knowledgeable audiences (Crockford et al. 2012; Schel et al. 2013a). However, more research is needed to confirm these initial results, as well as to test whether knowledge-based audience awareness is also available to other primates.

7.6 Communication as Cooperation

7.6.1 The "Interaction Engine"

As discussed in Sect. 7.4, language requires capacities to extract meaning by inference and to perceive others as governed by mental states with an intention to inform (Tomasello 2008). But it also requires a basic cooperative mind and a desire to engage in joint activities, which is often initiated, maintained, and terminated linguistically (Tomasello 2014a). Thus, another possible key ingredient of the linguistic mind is what has been termed the *interaction engine* in humans (Levinson 2006).

During social interactions, humans display a set of inherited tendencies, capacities, and motivations that produce characteristic outputs that enable them to interact effectively with or without language and across cultures. Interactions are structured into sequences with highly synchronized individual actions governed by mutually shared expectations, which are understood as a function of social roles and coordinated via multimodal signals.

This highly structured way of interacting is particularly visible in linguistic discourse, characterized by rapid exchanges of short syntactic units (Levinson 2016). Humans produce about 1,500 such short turns per day when talking to each other as a major part of their daily linguistic activity. Length of turns can be flexible, as can the number of speakers, but one important principle is that participants always avoid overlap and observe gaps of about 200 ms between turns. This requires participants to predict when a partner's turn comes to an end, and these predictions must be based on assessing the semantic content while the forthcoming production unit is already constructed, which is a considerable cognitive challenge. As this exchange principle is found across languages, the hypothesis is that this kind of turn taking is based on an ancient biological predisposition with possibly deep evolutionary roots. In fact, Levinson (2016) has found vocal and gestural turn taking across the entire primate order, particularly in pair-bonded species where males and females engage in duetting behavior and in apes that engage in gesturing. The fact that the content of most human dialogue is "gossip" (Dunbar 1996) is a vivid demonstration of the social nature of language use that is governed by an interaction engine likely to be evolutionarily older than language itself.

7.6.2 Joint Intentions

One of the most remarkable features of human sociality is the high degree of cooperation visible at every level of human behavior. Compared to nonhuman primates, humans are much more collaborative, prosocial, and willing to conform to and impose social norms (Tomasello 2014b). This cooperative predisposition can explain much of what makes our species unique, from language to social norms and institutions to other aspects of culture. For example, although cultural (i.e., shared, learned) behavior has been described in nonhuman primates, particularly in chimpanzees (Whiten et al. 1999), human cultures go much beyond what is usually seen as group-specific foraging or tool use traditions in animals. Human cooperation allows for the division of labor, which gave early humans a distinct competitive advantage over other species. So how did humans evolve their cooperative minds?

A first hypothesis is that cooperative motivation has initially evolved in the context of parental behavior. Unlike most other primates, humans are highly cooperative breeders where offspring are regularly tended to by relatives and genetically unrelated individuals, in both traditional hunter-gatherer and modern industrialized societies (Hrdy 2009). Probably related to this is the fact that humans exhibit stronger prosocial behavior (defined as voluntary behavior intended to benefit another) in standardized behavioral tests than most nonhuman primates. Cooperative breeding, according to this hypothesis, has had consequences for human social cognition to the subsequent benefit of other domains of human behavior (Burkart et al. 2009). In particular, Tomasello et al. (2005) have argued that humans are unique in aligning their own intentions with the intentions of their social partners, due to high levels of awareness of others' mental states, which is a most powerful facilitator of joint activities. Unlike other great apes, none of which have a cooperative breeding system, early humans may have experienced a unique addition to their already available cognitive tool kit that fundamentally changed the way individuals could interact with each other and their environment (Burkart et al. 2009).

A second hypothesis is in terms of cooperative foraging, which centers on the idea that survival in open savanna habitats has been possible only if group members cooperated during foraging (Tomasello et al. 2012). Although chimpanzees also cooperate during hunting, this is more on an ad hoc basis, whereas human cooperation during hunting and other foraging activities appears to have become mandatory for survival. According to Tomasello et al. (2012), this led to high levels of interdependency between group members and social partners became essential assets for survival. Helping others thus became a self-serving activity, a stable system provided there are reliable mechanisms to prevent cheaters from exploiting such behavior. Here, again, the hypothesis is that this generated additional evolutionary selective pressure toward strengthening social cognition, in particular, the ability to perceive others' mental states, enabling individuals to perceive and share each other's intentions.

Third, human cooperation may have evolved in the context of intergroup conflict. Early humans may have experienced severe selection pressure to defend their home ranges from neighboring groups, which is likely to have favored cooperative motivation among group members. In ants there is a relationship between degree of cooperation and levels of intergroup conflict (Hölldobler and Wilson 1990), and it has been suggested that similar patterns apply to humans (Sterelny 2016). However, intergroup conflict is quite common in most species of nonhuman primates and can be a source of intense selection pressure in chimpanzees (Wilson et al. 2014). But in humans, intergroup conflict often goes beyond mere resource defense and can additionally be based on cultural practices.

In the two-step evolutionary scenario proposed by Tomasello et al. (2012), humans first evolved mandatory collaboration in the context of foraging, which then paved the way for a second step, the formation of cultural groups (Tomasello et al. 2012). This may have been facilitated by increases in population size and increased competition, but if shared skills and values are a prerequisite for the trust needed in collaboration, it would have created an additional need to protect social groups from invasions of other ways of life (i.e., cultural practices).

Whatever the origin, humans are intrinsically more cooperative than any other primate species. It is reasonable to assume that hypercooperation evolved early during human evolution and that this led to the coevolution of advanced communication skills. For example, across primates, vocal repertoire size is correlated with social complexity (McComb and Semple 2005), suggesting that in primates and perhaps other groups of animals, the number of social interactions and the complexity of social relationships are main driving forces behind the evolution of complex communication.

Human social behavior is unique in terms of its high levels of cooperation with an associated cognitive capacity of taking the mental perspectives of others into account. Genetically fixed signal repertoires may simply be insufficient to manage social interactions at this level. Understanding others' mental states, including their intentions, goes beyond simple representations of events and requires the ability to think in compositional recursive terms, that is, mentally representing someone else's mental representations. Compositional thought, in other words, may have paved the way for the evolution of compositional communication, which enables individuals, by means of convention, to construct and understand an unlimited number of complex expressions assembled from simpler ones.

7.7 Conclusions

Humans split from the last common ancestor with nonhuman primates approximately 6–7 million years ago (Langergraber et al. 2012). An important scientific problem is to determine the evolutionary changes that took place before and after this split and how these changes gave rise to complex behaviors such as language. There is considerable support for the idea that the basic components of human cognition, including the ones required for the language faculty, did not emerge de novo but have deep evolutionary roots in the primate lineage. The depth of these evolutionary roots can only be determined by comparative studies of nonhuman primates in the wild and in laboratory settings. This research program has been widely adopted and continues to produce valuable data, in sharp contrast with the view of Hauser et al. (2014) that nothing can be learned from comparative primate research on the grounds that nonhuman primates (a) show no signs of vocal learning with only poor voluntary control over call structure; (b) produce innately specified signals that refer to observable external events only; (c) do not have abstract concepts; and (d) do not combine smaller units into larger ones to create new meanings.

The findings reviewed here paint a rather different picture with evolutionary continuity visible at all levels. In particular, the evidence suggests that great apes, and possibly other groups of primates, appear to have reasonable control over their supralaryngeal vocal tract, although they struggle to control the larynx voluntarily. What selective forces enabled humans to evolve the required enhanced motor control for speech is unclear, but the result was a gradual transition from a genetically fixed vocal repertoire, as in nonhuman primates (remnants of which are still present in modern humans), to an additional ability to produce the flexible, compositional, conventionalized, and socially learned acoustic structures seen in modern language. Some nonhuman primate species have partially overcome the constraints of limited vocal repertoires by combinatorial organization of some call types, although they still grossly underutilize the communicative potential of this capacity.

There is also overwhelming evidence that nonhuman primates can extract meaning from others' calls, sometimes by taking into account the pragmatic context and likely cause of the calls or call sequences. A basic pattern here is that specific events trigger specific vocalizations, which are recognized and interpreted by listeners. Yet there is still a profound discrepancy between signalers generally uttering vocalizations to serve specific functions— such as to interfere with the hunting tactic of predators, to recruit nearby allies, or to incite male-male competition—and recipients able to infer detailed representations of the causative events in the absence of any direct experience. Hence, although animals may not produce signals with a targeted intention to inform others, receivers are capable of inferring the likely cause of the events associated with vocal signals, sometimes by taking contextual information into account.

At the same time, there is evidence in the great apes that signalers are aware of the social consequences of some of their signals, but there is no strong evidence that they also communicate to actively inform others or take their mental states into account during call production. It is possible that (socially aware) signal outputs are based on simple associations of social categories, such as dominance or affiliation, rather than the shared history with their recipients and the mental states induced by this, such as knowledge or ignorance. Although primates can perceive intention in others, they do not appear to routinely consider higher mental status, such as beliefs or knowledge, when addressing others. Similarly, there is no good evidence that primates are generally motivated to use their communication signals in a cooperative way to inform others about facts or events that are relevant to others but not themselves. Noteworthy exceptions are in the domains of predator encounters when signalers apparently can be concerned about the well-being of others, whereas other forms of seemingly altruistic signaling, such as when encountering food, can be explained by callers trying to avoid negative consequences if they fail to advertise their finds. Although the differences are vast, human linguistic communication shows a clear evolutionary continuity with precursors seen in nonhuman primates in almost every relevant capacity.

Acknowledgments Much of the research reviewed here has been funded by the European Research Council (PRILANG 283871) and by the Swiss National Science Foundation. I thank the Royal Zoological Society of Scotland for providing core support for the Budongo Conservation Field Station and the editors for their valuable comments.

Compliance with Ethics Requirements
Klaus Zuberbühler declares that he has no conflict of interest.

References

Ackermann, H., Hage, S. R., & Ziegler, W. (2014). Brain mechanisms of acoustic communication in humans and nonhuman primates: An evolutionary perspective. *Behavioral and Brain Sciences*, 37(6), 529–546.

Arnold, K., & Zuberbühler, K. (2006). Semantic combinations in primate calls. *Nature*, 441(7091), 303.

Arnold, K., & Zuberbühler, K. (2008). Meaningful call combinations in a non-human primate. *Current Biology*, 18(5), R202–R203.

Arnold, K., & Zuberbühler, K. (2013). Female putty-nosed monkeys use experimentally altered contextual information to disambiguate the cause of male alarm calls. *PLoS One*, 8(6), e65660.

Bergman, T. J. (2010). Experimental evidence for limited vocal recognition in a wild primate: Implications for the social complexity hypothesis. *Proceedings of the Royal Society of London B: Biological Sciences*, 277(1696), 3045–3053.

Burkart, J. M., Hrdy, S. B., & Van Schaik, C. P. (2009). Cooperative breeding and human cognitive evolution. *Evolutionary Anthropology*, 18(5), 175–186.

Call, J., & Tomasello, M. (2008). Does the chimpanzee have a theory of mind? 30 years later. *Trends in Cognitive Sciences*, 12(5), 187–192.

Call, J., Hare, B., Carpenter, M., & Tomasello, M. (2004). 'Unwilling' versus 'unable': Chimpanzees' understanding of human intentional action. *Developmental Science*, 7(4), 488–498.

Candiotti, A., Zuberbühler, K., & Lemasson, A. (2012). Context-related call combinations in female Diana monkeys. *Animal Cognition*, 15(3), 327–339.

Cartmill, E. A., & Byrne, R. W. (2007). Orangutans modify their gestural signaling according to their audience's comprehension. *Current Biology*, 17(15), 1345–1348.

Cäsar, C., Zuberbühler, K., Young, R. J., & Byrne, R. W. (2013). Titi monkey call sequences vary with predator location and type. *Biology Letters*, 9(5), 20130535. doi:10.1098/rsbl.2013.0535.

Chomsky, N. (1995). *The minimalist program*. Cambridge, MA: MIT Press.

Clarke, E., Reichard, U. H., & Zuberbühler, K. (2006). The syntax and meaning of wild gibbon songs. *PLoS One*, 1(1), e73.

Clay, Z., & Zuberbühler, K. (2011). Bonobos extract meaning from call sequences. *PLoS One*, 6(4), e18786.

Clay, Z., & Zuberbühler, K. (2012). Communication during sex among female bonobos: Effects of dominance, solicitation and audience. *Scientific Reports*, 2.

Clay, Z., Pika, S., Gruber, T., & Zuberbühler, K. (2011). Female bonobos use copulation calls as social signals. *Biology Letters*, 7(4), 513–516.

Clay, Z., Archbold, J., & Zuberbühler, K. (2015). Functional flexibility in wild bonobo vocal behaviour. *PeerJ*, 3, e1124.

Collier, K., Bickel, B., van Schaik, C. P., Manser, M. B., & Townsend, S. W. (2014). Language evolution: Syntax before phonology? *Proceedings of the Royal Society of London B: Biological Sciences*, 281(1788), DOI:10.1098/rspb.2014.0263.

Coye, C., Ouattara, K., Zuberbühler, K., & Lemasson, A. (2015). Suffixation influences receivers' behaviour in non-human primates. *Proceedings of the Royal Society of London B: Biological Sciences*, 282(1807), DOI:10.1098/rspb.2015.0265.

Coye, C., Zuberbühler, K., & Lemasson, A. (2016). Morphologically structured vocalizations in female Diana monkeys. *Animal Behaviour*, 115, 97–105.

Cramer, I. (2005). The parameters of the Altmann-Menzerath Law. *Journal of Quantitative Linguistics*, 12(1), 41–52. doi:10.1080/09296170500055301.

Crockford, C., Herbinger, I., Vigilant, L., & Boesch, C. (2004). Wild chimpanzees produce group-specific calls: A case for vocal learning? *Ethology*, 110(3), 221–243.

Crockford, C., Wittig, R. M., Mundry, R., & Zuberbühler, K. (2012). Wild chimpanzees inform ignorant group members of danger. *Current Biology*, 22(2), 142–146.

Crockford, C., Wittig, R. M., & Zuberbühler, K. (2015). An intentional vocalization draws others' attention: A playback experiment with wild chimpanzees. *Animal Cognition*, 18(3), 581–591.

Dennett, D. C. (1983). Intentional systems in cognitive ethology: The Panglossian Paradigm defended. *Behavioral and Brain Sciences*, 6(3), 343–355.

Dunbar, R. I. M. (1996). *Grooming, gossip and the evolution of language*. Cambridge, MA: Harvard University Press.

Enard, W., Przeworski, M., Fisher, S. E., Lai, C. S. L., et al. (2002). Molecular evolution of FOXP2, a gene involved in speech and language. *Nature*, 418(6900), 869–872.

Evans, N., & Levinson, S. C. (2009). The myth of language universals: Language diversity and its importance for cognitive science. *Behavioral and Brain Sciences*, 32(5), 429.

Fant, G. (1960). *Acoustic theory of speech production*. The Hague: Mouton.

Fedurek, P., & Slocombe, K. E. (2011). Primate vocal communication: A useful tool for understanding human speech and language evolution? *Human Biology*, 83(2), 153–173.

Fedurek, P., & Slocombe, K. E. (2013). The social function of food-associated calls in male chimpanzees. *American Journal of Primatology*, 75(7), 726–739.
Fedurek, P., Slocombe, K. E., & Zuberbühler, K. (2015). Chimpanzees communicate to two different audiences during aggressive interactions. *Animal Behaviour*, 110, 21–28.
Fedurek, P., Zuberbühler, K., & Dahl, C. D. (2016). Sequential information in a great ape utterance. *Scientific Reports*, 6, 38226.
Fichtel, C., & Kappeler, P. M. (2002). Anti-predator behavior of group-living Malagasy primates: Mixed evidence for a referential alarm call system. *Behavioral Ecology and Sociobiology*, 51(3), 262–275.
Fisher, S. E., & Scharff, C. (2009). FOXP2 as a molecular window into speech and language. *Trends in Genetics*, 25(4), 166–177.
Fitch, W. T., & Hauser, M. D. (1995). Vocal production in nonhuman primates: Acoustics, physiology, and functional constraints on "honest" advertisement. *American Journal of Primatology*, 37, 191–219.
Fitch, W. T., & Reby, D. (2001). The descended larynx is not uniquely human. *Proceedings of the Royal Society of London B: Biological Sciences*, 268(1477), 1669–1675.
Fitch, W. T., & Zuberbühler, K. (2013). Primate precursors to human language: Beyond discontinuity. In E. Altenmuller, S. Schmidt, & E. Zimmerman (Eds.), *Evolution of emotional communication: From sounds in nonhuman mammals to speech and music in man*. Oxford: Oxford University Press.
Fitch, W. T., de Boer, B., Mathur, N., & Ghazanfar, A. A. (2016). Monkey vocal tracts are speech-ready. *Science Advances, 2*.
Ford, S. M. (2009). Dialects. In W. Perrin, B. Würsig, & J. G. M. Thewissen (Eds.), *Encyclopedia of marine mammals* (2nd ed., pp. 310–311). New York: Academic Press.
Gautier, J.-P., & Gautier, A. (1977). Communication in old world monkeys. In T. A. Sebeok (Ed.), *How animals communicate* (pp. 890–964). Bloomington, IN: Indiana University Press.
Geissmann, T. (2002). Duet-splitting and the evolution of gibbon songs. *Biological Reviews*, 77(1), 57–76.
Geissmann, T., & Orgeldinger, M. (2000). The relationship between duet songs and pair bonds in siamangs, *Hylobates syndactylus*. *Animal Behaviour*, 60(6), 805–809.
Genty, E., Clay, Z., Hobaiter, C., & Zuberbühler, K. (2014). Multi-modal use of a socially directed call in bonobos. *PLoS One*, 9(1), e84738.
Goodall, J. (1986). *The chimpanzees of Gombe: Patterns of behavior*. Cambridge, MA: Harvard University Press.
Gouzoules, S., Gouzoules, H., & Marler, P. (1984). Rhesus monkey (*Macaca mulatta*) screams: Representational signalling in the recruitment of agonistic aid. *Animal Behaviour*, 32(1), 182–193.
Grice, H. P. (1969). Utterers' meaning and intentions. *Philosophical Review*, 78, 147–177.
Gustison, M. L., Semple, S., Ferrer-I-Cancho, R., & Bergman, T. J. (2016). Gelada vocal sequences follow Menzerath's linguistic law. *Proceedings of the National Academy of Sciences of the United States of America*, 113, E2750–E2758.
Hauser, M. D., Yang, C., Berwick, R. C., Tattersall, I., et al. (2014). The mystery of language evolution. *Frontiers in Psychology*, 5, 401.
Hayes, K. J., & Hayes, C. (1951). The intellectual development of a home-raised chimpanzee. *Proceedings of the American Philosophical Society*, 95, 105–109.
Herbinger, I., Papworth, S., Boesch, C., & Zuberbühler, K. (2009). Vocal, gestural and locomotor responses of wild chimpanzees to familiar and unfamiliar intruders: A playback study. *Animal Behaviour*, 78(6), 1389–1396.
Hockett, C. F. (1960). The origin of speech. *Scientific American*, 203(3), 89–96.
Hölldobler, B., & Wilson, E. O. (1990). *The ants*. Cambridge, MA: Harvard University Press.
Hostetter, A. B., Cantero, M., & Hopkins, W. D. (2001). Differential use of vocal and gestural communication by chimpanzees (*Pan troglodytes*) in response to the attentional status of a human (*Homo sapiens*). *Journal of Comparative Psychology*, 115(4), 337–343.
Hrdy, S. B. (2009). *Mothers and others: The evolutionary origins of mutual understanding*. Cambridge, MA: Belknap Press, Harvard University Press.

Janik, V. M. (2014). Cetacean vocal learning and communication. *Current Opinion in Neurobiology*, 28, 60–65.
Janik, V. M., & Slater, P. J. B. (1997). Vocal learning in mammals. *Advances in the Study of Behavior*, 26, 59–99.
Jürgens, U. (2002). Neural pathways underlying vocal control. *Neuroscience and Biobehavioral Reviews*, 26(2), 235–258.
Kaplan, G. (2014). Animal communication. *Wiley Interdisciplinary Reviews: Cognitive Science*, 5(6), 661–677.
Kershenbaum, A., Blumstein, D. T., Roch, M. A., Akcay, C., et al. (2016). Acoustic sequences in non-human animals: A tutorial review and prospectus. *Biological Reviews*, 91(1), 13–52.
Krause, J., Lalueza-Fox, C., Orlando, L., Enard, W., et al. (2007). The derived FOXP2 variant of modern humans was shared with neandertals. *Current Biology*, 17(21), 1908–1912.
Lameira, A. R., Hardus, M. E., Kowalsky, B., de Vries, H., et al. (2013). Orangutan (*Pongo* spp.) whistling and implications for the emergence of an open-ended call repertoire: A replication and extension. *The Journal of the Acoustical Society of America*, 134(3), 2326–2335.
Lameira, A. R., Maddieson, I., & Zuberbühler, K. (2014). Primate feedstock for the evolution of consonants. *Trends in Cognitive Sciences*, 18(2), 60–62.
Langergraber, K. E., Pruefer, K., Rowney, C., Boesch, C., et al. (2012). Generation times in wild chimpanzees and gorillas suggest earlier divergence times in great ape and human evolution. *Proceedings of the National Academy of Sciences of the United States of America*, 109(39), 15716–15721.
Laporte, M. N. C., & Zuberbühler, K. (2010). Vocal greeting behaviour in wild chimpanzee females. *Animal Behaviour*, 80(3), 467–473.
Laporte, M. N. C., & Zuberbühler, K. (2011). The development of a greeting signal in wild chimpanzees. *Developmental Science*, 14(5), 1220–1234.
Lemasson, A., & Hausberger, M. (2004). Patterns of vocal sharing and social dynamics in a captive group of Campbell's monkeys (*Cercopithecus campbelli campbelli*). *Journal of Comparative Psychology*, 118(3), 347–359.
Lemasson, A., Hausberger, M., & Zuberbühler, K. (2005). Socially meaningful vocal plasticity in adult Campbell's monkeys (*Cercopithecus campbelli*). *Journal of Comparative Psychology*, 119, 220–229.
Levinson, S. C. (2006). On the human "interaction engine." In N. J. Enfield & S. C. Levinson (Eds.), *Roots of human sociality: Culture, cognition and interaction* (pp. 39–69). Oxford: Berg Publishers.
Levinson, S. C. (2016). Turn-taking in human communication—Origins and implications for language processing. *Trends in Cognitive Sciences*, 20(1), 6–14.
Lieberman, P. (2012). Vocal tract anatomy and the neural bases of talking. *Journal of Phonetics*, 40(4), 608–622.
Marler, P. (1998). Animal communication and human language. In N. G. Jablonski & L. C. Aiello (Eds.), *Origin and diversification of language* (pp. 1–19). San Francisco, CA: California Academy of Sciences.
Marshall, A. J., Wrangham, R. W., & Arcadi, A. C. (1999). Does learning affect the structure of vocalizations in chimpanzees? *Animal Behaviour*, 58, 825–830.
McComb, K., & Semple, S. (2005). Coevolution of vocal communication and sociality in primates. *Biology Letters*, 1(4), 381–385.
Nishimura, T. (2005). Developmental changes in the shape of the supralaryngeal vocal tract in chimpanzees. *American Journal of Physical Anthropology*, 126(2), 193–204.
Nishimura, T., Mikami, N., Suzuki, J., & Matsuzawa, T. (2003). Descent of the larynx in chimpanzee infants. *Proceedings of the National Academy of Sciences of the United States of America*, 100, 6930–6933.
Ouattara, K., Lemasson, A., & Zuberbühler, K. (2009a). Campbell's monkeys use affixation to alter call meaning. *PLoS One*, 4(11), e7808.

Ouattara, K., Lemasson, A., & Zuberbühler, K. (2009b). Campbell's monkeys concatenate vocalizations into context-specific call sequences. *Proceedings of the National Academy of Sciences of the United States of America*, 106(51), 22026–22031.
Owren, M. J., & Rendall, D. (2001). Sound on the rebound: Bringing form and function back to the forefront in understanding nonhuman primate vocal signaling. *Evolutionary Anthropology*, 10, 58–71.
Papworth, S., Bose, A. S., Barker, J., Schel, A. M., & Zuberbühler, K. (2008). Male blue monkeys alarm call in response to danger experienced by others. *Biology Letters*, 4(5), 472–475.
Pereira, M. E., & Macedonia, J. M. (1991). Ringtailed lemur anti-predator calls denote predator class, not response urgency. *Animal Behaviour*, 41, 543–544.
Rainey, H. J., Zuberbuhler, K., & Slater, P. J. B. (2004). Hornbills can distinguish between primate alarm calls. *Proceedings of the Royal Society of London B: Biological Sciences*, 271(1540), 755–759.
Riede, T., Bronson, E., Hatzikirou, H., & Zuberbühler, K. (2005). Vocal production mechanisms in a non-human primate: Morphological data and a model. *Journal of Human Evolution*, 48(1), 85–96.
Scarantino, A., & Clay, Z. (2015). Contextually variable signals can be functionally referential. *Animal Behaviour*, 100, E1–E8.
Schel, A. M., Tranquilli, S., & Zuberbühler, K. (2009). The alarm call system of two species of black-and-white colobus monkeys (*Colobus polykomos* and *Colobus guereza*). *Journal of Comparative Psychology*, 123(2), 136–150.
Schel, A. M., Candiotti, A., & Zuberbühler, K. (2010). Predator-deterring alarm call sequences in Guereza colobus monkeys are meaningful to conspecifics. *Animal Behaviour*, 80(5), 799–808.
Schel, A. M., Townsend, S. W., Machanda, Z., Zuberbühler, K., & Slocombe, K. E. (2013a). Chimpanzee alarm call production meets key criteria for intentionality. *PLoS One*, 8(10), e76674.
Schel, A. M., Machanda, Z., Townsend, S. W., Zuberbühler, K., & Slocombe, K. E. (2013b). Chimpanzee food calls are directed at specific individuals. *Animal Behaviour*, 86(5), 955–965.
Schlenker, P., Chemla, E., Arnold, K., Lemasson, A., et al. (2014). Monkey semantics: Two 'dialects' of Campbell's monkey alarm calls. *Linguistics and Philosophy*, 37(6), 439–501.
Schlenker, P., Chemla, E., Schel, A. M., Fuller, J., et al. (2016). Formal monkey linguistics. *Theoretical Linguistics*, 42(1-2), 1–90. doi:10.1515/tl-2016-0001.
Sebeok, T. A. (1977). Zoosemiotic components of human communication. In T. A. Sebeok (Ed.), *How animals communicate*. Bloomington, IN: Indiana University Press.
Semple, S., Gerald, M. S., & Suggs, D. N. (2009). Bystanders affect the outcome of mother-infant interactions in rhesus macaques. *Proceedings of the Royal Society of London B: Biological Sciences*, 276(1665), 2257–2262.
Seyfarth, R. M., & Cheney, D. L. (1986). Vocal development in vervet monkeys. *Animal Behaviour*, 34, 1640–1658.
Seyfarth, R. M., Cheney, D. L., & Marler, P. (1980). Monkey responses to three different alarm calls: Evidence of predator classification and semantic communication. *Science*, 210, 801–803.
Slocombe, K. E., & Zuberbühler, K. (2005). Functionally referential communication in a chimpanzee. *Current Biology*, 15(19), 1779–1784.
Slocombe, K. E., & Zuberbühler, K. (2007). Chimpanzees modify recruitment screams as a function of audience composition. *Proceedings of the National Academy of Sciences of the United States of America*, 104, 17228–17233.
Slocombe, K. E., & Zuberbühler, K. (2010). Vocal communication in chimpanzees. In E. V. Lonsdorf, S. Ross, & T. Matsuzawa (Eds.), *The mind of the chimpanzee: Ecological and empirical perspectives* (pp. 192–207). Chicago: University of Chicago Press.
Slocombe, K. E., Kaller, T., Call, J., & Zuberbühler, K. (2010a). Chimpanzees extract social information from agonistic screams. *PLoS One*, 5(7), e11473.

Slocombe, K. E., Kaller, T., Turman, L., Townsend, S., et al. (2010b). Production of food-associated calls in wild male chimpanzees is dependent on the composition of the audience. *Behavioral Ecology and Sociobiology*, 64(12), 1959–1966.

Snowdon, C. T., Brown, C. H., & Petersen, M. R. (Eds.). (1982). *Primate communication*. Cambridge, UK: Cambridge University Press.

Sterelny, K. (2016). Cooperation, culture, and conflict. *British Journal for the Philosophy of Science*, 67(1), 31–58.

Taylor, A. M., & Reby, D. (2010). The contribution of source-filter theory to mammal vocal communication research. *Journal of Zoology*, 280(3), 221–236.

Tomasello, M. (2008). *Origins of human communication*. Cambridge, MA: MIT Press.

Tomasello, M. (2014a). *A natural history of human thinking*. Cambridge, MA: Harvard University Press.

Tomasello, M. (2014b). The ultra-social animal. *European Journal of Social Psychology*, 44(3), 187–194.

Tomasello, M., Call, J., & Hare, B. (2003). Chimpanzees understand psychological states—The question is which ones and to what extent. *Trends in Cognitive Sciences*, 7(4), 153–156.

Tomasello, M., Carpenter, M., Call, J., Behne, T., & Moll, H. (2005). Understanding and sharing intentions: The origins of cultural cognition. *Behavioral and Brain Sciences*, 28(5), 675–691.

Tomasello, M., Melis, A. P., Tennie, C., Wyman, E., & Herrmann, E. (2012). Two key steps in the evolution of human cooperation. The interdependence hypothesis. *Current Anthropology*, 53(6), 673–692.

Townsend, S. W., Deschner, T., & Zuberbühler, K. (2008). Female chimpanzees use copulation calls flexibly to prevent social competition. *PLoS One*, 3(6), e2431.

Watson, S. K., Townsend, S. W., Schel, A. M., & Wilke, C. (2015). Vocal learning in the functionally referential food grunts of chimpanzees. *Current Biology*, 25(4), 495–499.

Wheeler, B. C., & Fischer, J. (2012). Functionally referential signals: A promising paradigm whose time has passed. *Evolutionary Anthropology*, 21(5), 195–205.

Whiten, A., Goodall, J., McGrew, W. C., Nishida, T., et al. (1999). Cultures in chimpanzees. *Nature*, 399(6737), 682–685.

Wich, S. A., & de Vries, H. (2006). Male monkeys remember which group members have given alarm calls. *Proceedings of the Royal Society of London B: Biological Sciences*, 273(1587), 735–740.

Wilson, M. L., Boesch, C., Fruth, B., Furuichi, T., et al. (2014). Lethal aggression in Pan is better explained by adaptive strategies than human impacts. *Nature*, 513(7518), 414–417.

Zuberbühler, K. (2000). Interspecific semantic communication in two forest monkeys. *Proceedings of the Royal Society of London B: Biological Sciences*, 267(1444), 713–718.

Zuberbühler, K. (2001). Predator-specific alarm calls in Campbell's monkeys, *Cercopithecus campbelli*. *Behavioral Ecology and Sociobiology*, 50(5), 414–422.

Zuberbühler, K. (2008). Audience effects. *Current Biology*, 18(5), R189–R190.

Zuberbühler, K., Cheney, D. L., & Seyfarth, R. M. (1999). Conceptual semantics in a nonhuman primate. *Journal of Comparative Psychology*, 113(1), 33–42.

Chapter 8
Evolution of Hearing and Language in Fossil Hominins

Rolf M. Quam, Ignacio Martínez, Manuel Rosa, and Juan Luis Arsuaga

Abstract This chapter outlines the evolution of auditory capacities during the course of human evolution and the implications for understanding when human language may have evolved. These findings are considered within the context of habitat acoustics, the mathematical theory of communication, and the frame/content theory of speech production. Compared to chimpanzees, the auditory pattern in the early hominin taxa *Australopithecus* and *Paranthropus* show a heightened sensitivity between 1.0 and 3.5 kHz but a similarly narrow bandwidth of maximum sensitivity. The early hominin auditory pattern may have facilitated short-range communication in open habitats, but their communication pattern apparently did not involve transmission of information beyond that of a chimpanzee. The early hominins likely were restricted to the frame stage of speech production, a phoneme-based, presyntactic form of communication with only limited word formation. In contrast, the Middle Pleistocene Atapuerca Sima de los Huesos (SH) hominins resemble *H. sapiens* in showing a broad region of heightened sensitivity between 1 and 5 kHz and a wider bandwidth of maximum sensitivity that is extended toward higher frequencies. The wider bandwidth in the Atapuerca (SH) hominins facilitated specialization in the use of complex, short-range vocal communication, including

---

R.M. Quam (✉)
Department of Anthropology, Binghamton University (SUNY),
Binghamton, NY 13902-6000, USA
e-mail: rquam@binghamton.edu

I. Martínez
Departamento de Ciencias de la Vida, Universidad de Alcalá,
Edificio de Ciencias, Campus Universitario, Alcalá de Henares 28805, Spain
e-mail: ignacio.martinezm@uah.es

M. Rosa
Departamento de Teoría de la Señal y Comunicaciones, Universidad de Alcalá,
Escuela Politécnica, Campus Universitario, Alcalá de Henares 28871, Spain
e-mail: manuel.rosa@uah.es

J.L. Arsuaga
Centro de Investigación (UCM-ISCIII) sobre la Evolución y Comportamiento Humanos,
Avda. Monforte de Lemos 5, Madrid 28029, Spain
e-mail: jlarsuaga@isciii.es

R.M. Quam et al. (eds.), *Primate Hearing and Communication*, Springer
Handbook of Auditory Research 63, DOI 10.1007/978-3-319-59478-1_8

an emphasis on high-frequency consonant production and increased word formation. The Atapuerca (SH) hominins, then, may have been on the threshold of passing into the frame/content stage of speech production. The evolution of auditory capacities is consistent with the presence of some form of spoken language in the genus *Homo* prior to the appearance of *H. sapiens*.

Keywords Atapuerca • *Australopithecus* • Communication • Consonant production • Ear ossicles • Frame/content theory • Habitat acoustics • Inner ear • Middle ear • Neandertal • Occupied band • Outer ear • *Paranthropus*

8.1 Introduction

The human species, *Homo sapiens*, is today a cosmopolitan, culture-bearing linguistic entity capable of sophisticated technological endeavors. Yet the evolutionary origins of humans can be traced back millions of years to when the earliest human ancestors arose in Africa. While disagreement exists among scholars regarding the number of taxa that can currently be identified in the available fossil record, the view that human evolution follows a cladogenetic branching pattern is widely accepted (Wood and Lonergan 2008). This view emphasizes the multiplication of taxa, interspecific diversity, and competition between different human species as primary features of the evolutionary process (Fig. 8.1).

The earliest fossils tentatively considered to be members of the human family date to around 6–7 million years ago (Ma) (Senut et al. 2001; Brunet et al. 2002), which is close in time to the last common ancestor of humans and chimpanzees (*Pan troglodytes*) (Patterson et al. 2006). These fragmentary fossils are primarily identified as hominins, the group that includes living humans and fossil human species that arose after the split with chimpanzees, based on purported anatomical indicators of bipedalism (i.e., walking on two feet). The genus *Australopithecus* emerged around 4 Ma (Fig. 8.1) and is found in both eastern and southern Africa (White et al. 2006; Berger et al. 2010). Species attributed to this genus combine clear evidence of bipedalism with some degree of reduction in the canine teeth but retain many primitive features, including small brain size, curved fingers, and a projecting face. Evidence for tool use and manufacture in *Australopithecus* comes from the presence of a human-like elongated thumb in the hand skeleton (Alba et al. 2003), the recognition of cut marks on animal bones indicative of defleshing (McPherron et al. 2010), and the discovery of stone tools that date to around 3.3 Ma (Harmand et al. 2015), which predate the emergence of the genus *Homo*. A second genus of early human ancestor, *Paranthropus*, is also found in both eastern and southern Africa and can be distinguished from *Australopithecus* mainly by the presence of enlarged cheek teeth and robust cranial morphology to support more powerful chewing muscles (Constantino and Wood 2004). Species attributed to this genus are sometimes referred to as "robust australopithecines," and taxa that are evolutionarily more

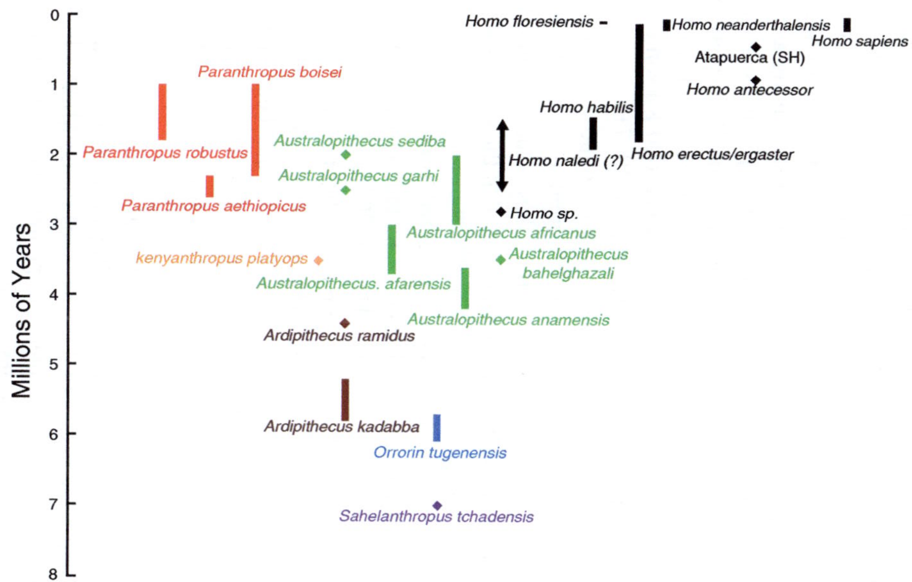

Fig. 8.1 The human evolutionary tree, indicating the temporal span of existence for different fossil hominin taxa. Each genus is represented by a different color. Species in genera other than *Homo* are generally referred to as "early hominins" by paleoanthropologists.

primitive than the genus *Homo*, including both *Australopithecus* and *Paranthropus*, are generally referred to as “early hominins” by anthropologists.

All species of the genus *Homo* are generally considered tool users that consumed meat through either scavenging or hunting. The earliest fossils attributed to the genus *Homo* date to more than 2.5 Ma and are found in eastern Africa (Villmoare et al. 2015). In addition, a large fossil assemblage has been excavated in South Africa and attributed to the species *Homo naledi* (Fig. 8.1), which may represent an early member of the genus *Homo* (Berger et al. 2015). These fossils retain some primitive features seen in the early hominins, including a fairly small brain and curved fingers, but they also seem to depart toward *H. sapiens* in a number of aspects, including dental reduction, bipedal foot anatomy, and longer legs relative to arms. Subsequently, body size increased dramatically, with species such as *Homo ergaster* being similar in stature to *H. sapiens* but likely having a considerably heavier body mass (Ruff et al. 1997). These body proportions were largely maintained throughout the Pleistocene until the emergence of *H. sapiens* around 200 thousand years ago (Ka) (Arsuaga et al. 2015). Brain size increased during the course of evolution of the genus *Homo*, and after about 600 Ka some fossil specimens show brain sizes similar to *H. sapiens* (Ruff et al. 1997).

In Europe, the earliest hominin fossils date to around 1.2–1.4 Ma and are attributed to the species *Homo antecessor* (Bermúdez de Castro et al. 1997; Carbonell et al. 2008). The earliest fossils that show clear Neandertal features date to somewhat later, around 500 Ka, and are found at the site of the Sima de los Huesos (SH) in the Sierra de Atapuerca in northern Spain (Arsuaga et al. 2014). The Neandertals, *Homo neanderthalensis*, emerged sometime around 200 Ka (Fig. 8.1) and show a considerably enlarged brain and perhaps some evidence of cold adaptation in their limb proportions (Harvati 2007). Neandertals survived in Europe until around 40 Ka and were adept hunters that buried their dead and had a controlled use of fire. Some disputed evidence of symbolism, in the form of personal adornments and engravings on cave walls, has been found at a few Neandertal sites (Caron et al. 2011; Rodríguez-Vidal et al. 2014). While the Neandertals were not ancestral to *H. sapiens*, some evidence of hybridization has been found (Fu et al. 2015).

H. sapiens emerged in Africa around 200 Ka, and the earliest *H. sapiens* fossils have been found in eastern Africa (White et al. 2003; McDougall et al. 2005). While early *H. sapiens* fossils are also found at two sites in the Middle East, dating to around 100 Ka, *H. sapiens* did not enter Europe until much later (around 40 Ka). Evidence of symbolic expression is present at a few sites in southern Africa in the form of personal adornments around 70–80 Ka (Henshilwood et al. 2004) and in Europe around 35–40 Ka in the form of cave paintings (Pike et al. 2012).

The evolution of language is a long-standing controversial topic in evolutionary studies, in part because language is often considered a defining feature of what it means to be human. Nevertheless, the study of language in fossil specimens has been a frustrating endeavor, partly due to the fragmentary nature of the fossil record but even more due to the difficulty in establishing a correspondence between anatomical features of the skull and skeleton and language capacities.

8.2 Language Studies in Hominin Fossils

Given the degree of neural processing involved in producing and perceiving auditory stimuli and decoding symbolic signals (Ramsier and Rauschecker, Chap. 3), brain morphology represents an obvious first approach to studying language evolution. Some attempts have been made to identify the development of language centers in particular Broca's area, in fossil hominin endocasts (fossilized impressions of the brain surface) (Tobias 1987; Holloway et al. 2004). These studies have generally concluded that language capacities were not present in the early hominin genera *Australopithecus* and *Paranthropus* and that language capacity first appeared during the evolution of the genus *Homo*. However, drawing clear functional inferences from the gross morphology and contours of the brain surface is problematic (Sherwood et al. 2003), and it is well-known that linguistic processing in the brain is not restricted to Broca's area (Grodzinksy 2000).

Because of its direct role in the production of vocalizations, the anatomy of the supralaryngeal vocal tract (SVT) has been considered fundamental by many researchers for establishing the presence of spoken language in fossil hominins, particularly Neandertals (Laitman et al. 1979). The SVT in mammals is composed of a horizontal segment (the oral cavity or mouth) and a vertical segment (the pharynx or throat). In order to produce the quantal vowels /a/, /i/, and /u/, the two segments of the SVT should be approximately the same length (i.e., a 1:1 ratio). In adult *H. sapiens*, both segments are of a similar length due to the combination of a short oral cavity and a low position of the larynx in the neck. This latter feature, often called "the descent of the larynx," is considered to be exclusive to *H. sapiens* and a direct adaptation for the production of spoken language (Lieberman et al. 1992).

In contrast, Lieberman (2007a) argued that Neandertals depart from a 1:1 ratio for the two segments of the SVT due to the combination of a shorter vertical segment, implying a higher placement for the larynx, and a longer horizontal segment related to a higher degree of facial projection. While the Neandertal ratio likely does depart from the 1:1 ratio in adult *H. sapiens*, it is similar to that of a ten-year-old *H. sapiens* child who is capable of producing the quantal vowels (Boë et al. 2007).

Other researchers used a prediction model for larynx height based on correlations with various dimensions of the skull and mandible in *H. sapiens* (Boë et al. 2002) and found Neandertals were characterized by a low-placed larynx. In fact, a low-placed larynx is not an exclusively *H. sapiens* feature since it is also found in other mammals in which it is related to biological functions other than speech production (Fitch and Reby 2001). Furthermore, Nishimura et al. (2006) have shown that the larynx in chimpanzees descends during the growth process. Although its final placement is not as low as in humans, chimpanzees appear to represent an initial stage in the descent of the larynx, and this feature was likely already present in the last common ancestor of chimpanzees and humans.

Another line of evidence into reconstructing the SVT in human fossils is the study of the morphology and dimensions of the hyoid bone, which has insertions for the muscles of the tongue and of the pharynx. The *H. sapiens* hyoid bone differs from chimpanzees in which the body of the hyoid bone is expanded and box shaped and houses an extension of the laryngeal air sacs on the ventral face. In contrast, the hyoid body in *H. sapiens* is not expanded, takes on a horseshoe shape, and lacks laryngeal air sacs (Aiello and Dean 1990). The absence of laryngeal air sacs is an important anatomical feature for creating subtle, timed, and distinct sounds, which are necessary for human speech.

The hyoid is among the rarest bones preserved in the hominin fossil record. The single *Australopithecus afarensis* specimen shows a hyoid body with morphology and dimensions similar to those of chimpanzees (Alemseged et al. 2006), while the hyoid in Neandertals and the Atapuerca (SH) hominins is human-like (Arensburg et al. 1989; Martínez et al. 2008). This suggests that laryngeal air sacs were still retained in *A. afarensis* but were absent in Neandertals and the Atapuerca (SH) hominins (de Boer 2012). Thus, although the vocal tract proportions and their functional significance in Neandertals continue to be a subject of debate (due in no small part to the amount of soft-tissue reconstruction necessary), it seems likely that middle (780–128 Ka) and late Pleistocene (128–10 Ka) hominins possessed the capacity for speech production.

Ancient DNA analysis also offers some insights into the evolution of human language. Neandertals share with *H. sapiens* two derived mutations in the *FOXP2* gene that are absent in chimpanzees (Krause et al. 2007). Although the precise functional implications of these two derived mutations for language ability are not well-understood, certain mutations in this gene are known to produce speech and language disorders in living humans (Lai et al. 2001). These results offer tantalizing evidence for the possible presence of spoken language in Neandertals.

In contrast, there is a movement to limit language capacities solely to *H. sapiens* based on evidence for symbolic behavior in the archaeological record. Several researchers argue that symbolic objects and activities such as cave paintings, "Venus" figurines, grave goods, and use of body ornaments and/or pigmentation should be taken as clear and direct indicators of symbolic mental representations and language (Henshilwood et al. 2001; Tattersall 2009). Indeed, the emergence of these symbolic behaviors in the archaeological record has been explicitly linked with the beginning of human spoken language (Tattersall 2004). Such evidence for symbolic activities has historically focused on European Upper Paleolithic sites, which are relatively late in time (generally younger than 40 Ka), and has been attributed to *H. sapiens*. Earlier evidence for body ornaments and abstract engravings has been documented at a few sites in northern and southern Africa, dating to approximately 80 Ka (Henshilwood et al. 2002; Bouzouggar et al. 2007). Although no diagnostic skeletal remains have been recovered from these sites, it is generally held that the activities were carried out by *H. sapiens* individuals. Evidence for Neandertal symbolic behavior is more limited (Jaubert et al. 2016), coming mainly from a few sites in western Europe, and remains controversial (d'Errico et al. 1998; Mellars 2010). While symbolic activities in the archaeological record likely do reflect the presence of human spoken language, there is no reason to suspect that the absence of archaeologically detectable symbolic behaviors precludes the possibility of some form of spoken language.

This review suggests that the balance of evidence based on the preserved anatomical features of the skeleton, as well as ancient DNA, is most consistent with some speech production capability characterizing the human evolutionary lineage prior to the emergence of *H. sapiens*. With this in mind, the chapter reviews research into the evolution of audition in fossil hominins and discusses the potential of this avenue of inquiry for providing new insights into this topic.

8.3 Hearing in Living Primates

The auditory capacities of numerous primate species have been measured in the laboratory by a variety of techniques (Ramsier and Rauschecker, Chap. 3). A review of primate hearing has stressed the fundamentally mammalian nature of the audiograms reported among primate species, including humans, in terms of high-frequency cutoff and low-frequency sensitivity, as well as sound localization abilities (Heffner 2004). However, a detailed study of primate audiograms does reveal some important distinctions between the major subgroups within the primate order (Coleman 2009).

The majority of Old World monkey species are characterized by a W-shaped audiogram with two peaks of heightened sensitivity, often at around 1 and 8 kHz, and a region of reduced sensitivity in the midrange frequencies, usually around 4 kHz (Coleman 2009). Despite some variation in a few species (Owren et al. 1988), the widespread distribution of a W-shaped audiogram among Old World monkeys suggests that this represents the primitive condition from which the human audiogram evolved.

Hearing in chimpanzees was first studied by Elder (1934) and subsequently studied by Kojima (1990). The results of these two studies coincide in showing a W-shaped audiogram with two peaks in heightened sensitivity at 1–2 kHz and 8 kHz and a relative reduction in sensitivity at around 4 kHz (Fig. 8.2). In addition, the high-frequency cutoff was around 33 kHz (Elder 1935), which is substantially higher than in humans but lower than in other primate species tested subsequently.

The *H. sapiens* audiogram generated under free-field conditions (Fig. 8.2) suggests good low-frequency sensitivity (below 1 kHz) compared to that in most primate species and a broad range of heightened sensitivity in the midrange frequencies (1–4 kHz) with the best sensitivity around 3 kHz (Sivian and White 1933). This is followed by a subsequent decrease in sensitivity to higher frequencies with the lowest high-frequency cutoff of any primate species (Heffner 2004). Although there is some variation in the published human audiograms, in non-hearing-impaired people, the best sensitivity is consistently between 2 and 4 kHz and the high-frequency limit is around 20 kHz.

The greatest contrast between the chimpanzee and human audiograms, then, is in the midrange frequencies where humans appear to have widened the region of maximum sensitivity. Despite assertions to the contrary (Heffner 2004), the human audiogram appears to be unique and evolutionarily derived (i.e., differing from the ancestral condition) within the living primates, suggesting the possibility of an auditory specialization. Given this difference in the audiogram between chimpanzees and humans, the study of auditory capacities in fossil hominins has the potential to reveal when the *H. sapiens* auditory pattern first emerged during evolutionary history. The region of heightened auditory sensitivity in humans also coincides with a portion of the frequency range of spoken language, which reaches up to around 6 kHz (Fig. 8.3). Although the majority of the sound power of vowels is concentrated below about 2.5 kHz, many consonants are concentrated at the higher

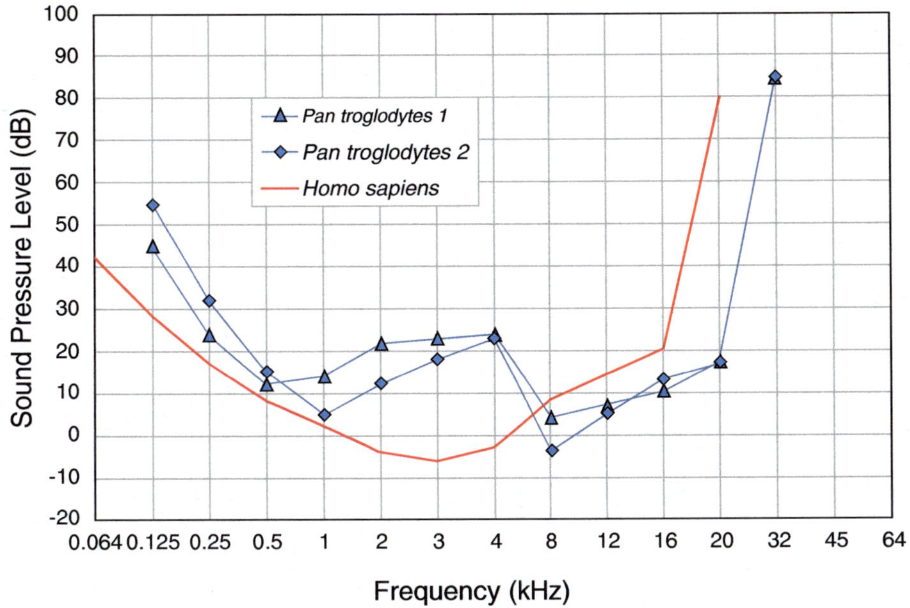

Fig. 8.2 Behavioral audiograms of humans and two chimpanzees (*Pan troglodytes*: 1, 2). The solid line for each species/individual represents the minimum sound level at a given frequency that a subject can perceive correctly at least 50% of the time. Points lower on the curve indicate a greater auditory sensitivity, and any sounds that fall above the curve are audible. Humans show a U-shaped audiogram with a broad region of heightened sensitivity in the midrange frequencies and a best frequency around 3 kHz. Chimpanzees show a W-shaped audiogram with two points of maximum sensitivity around 1 and 8 kHz and a loss in sensitivity around 4 kHz. The human audiogram presented here corresponds to the M. A. P. function from Sivian and White (1933, p. 313), and the chimpanzee data are from Kojima (1990).

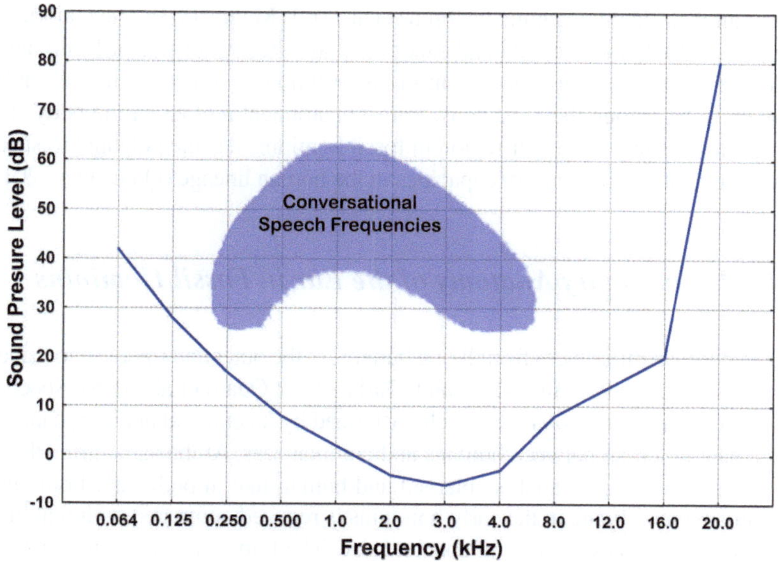

Fig. 8.3 The human audiogram and speech frequencies. The frequency range of conversation-level human spoken language, the "speech banana," is indicated by the shaded region. The region of heightened sensitivity in the human audiogram overlaps with a portion of the frequency range of sounds emitted during spoken language. (Reprinted from Quam et al. 2012; used with permission from Nova Science Publishers, Inc.)

frequencies (Hughes and Halle 1956). Thus, the appearance of the human auditory pattern also may have implications for communicative capacities.

8.4 Studying Hearing in Fossils

The appearance of the ossicular chain (malleus, incus, and stapes) of the mammalian middle ear had clear auditory consequences, most notably the emergence of high-frequency hearing in early mammalian forms, including primates (Masterton et al. 1969; Rosowski and Graybeal 1991). The functional implications of size and shape variation in ear anatomy underlie much of modern audiological research and have made it possible to predict aspects of the known hearing capacities in living primate species from their ear dimensions (Coleman and Ross 2004; Coleman and Colbert 2010) (Nummela, Chap. 2). There are clear anatomical differences in the outer, middle, and inner ears between humans and chimpanzees (House 1972; Quam 2006), and auditory differences also have been reported (Sect. 8.3). The correlations between structure and function established in these studies can be used to predict aspects of the hearing capacities in fossil taxa (Coleman et al. 2010).

Comparative genomic studies have revealed changes during the course of human evolutionary history in several genes related to the development of the auditory structures (Clark et al. 2003) and hearing (Scally et al. 2012). Importantly, both the functional morphology (Rosowski 1994) and developmental biology of the ear have been studied across mammals, including humans (Frenz et al. 2001; Mallo 2003). The majority of the structures of the human middle and inner ears are already formed at birth, and their development is under tight genetic control. In evolutionary terms, the middle and inner ears seem to be among the most conservative anatomical regions of the body. Thus, a comprehensive study of the ear region in fossil hominins has the potential to shed new light on the evolution of sensory capacities in the human lineage (Quam et al. 2012).

8.4.1 Evolutionary Anatomy of the Ear in Fossil Hominins

Some anatomical differences from living humans in the outer and middle ears have been reported in the early hominins (Rak and Clarke 1979; Quam et al. 2015). Specifically, both *A. africanus* and *P. robustus* are characterized by an external auditory canal that is intermediate in length between humans and chimpanzees. At the same time, the cross-sectional area of the ear canal is enlarged and human-like in both early hominin taxa. The tympanic membrane in the early hominins is reduced compared with that in chimpanzees and resembles humans in absolute size. The tympanic cavity in *P. robustus* is similar in size to the mean size in both the chimpanzee and human, but the tympanic cavity in *A. africanus* is somewhat smaller (Quam et al. 2015). The mastoid region in chimpanzees shows a more extensive network of air-filled cavities (i.e., pneumatization) than in *H. sapiens* (Sherwood 1999), but both species show considerable variation, which makes it difficult to interpret the possible significance of differences in fossil hominin taxa.

The outer and middle ears in the genus *Homo* subsequently changed in several aspects compared with those in the early hominins (Martínez et al. 2004; Quam 2006). In particular, fossil specimens attributed to the genus *Homo* are characterized by a mediolaterally shortened ear canal, resembling *H. sapiens*. The tympanic membrane is enlarged and more similar in size to the larger tympanic membranes in chimpanzees than to the smaller values in *H. sapiens*. The volumes of the tympanic cavity and mastoid air cells are also enlarged in the genus *Homo*. All of these features are present in the Middle Pleistocene hominins from the Atapuerca (SH) site (Martínez et al. 2013) and also appear to characterize Neandertals (Quam 2006; Balzeau and Radovčić 2008). Although the data are more limited, some of these features also seem to be present in *H. erectus* fossils from Asia (Balzeau and Grimaud-Hervé 2006) and at least one early Pleistocene *Homo* fossil (SK 847) from South Africa; thus, they appear to have emerged very early in the evolution of the genus *Homo*.

Regarding the ear ossicles, the malleus in both *A. africanus* and *P. robustus* is derived and human-like in its proportions, and it is clearly distinct from chimpanzees (Quam et al. 2013b). While no incus is known for *A. africanus*, *P. robustus* shows a unique combination of a human-like malleus and ape-like incus. The resulting malleus/incus lever ratio in *P. robustus* is intermediate between the higher values in chimpanzees and the lower values in humans. The stapes in *A. africanus* resembles the chimpanzee stapes in its overall size, including the small size of the footplate (Moggi-Cecchi and Collard 2002). Although a different study suggested larger stapes footplates in early hominins (Braga et al. 2015), the discovery of additional stapes from both *A. africanus* and *P. robustus* has confirmed their small dimensions (Quam et al. 2013b).

Thus, the early hominins seem to be characterized by a human-like malleus, while the incus and stapes are primitive and most similar to chimpanzees in their size and shape.

The ear ossicles in fossils attributed to the genus *Homo* show some changes from the early hominins but also preserve some primitive features. Specifically, the malleus/incus lever ratio in the Middle Pleistocene Atapuerca (SH) specimens is similar to *H. sapiens* (Martínez et al. 2004), indicating that a lengthening of the incus long process occurred. At the same time, the stapes remains small in overall size, including the footplate (Quam et al. 2006). The few known Neandertal ear ossicles show similar morphology and dimensions as in the Atapuerca (SH) specimens, while *H. sapiens* differs mainly in showing a larger stapes, including the footplate (Quam and Rak 2008; Quam et al. 2013a).

Studies of the inner ear in early hominin taxa have provided insights into their taxonomic relationships and hominin locomotion (Spoor et al. 1994). Although most analyses have focused on the semicircular canals, the cochlear basal turn is similar in size in chimpanzees and early hominins but slightly larger in humans. Nevertheless, little inference regarding hearing abilities can be drawn from these limited data. Perhaps more relevant, the length of the cochlea along the outer surface has been measured in several early hominin specimens and is shorter than in *H. sapiens* (Braga et al. 2015). If this is taken as a proxy measure for the length of the bony spiral lamina, it may indicate a shorter basilar membrane length in the early hominins. This shorter length would be consistent with a higher high-frequency cutoff (West 1985) above the approximate 20 kHz cutoff in humans, perhaps more closely resembling chimpanzees (about 30 kHz cutoff).

The cochlea in the genus *Homo* is generally larger than in the early hominins. The radius of the basal turn is similar in most members of the genus *Homo* (Quam et al. 2016), and the length along the outer surface of the cochlea is similar in Neandertals and *H. sapiens* (Braga et al. 2015; Beals et al. 2016). While more data are needed, there currently appears to be little change in cochlear dimensions within the genus *Homo*.

8.4.2 Modeling Auditory Capacities

These differences in ear anatomy have clear auditory implications, but few attempts have been made to reconstruct auditory capacities in fossil hominins (Masali et al. 1991; Moggi-Cecchi and Collard 2002). Results from these studies should be viewed cautiously since they are based on only a few ear dimensions. In contrast, a comprehensive model has been used to evaluate the auditory capacities in the early hominin taxa *A. africanus* and *P. robustus* from South Africa (Quam et al. 2015) and the Middle Pleistocene Atapuerca (SH) hominins (Martínez et al. 2004; Martínez et al. 2013). The use of models is a common practice in auditory research (Rosowski 1996; O'Connor and Puria 2008) and makes it possible to theoretically predict the flow of sound power through the outer and middle ears.

The model incorporates nearly 30 variables related to head size and the dimensions and physical properties of the anatomical structures of the outer and middle ears (Table 8.1; Fig. 8.4). In general, variables of the outer ear and ear ossicles have

Table 8.1 Modeling auditory capacities: definition of the model variables, the source of the value used, and the sensitivity analysis for frequencies above 2 kHz in the model

Electrical Parameters[a]			Related Anatomical Variables		
			Definition	Value Used	Sensitivity[b] (≥2 kHz)
Outer ear	Two-port network that models the concha horn		Concha length	Rosowski (1996)	High (A)
			Cross-sectional area of wide end	Rosowski (1996)	High (A)
			Cross-sectional area of narrow end	Measured as A_{EAC}	High (A)
	Two-port network that models the ear canal tube		Ear canal length	Measured as L_{EAC} complete	High (A)
			Cross-sectional area of the ear canal	Measured as A_{EAC}	High (A)
Middle ear	Middle ear cavity	C_{TC}	Volume of the middle ear cavity	Measured as V_{MEC}	Low (A)
		C_{MC}	Volume of the mastoid air spaces	Measured as V_{MA}	Low (A)
		R_A	Surface area of the *aditus ad antrum* and mastoid air spaces	Rosowski (1996)	Low (E)
		L_A	Length and radius of the *aditus ad antrum*	Measured as L_{AD} R_{AD1} and R_{AD2}	Low (E)
	Tympanic membrane and mallear attachment	A_{TM}	Area of the tympanic membrane	Measured as A_{TM}	High (A)
		L_{T1}	Mass of the tympanic membrane	Estimated from A_{TM}[c]	High (A)
		C_T	Structural properties of the tympanic membrane and mallear attachment	Rosowski (1996)	Low (E)
		R_T			Low (E)
		L_T			Low (E)
		C_{T2}			Low (E)
		R_{T2}			Medium (E)
		C_{TS}^M			High (E)
		R_{TS}^M			Medium (E)
	Malleus, incus, ligaments, and stapes	l_M:l_I	Functional lengths of the malleus and incus	Measured as L_M/L_I	High (A)
		L_{MI}^M	Masses of the malleus and incus	Measured as $M_M + M_I$	Medium (A)
		R_{MI}^M	Non-articular surface area of the malleus and incus	Rosowski (1996)	Low (E)
		C_{MI}^M	Structural properties of the malleus and incus	Rosowski (1996)	Low (E)
		L_S^M	Mass of the stapes	Measured as M_S	Low (A)
		R_J^M	Structural properties of the ossicular joints	Rosowski (1996)	Low (E)
		C_J^M			Low (E)
		A_{FP}	Area of the stapes footplate	Measured as A_{FP}	Medium (A)

Inner ear	Annular ligament	C_{AL}	Structural properties of the annular ligament	Rosowski (1996)	–[d]
		R_{AL}			High (E)
	Cochlea	Z_C	Structural properties of the cochlea	Aibara et al. (2001)	High (E)

[a]Definitions and abbreviations of the electrical parameters generally follow Rosowski (1996), except Zc (cochlear input impedance), which follows Aibara et al. (2001). Anatomical variables (Value Used) are defined in Fig. 8.4

[b]Sensitivity is related to the difference in the value for sound power at the entrance to the cochlea (in dB) obtained by increasing and decreasing the individual anatomical variable (A) or electrical parameter (E) by 50%. Sensitivity has been classified into three broad groupings: Low (≤1 dB difference), Medium (>1 to ≤3 dB difference), and High (>3 dB difference)

[c]Mass of the tympanic membrane was estimated based on its area, extrapolating from the values for *H. sapiens* provided by Rosowski (1996)

[d]The value provided for this variable in Rosowski (1996) is infinite, and it is not included in the sensitivity analysis

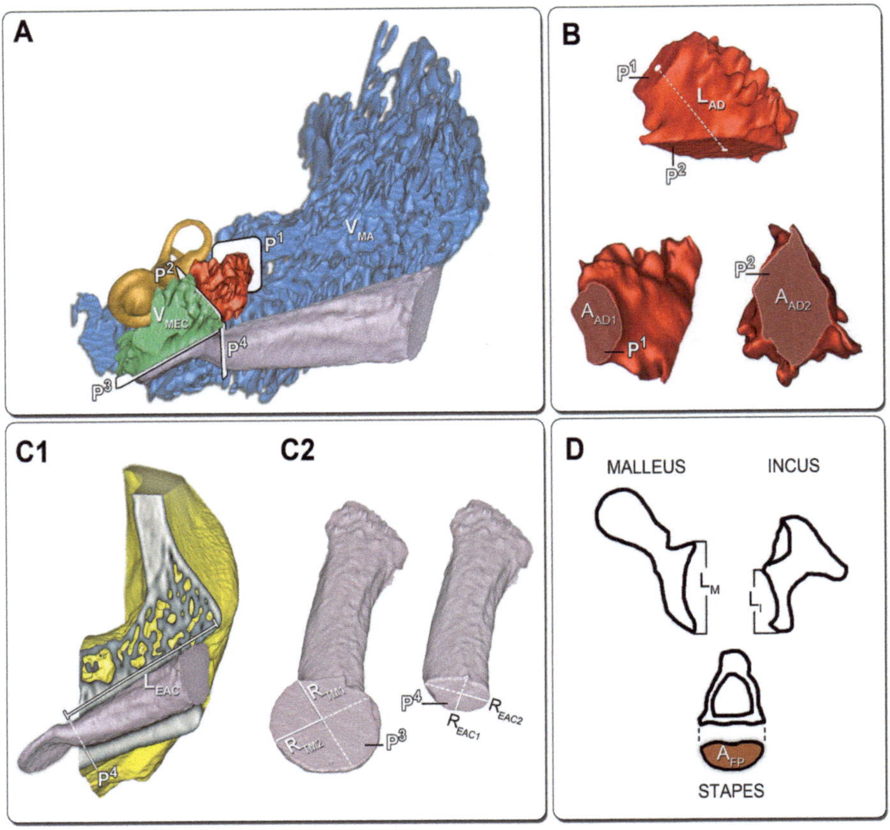

Fig. 8.4 Measurements of the middle and external ears (**A-C**) and ear ossicles (**D**). A, B, C1, C2, and D are not drawn to the same scale. (**A-C**) are based on the 3-D reconstruction of the left side of HTB 1769 (*Pan troglodytes*), showing the external auditory canal (*gray*), the middle ear cavity (*green*), the *aditus ad antrum* (*red*), the mastoid antrum and connected mastoid air cells (*blue*), the inner ear (*orange*), and the temporal bone (*yellow*). (**D**) is based on the profiles of the Atapuerca (SH) ear ossicles. Abbreviations: A_{AD1}, measured area of the exit from the aditus ad antrum (P^1); A_{AD2}, measured area of the entrance to the aditus ad antrum (P^2); A_{FP}, measured area of the footplate of the stapes (*brown*, in D); L_{AD}, length of the aditus ad antrum; L_{EAC}, length of the external auditory canal; L_I, functional length of the incus; L_M, functional length of the malleus; P^1, plane of the exit from the aditus ad antrum; P^2, plane of the entrance to the aditus ad antrum; P^3, plane of the tympanic membrane; P^4, plane marking the cross-section of the external auditory canal; R_{EAC1} and R_{EAC2}, half of the measured diameters of the two major perpendicular axes of the external auditory canal; R_{TM1}, half of the measured greater diameter of the tympanic membrane; R_{TM2}, half of the measured lesser diameter of the tympanic membrane; V_{AD}, volume of the aditus ad antrum; V_{MA}, volume of the *mastoid antrum* and connected mastoid air cells; V_{MEC}, volume of the middle ear cavity. (Reprinted from Quam et al. 2012; used with permission from Nova Science Publishers, Inc.)

a stronger influence on the model results (labeled as *medium* and *high* in Table 8.1), while the middle ear spaces (tympanic cavity, mastoid antrum and air cells, and *aditus ad antrum*) have a much weaker influence on the results (labeled as *low* in Table 8.1) (Fig. 8.4). Given the important roles of both the cochlea and the brain in sound perception, the model results are not a true audiogram. Nevertheless, there is a strong correlation between sound power transmission through the outer and middle ears and auditory sensitivity to different frequencies (Rosowski 1991), and the model predicts auditory sensitivity in decibels (dB) up to 5 kHz.

The model results also make it possible to analyze the frequency range of maximum auditory sensitivity as a communication channel. The capacity of the channel is an indication of the maximum rate at which information can be transmitted (Shannon 1948). Channel bandwidth is a proxy for channel capacity. A wider channel bandwidth corresponds to a greater channel capacity, which allows for more rapid transmission of information. The model results can provide an estimate of the occupied band, defined as the range of frequencies across which at least 90% of the sound power of an incident plane wave with flat spectrum is transmitted to the inner ear (ITUR 2000). This approach can help place the relationship between sound perception and communication on a firmer theoretical basis (Sect. 8.6.4).

8.5 Evolution of Auditory Capacities

The model results for humans and chimpanzees (Fig. 8.5) agree with published audiograms for these species (Fig. 8.4). In particular, humans are characterized by a broad region of heightened sensitivity between approximately 1.0 and 4.5 kHz, while chimpanzees show a decrease in sensitivity above 3.0 kHz and reach a minimum between 4.0 and 5.0 kHz. The mean separation between the human and chimpanzee curves at 4.0 kHz is 15 dB. Importantly, this mean difference is considerably greater than the average intersubject variability (8.4 dB) in hearing sensitivity reported across a large number of primate audiogram studies (Coleman 2009) and is clearly more consistent with interspecific variation in hearing sensitivity. Between 1.0 and 5.0 kHz, humans are more sensitive than chimpanzees at most frequencies. Similarly, the occupied band in humans is significantly wider (about 40%) than in chimpanzees and is extended toward higher frequencies (Fig. 8.5). This evolutionary transformation of the hominin audiogram seems to have gone through two main stages.

8.5.1 Early Hominins

The model results for *A. africanus* and *P. robustus* reveal similar patterns of auditory sensitivity and occupied bandwidth, suggesting both early hominin taxa have similar auditory capacities. Compared with chimpanzees, early hominins show a greater

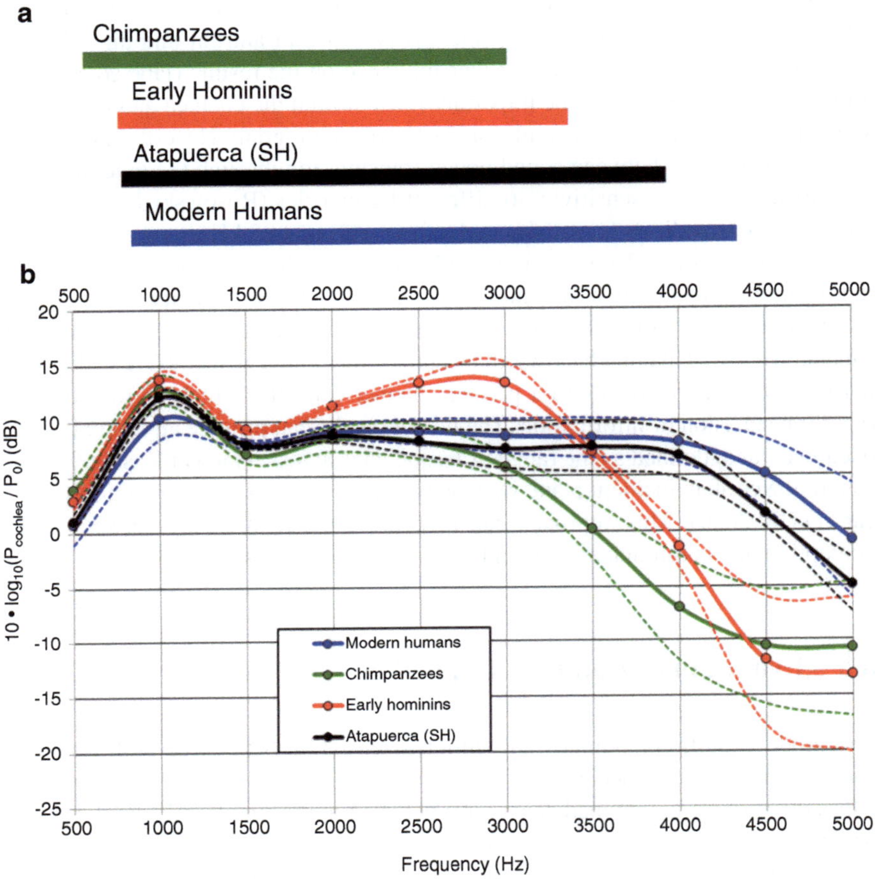

Fig. 8.5 Auditory capacities from 0.5 to 5.0 kHz in chimpanzees, fossil hominins, and *H. sapiens*. **(a)** The occupied band is similar in chimpanzees and early hominins but is shifted toward slightly higher frequencies in the latter. The Atapuerca (SH) hominins and *H. sapiens* show a widened occupied band that is further extended toward higher frequencies. **(b)** The sound power transmission curves correspond to dB at the entrance to the cochlea relative to $P_0 = 10^{-18}$ W for an incident plane wave intensity of 10^{-12} W/m^2. The mean value ± 1.0 standard deviation for each frequency position is shown for each group, except the early hominins, which encompasses the entire range of variation within the small sample. Points higher along the curve indicate better sound power transmission and heightened auditory sensitivity. (Adapted from Quam et al. 2015)

sensitivity between 1.5 and 3.5 kHz and an occupied band of maximum sensitivity that is slightly shifted toward higher frequencies (Fig. 8.5). In both these aspects, the early hominins are evolutionarily derived compared with chimpanzees, and the homogeneity in auditory capacities in both early hominin taxa is compatible with the similarity documented in most dimensions of their outer and middle ears. The increased sensitivity between 1.5 and 3.0 kHz in early hominins, as compared with chimpanzees and humans, may be explained by variation in the skeletal structures that have a strong influence on the model results (Table 8.1). These include the shortened length and enlarged cross-section of the ear canal, the size reduction of the tympanic membrane, and the lower lever ratio of the ear ossicles when compared with chimpanzees. Since the occupied bandwidth in the early hominins is essentially similar to that in chimpanzees, this first stage in the evolution of human hearing mainly involved an increased auditory sensitivity to the midrange frequencies.

The shorter ear canal in the early hominins (compared with chimpanzees) likely reflects the mediolateral expansion of the central cranial base and a reduction of the tympanic length in early hominins that is associated with the forward migration of the foramen magnum (Dean and Wood 1982; Kimbel et al. 2004). This suite of changes in the cranial base in early hominins is likely related to the emergence of bipedalism. In addition, the size reduction of the tympanic membrane and changes in the malleus proportions in early hominins might plausibly be explained as pleiotropic effects of these changes in the external ear canal. This hypothesis finds some support in the strong developmental integration of the external ear canal, the tympanic membrane, and malleus manubrium (Mallo et al. 2000). Perhaps relevant in this regard, morphological integration has been demonstrated between the cranial base and the inner ear in humans as changes in cranial base width affect certain dimensions of the bony labyrinth (Gunz et al. 2013). The auditory consequences of these anatomical changes in the early hominin outer and middle ears are a greater sensitivity from 1.5 to 3.0 kHz compared with that in chimpanzees and humans.

8.5.2 *Genus* Homo

Compared with the early hominins, the genus *Homo* is characterized by a slight reduction in sensitivity between 1.5 and 3.0 kHz and a considerable widening of the occupied bandwidth of maximum sensitivity, which is extended toward higher frequencies (>3.5 kHz) (Fig. 8.5). This second stage in the evolution of human hearing occurred by at least the Middle Pleistocene since the hearing pattern in the Atapuerca (SH) hominins is already largely human-like. Because the Atapuerca (SH) hominins are not on the direct evolutionary line that gave rise to *H. sapiens* but form part of the Neandertal evolutionary clade (Arsuaga et al. 2014), it is likely that this human-like hearing pattern was already present in the last common ancestor of *H. sapiens* and Neandertals and may characterize the entire genus *Homo*. The slight reduction in sensitivity between 1.5 and 3.0 kHz is likely to be related to a narrowing of the ear canal, an expansion in the size of the tympanic membrane, and a lowering of the ossicular lever ratio compared to the early hominins. The extension of the occupied band toward higher frequencies is mainly related to the shortened ear canal, which leads to higher resonance frequencies.

Compared with the early hominins, the shorter ear canal length in the genus *Homo* may reflect further reorganization of the cranial base related to increases in brain size. Brain expansion within the genus *Homo* is associated with changes in the cranial base (Dean and Wood 1981), including a relative stability of midline structures and more variability in the lateral cranial base (Bastir and Rosas 2009). Perhaps reflecting this, the tympanic plate in *H. sapiens* is oriented in a more sagittal (antero-posterior) direction than in Neandertals or *Homo erectus* (Martínez and Arsuaga 1997). Thus, changes in the outer and middle ears can be seen as pleiotropic effects of this reorganization of the cranial base, the auditory consequences of which include a considerable widening of the occupied band toward higher frequencies.

8.6 Hearing and Language

While the correlation between hearing and vocalizing is intuitive, the precision of this correlation is complicated by several factors. Clearly, there are many other auditory stimuli in the soundscape, in addition to conspecifics, that are biologically relevant to an organism, including predator and prey movements and ambient noise (Brown and Waser 1988; Seyfarth and Cheney 2003; Ramsier and Rauschecker, Chap. 3; Brown and Waser, Chap. 4). Furthermore, all organisms can both hear and vocalize within a wide range of frequencies, and the vocalizations of numerous sympatric primate species can often be heard and understood interspecifically (Zuberbühler 2000b). In fact, many similar vocalizations are shared by numerous species, and alarm calls tend to be generalized and understood by different species since there is a mutual benefit to be had. Thus, limiting auditory perception to frequencies emitted by conspecifics would represent a serious liability for any organism, making any precise correlation that may exist between auditory perception and vocalization all the more difficult to identify.

Nevertheless, intraspecific communication with other group members is likely to represent the most frequent auditory stimulus among primates. Given this, there might be a correlation between the best hearing frequency of an organism (i.e., the frequency at which the animal is most sensitive) and the range of frequencies emitted most frequently during vocalization. However, auditory capacities and vocal behavior are considered separately in most primate studies with, at most, only a passing reference made to their correspondence or lack thereof. One notable study documented heightened low-frequency auditory acuity in blue monkeys (*Cercopithecus mitis*) that coincided with the dominant frequency (around 250 Hz) contained in their long-distance *boom calls* (Brown and Waser 1984; Brown and Waser, Chap. 4). Similarly, there does appear to be a broad correspondence between the frequency ranges of heightened hearing sensitivity and vocalizations in the vervet monkey (*Chlorocebus aethiops*) (Owren et al. 1988).

Given the strong genetic component to the development of the ear structures (Moore and Linthicum 2007) and the remarkable vocal flexibility exhibited in human spoken language (Ladefoged and Maddieson 1996), it seems reasonable to conclude that the mechanisms of speech production have probably adapted to the audible hearing range, rather than vice versa. Although much of the acoustic information in spoken language is concentrated in the region up to around 2.5 kHz (e.g., the first two formant frequencies of the vowels), the region between 3 and 5 kHz also contains relevant acoustic information in human speech (Fant 1973). Given this, it is tempting to draw some tentative conclusions on speech capabilities in fossil hominins based on their auditory capacities.

8.6.1 Habitat Acoustics

Studies of habitat acoustics have suggested that the structural properties of primate vocalizations are related to environmental characteristics (Waser and Brown 1986; Brown and Waser, Chap. 4). Lower frequency calls, generally ≤1 kHz, emitted by arboreal primates travel far in the rain forest canopy and are important for maintaining intergroup spacing. However, attenuation of the sound signal at distances ≥25 m is considerably greater close to the ground in open environments, such as the savanna, and the presence of the visual channel further limits the utility of long-range intergroup calls by terrestrial primates (Waser and Brown 1986). In contrast, short-range communication (generally ≤25 m) is more strongly dependent on ambient background noise, which is lower across most frequencies up to 10 kHz in the savanna compared to the rain forest. This combination of higher signal attenuation and lower ambient background noise means that short-range intragroup vocal communication is favored in open habitats, and the fine structure of the signal acoustics may be expected to be more elaborate and varied (Waser and Brown 1986).

In fact, this relationship has been used for the vervet monkey, a species that regularly occupies open habitats, to explain both their vocal repertoire and their auditory pattern. Their vocal repertoire includes short-range sounds, some of which reach up to 5 kHz (Owren and Bernacki 1988) and have complex acoustic features (Seyfarth and Cheney 1984), and their auditory pattern shows a heightened sensitivity to frequencies between 1 and 8 kHz (Owren et al. 1988). Interestingly, both *A. africanus* and *P. robustus* are reported to have regularly consumed open habitat resources, perhaps comprising up to 50% of their diet (van der Merwe et al. 2003; Sponheimer et al. 2006). Thus, the early hominin auditory pattern, showing heightened sensitivity between 1.5 and 3.5 kHz, may have facilitated and reinforced an increased emphasis on short-range vocal communication in open habitats.

8.6.2 Vowel and Consonant Production

The extended occupied band of maximum sensitivity in both *H. sapiens* and the Atapuerca (SH) hominins reaches up to a frequency range that is mainly associated with consonant production in human spoken language. The frequency range from 3 to 5 kHz generally coincides with the "high consonant area" of the so-called "speech banana" (Fant 1973), which describes the frequencies emitted during spoken language (see Fig. 8.3). While other consonants do occur at lower frequencies where the human and chimpanzee audiograms do not differ significantly in their auditory sensitivity, the high-frequency consonants are particularly salient features in human spoken language. Consonants in this frequency range mainly consist of the voiceless plosives (stops), such as those associated with the letters /t/ and /k/, and the voiceless fricatives, including those associated with the letters /f/, /s/, and /th/. Plosives are the most common type of consonant, being found in 100% of the world's languages. Within this group of consonants, the voiceless plosives are present in over 90% of the world's languages with /t/ and /k/ being the most strongly represented (Maddieson 1984). Indeed, these two consonants were proposed to have been present in the first language (MacNeilage 1998). Similarly, fricative consonants are found in more than 90% of the world's languages, and among the fricatives, /s/ is the most common voiceless fricative, present in over 80% of languages, with /f/ also occurring at high frequencies (Maddieson 1984). Thus, an increased emphasis on high-frequency consonant production may have been an important component in the vocal repertoire of these Middle Pleistocene hominins, a feature that would have distinguished them from the pattern of vowel-based vocal communication in chimpanzees (Mitani et al. 1999) and, indeed, nearly all mammals (Lieberman 1975).

It has been suggested that communication in early hominins was likely largely restricted to just such a vowel-based system (Owren and Cardillo 2006). However, consonant-like sounds have been identified in the vocal repertoire of some great ape species, including *clicks*, *lip smacks*, *kisses*, and *whistles* (Lameira et al. 2014). Importantly, these consonant-like sounds are produced by the lips, teeth, and tongue independent of vocal fold vibration. Thus, these are voiceless or silent calls, and their production in living apes is compatible with the *frame/content theory* of speech evolution (MacNeilage 1998, 2008), which posits that the sounds that comprise human language initially evolved out of the cycles of mandible oscillation during feeding. Broadly speaking, vowels are produced when the mandible is depressed (i.e., the mouth is open), the vocal folds are vibrating, and the airflow is unobstructed through the vocal tract. In contrast, consonant production (including both voiced and voiceless consonants) is associated with an elevation of the mandible (i.e., closing of the mouth) and partial or total occlusion of the airflow through the vocal tract. Most forms of animal communication are limited to the *frame stage* and may include production of consonant-vowel (CV) combinations (i.e., syllables) and limited word formation. The subsequent *frame/content stage* characterizes human spoken language and involves sequential ordering of CV combinations, including new combinations, leading to an increase in vocabulary and sentence formation, syntax, and a developed symbolic component (MacNeilage 2008).

This distinction between the use of vowels and consonants seems to represent a fundamental step in the emergence of human speech, and there is evidence that vowels and consonants are processed as categorically distinct objects in the human brain (Caramazza et al. 2000). In addition, vowels in words have been demonstrated to play a more important role in determining the identity of the speaker while consonants are most important for determining word meaning (Owren and Cardillo 2006). The presence of voiceless calls in living ape species suggests the possibility that "… discrimination between voiceless protoconsonants and voiced protovowels would have already been possible at the earliest stages of consonant-vowel combinations" (p. 61 in Lameira et al. 2014). The development of voiceless consonants may have been a way to extend the vocal repertoire beyond the capabilities or limitations of the anatomy of the supralaryngeal vocal tract. The presence of these voiceless calls in living ape species suggests that early hominins were likely capable of producing a limited set of consonants and CV combinations.

Lindblom and Maddieson (1988) suggested that communication systems initially will preferentially include consonants that combine ease of articulatory complexity and acoustic distinctiveness. The voiceless consonants, including the high-frequency stop consonants, would seem to be an ideal category for early appearance since they are some of the most dissimilar to vowels (Caramazza et al. 2000), and they are among the easiest consonants to produce in terms of articulatory complexity (Lindblom and Maddieson 1988). Indeed, voiceless consonants may represent "…the evolutionarily oldest group of consonants" (p. 61 in Lameira et al. 2014). Such voiceless consonants are useful only in short-range communication (Sect. 8.6.1), and this emphasis on consonant production is consistent with the "…general perspective that speech evolved through imposing articulatory gestures on a vowel-based carrier signal" (p. 1736 in Owren and Cardillo 2006).

8.6.3 Modeling Language Evolution

While the beginnings of human speech may date to the origin of the genus *Homo* (Lieberman 2007b), the precise nature of vocal communication in fossil hominins is unknown. However, it need not be fully equivalent in syntax, linguistic flexibility, or cognitive complexity to present day human spoken language. Researchers who have taken a quantitative approach to the question of language evolution have focused on theoretical modeling of the emergence of words and syntax (Nowak and Krakauer 1999; Nowak 2000a). Results from these studies provide insight into the evolution of language, and some aspects find a parallel with the occupied bandwidth.

In particular, an error limit in communication has been identified as a general feature of the most basic (i.e., phoneme-based, presyntactic) communication systems, including those of most animals (Nowak et al. 1999). Individual phonemes (perceptually distinct units of sound) consist of single vowels or consonants. The presence of a finite number of phonemes in vocal communication places a limit on the information that can be transmitted successfully between individuals, since each phoneme refers to a distinct "object," including objects in the environment, other individuals, concepts, or actions. One obvious strategy to communicate additional information is to increase the number of phonemes to refer to more "objects," preferentially adding new phonemes that are as acoustically distinct as possible from previously existing phonemes. However, above a certain limit, increasing the repertoire of phonemes does not lead to an increase in the transmission of information since errors in communication become more frequent. As more and more phonemes are added, the acoustic distinctions between them become less and less clear, leading to confusion in interpreting the signal. Indeed, despite the spectacular linguistic diversity in *H. sapiens* today, of the 300+ living human languages for which a complete phoneme inventory exists, approximately 70% possess between twenty and thirty-seven phonemes (Nowak et al. 1999).

Human language has extended the error limit by combining phonemes into words. Generally, phonemes are combined into words as the amount of information to communicate increases beyond the error limit of phoneme-based communication (Nowak and Krakauer 1999). The relatively limited repertoire of phonemes in living languages suggests this threshold must be fairly low, and the combination of phonemes into words has been posited as an initial step in language evolution. Indeed, the first language has been suggested to have had about 100 words (MacNeilage 1998). Further word formation is achieved by combining more than two phonemes, forming longer words and leading to an increase in the information that can be transmitted as well as a reduction in errors in communication (Shannon and Weaver 1949). Word length, in turn, is limited by the number of "objects" it is necessary to communicate about (Nowak et al. 1999). Increases in the size of the vocabulary depend on the dynamics of social transmission of new words between individuals (Nowak 2000b). This leads to a potentially infinite number of words to describe a potentially infinite number of objects but also requires memorization of a large vocabulary.

Syntactic communication transmits even more information by combining two or more words into phrases, which Nowak and Krakuer (1999) argued evolved when the vocabulary reached a critical threshold. In particular, words describing objects and words describing actions can be combined into object-action pairings to describe an event. As long as the number of object-action pairings is small, nonsyntactic communication is sufficient to transmit the information. Indeed, in pre-syntactic communication, signals (i.e., phonemes or words) may refer to combined object-action pairings (holistic utterances), meaning there is often no clear distinction between nouns and verbs (Nowak et al. 2000). Syntactic communication is proposed to evolve when the number of events to be described equals or exceeds the number of nouns and verbs that can be used to describe the events, leading to errors in communication. As a minimum condition, the number of nouns and verbs need to equal or exceed three each, meaning more than nine possible combinations (Nowak et al. 2000). Thus, syntax evolves only when communication about a large number of events is necessary, and ultimately, a syntactic form of communication has more sentences than words (Nowak and Krakauer 1999).

The most obvious stimulus that increases the number of events to describe would appear to be increases in group size, and primate species that live in larger social groups do have larger vocal repertoires (McComb and Semple 2005). Limited evidence of possible syntax in the form of combination calls also has been reported for nonhuman primates (Zuberbühler 2002). Nevertheless, many species live in large, complex social groups yet, apparently, have not evolved syntactic communication. This implies that as socially complex as primate societies can be, the number of events to describe has not increased beyond the threshold for syntactic communication.

8.6.4 Hominin Communication

Given the context discussed in the previous section, changes in the occupied band can perhaps shed some additional light on the nature of hominin communication. The bandwidth is a direct reflection of the channel capacity with a wider band, indicating greater channel capacity and increased information transfer (Shannon 1948). The hearing pattern in the early hominins shows a narrow occupied band, suggesting that their communication pattern did not involve transmission of information beyond that of a chimpanzee, implying a lack of syntax. In addition, the brain size in early hominins is similar to that of chimpanzees, suggesting it is unlikely they possessed the cognitive abilities to master a large vocabulary or understand and use complex syntactic rules.

Nevertheless, there is evidence for referential signaling (Seyfarth et al. 1980; Zuberbühler 2000a) and for call (i.e., phoneme) combination in the vocalizations of some primate species (Zuberbühler and Lemasson 2014). Concatenation of calls into an ordered sequence with limited syntactic rules also has been documented in some species (Zuberbühler 2002). Since most primates have a restricted number of calls, these seem to be attempts to increase the information content of their limited vocal repertoire, a practice that approaches word formation. Indeed, individual phonemes could potentially function as words if they contain a semantic element. Thus, primate communication seems to be on the threshold, in some respects, of word formation. It is reasonable to infer, then, that early hominins likely had at least a phoneme-based presyntactic form of communication that likely involved some degree of CV syllable production, and their heightened auditory sensitivity between 1.5 and 3.0 kHz would have facilitated short-range communication, particularly in open environments. Thus, early hominin communication seems largely limited to the frame stage (MacNeilage 2008) and may have involved some form of "low-fidelity social transmission" beyond that of a chimpanzee (Morgan et al. 2015) but likely lacked a symbolic component.

Since human language exhibits word and sentence formation, syntax, and a developed symbolic component (MacNeilage 2008), *H. sapiens* has clearly passed into the frame/content stage. When hominin communication passed into the frame/content stage is not clear, partly because the distinction between phonemes and words is, at times, difficult to decipher. The hearing pattern in the Atapuerca (SH) hominins and *H. sapiens* shows a considerably wider occupied band, which extends toward higher frequencies and indicates a greater capacity for transmitting information. In addition, brain size has expanded considerably with some individuals from the Atapuerca (SH) site having brain sizes within the *H. sapiens* range of variation and showing cerebral asymmetries with inferred lateralization (Poza-Rey et al. 2015). Furthermore, the accumulation of bodies at the Atapuerca (SH) site has been argued to be a result of anthropic (i.e., hominin) activity and seems to document the earliest funerary practice (Arsuaga et al. 1997). This combination of anatomical and behavioral aspects indicates fairly complex cognitive abilities in the Atapuerca (SH) hominins.

If the linguistic capacities in the genus *Homo* evolved out of a phoneme-based, presyntactic form of communication in early hominins, then at some point the number of objects and events needing to be described accurately increased to the point that further word formation and syntax became necessary. It is reasonable to suggest that the wider occupied band of maximum sensitivity in the Atapuerca (SH) hominins facilitated specialization in the use of complex, short-range vocal communication, including an emphasis on high-frequency consonant production and increased word formation. Whether they had passed into the frame/content stage depends on the presence of syntax and a symbolic component, but it is plausible that their vocabulary included at least 100 words, the suggested initial corpus for the first language (MacNeilage 1998).

Aiello and Dunbar (1993) have argued that language evolved because humans needed to communicate a large amount of social information. These authors found a strong relationship between the relative size of the neocortex in the brain and social group size across primates, suggesting that larger social groups involve more complex social dynamics and cognitive abilities. They estimated social group size in fossil hominins based on brain size and found that group size, and hence social dynamics, would have reached similar levels as living humans, implying the appearance of language, sometime during the Middle Pleistocene. The extended occupied band in the Atapuerca (SH) hominins indicates this may have occurred prior to 430 Ka, and the presence of a shared, derived auditory pattern in both *H. sapiens* and the Neandertal clade suggests this represents an ancient feature of the genus *Homo*, one that was likely already present in their last common ancestor.

8.7 Conclusion

The strong similarity between the auditory capacities in the Atapuerca (SH) hominins and *H. sapiens* suggests that the skeletal characteristics of the outer and middle ears that support the perception of human spoken language were largely present in these hominins. This, in turn, might be taken to indicate that the anatomical structures related with speech production were already present in the human lineage by at least Middle Pleistocene times. Although the study of audition is an indirect approach to the question of speech capacity in fossil specimens, prior attempts to reconstruct the linguistic capacities in fossil hominins have led to conflicting results and often have not been based on sound anatomical relationships between skeletal structures and speech production (e.g., Arensburg et al. 1990; Kay et al. 1998).

In contrast, combining the study of auditory capacities with the frame/content theory of speech evolution (MacNeilage, 1998) provides meaningful insights into the emergence of language in fossil hominins. The first step in the evolutionary transformation of the hominin audiogram involved an increase in sensitivity to the midrange frequencies between 1.0 and 3.5 kHz, and the early hominins appear to have been restricted to the frame stage. The second step in the evolution of the human audiogram involved an extension of the occupied band of maximum sensitivity toward higher frequencies, indicating an increased capacity for transmitting information, and the Atapuerca (SH) hominins may be on the threshold of passing into the frame/content stage.

Promising future lines of inquiry include ancient DNA analysis and reconstructing the hearing pattern in an early member of the genus *Homo*. In particular, the recovery of nuclear DNA from the Sima de los Huesos hominins (Meyer et al. 2016) makes it possible to further refine our understanding of when the human version of the *FOXP2* gene may have first evolved. Several additional genes related to hearing that show evidence of accelerated evolution in humans should be targeted for retrieval in ancient DNA studies. Reconstruction of the auditory capacities in an early *Homo* species (e.g., *H. naledi*) would reveal whether the *H. sapiens* hearing pattern, including an extended occupied band, characterizes the entire genus or is a more recent development. The implications drawn from the auditory capacities in fossil hominins are consistent with other studies that have approached language evolution from different perspectives (Aiello and Dunbar 1993; Krause et al. 2007; Lieberman 2007b). Collectively, these different avenues of inquiry form a strong argument in favor of the presence of some form of spoken language in the genus *Homo* prior to the appearance of *H. sapiens*.

Acknowledgments Rolf Quam, Ignacio Martínez, and Manuel Rosa form part of the Evolutionary Bioacoustics and Paleoanthropology Research Group at the Universidad de Alcalá. The authors thank C. Lorenzo, A. Bonmatí, and P. Jarabo for their valuable contributions to the present work. The following individuals and institutions provided access to the fossil and extant specimens housed under their care: J.M. Carretero (Universidad de Burgos, Spain); Y. Haile-Selassie, B. Latimer, and L. Jellema (Cleveland Museum of Natural History, USA); J. Cabot (Estación Biológica Doñana, Spain); F. Thackeray and S. Potze (Ditsong National Museum of Natural History, South Africa); P.V. Tobias, B. Zipfel, and M. Raath (University of the Witwatersrand, South Africa). CT scanning of the early hominin specimens was carried out by D. De Ruiter, J. Moggi-Cecchi, and C. Menter. CT scanning of the *H. sapiens* and chimpanzee specimens was carried out at the Laboratorio de Evolucion Humana of the Universidad de Burgos (Spain) and the Cleveland Clinic (USA). B. Tatchell analyzed the CT scans and made the virtual reconstructions for a number of specimens in the present study. The authors thank the Atapuerca Research and Excavation Team for their work in the field, especially at the Sima de los Huesos site. The excavations at the Atapuerca sites are funded by the Junta de Castilla y León. Partial financial support for this project has been provided by the Leakey Foundation, the American Museum of Natural History (AMNH), Binghamton University (SUNY), the Junta de Castilla y León (project nos. BU032A06 and BU005A09), and the Spanish Ministry of Economía y Competitividad (project nos. CGL2009-12703-C03-03/02, CGL2012-38434-C03-01, and CGL2015-65387-C3-2-P-MINECO/FEDER).

Compliance with Ethics Requirements
Rolf Quam, Ignacio Martínez, Manuel Rosa, and Juan Luis Arsuaga declare that they have no conflicts of interest.

References

Aibara, R., Welsh, J., Puria, S., & Goode, R. (2001). Human middle-ear sound transfer function and cochlear input impedance. *Hearing Research*, 152, 100–109.

Aiello, L., & Dean, C. (1990). *An introduction to human evolutionary anatomy*. London: Academic.

Aiello, L., & Dunbar, R. (1993). Neocortex size, group size and the evolution of language. *Current Anthropology*, 34(2), 184–193.

Alba, D. M., Moyà-Solà, S., & Köhler, M. (2003). Morphological affinities of the *Australopithecus afarensis* hand on the basis of manual proportions and relative thumb length. *Journal of Human Evolution*, 44(2), 225–254.

Alemseged, Z., Spoor, F., Kimbel, W., Bobe, R., et al. (2006). A juvenile early hominin skeleton from Dikika, Ethiopia. *Nature*, 443, 296–301.

Arensburg, B., Tillier, A., Vandermeersch, B., Duday, H., et al. (1989). A Middle Paleolithic human hyoid bone. *Nature*, 338, 758–760.

Arensburg, B., Schepartz, L., Tillier, A., Vandermeersch, B., et al. (1990). A reappraisal of the anatomical basis for speech in Middle Paleolithic hominids. *American Journal of Physical Anthropology*, 83, 137–146.

Arsuaga, J. L., Martínez, I., Gracia, A., Carretero, J. M., et al. (1997). Sima de los Huesos (Sierra de Atapuerca, Spain). The site. *Journal of Human Evolution*, 33(2/3), 109–128.

Arsuaga, J. L., Martínez, I., Arnold, L., Aranburu, A., et al. (2014). Neandertal roots: Cranial and chronological evidence from Sima de los Huesos. *Science*, 344(6190), 1358–1363.

Arsuaga, J. L., Carretero, J.-M., Lorenzo, C., Gómez-Olivencia, A., et al. (2015). Postcranial morphology of the middle Pleistocene humans from Sima de los Huesos, Spain. *Proceedings of the National Academy of Sciences of the United States of America*, 112(37), 11524–11529.

Balzeau, A., & Grimaud-Hervé, D. (2006). Cranial base morphology and temporal bone pneumatization in Asian *Homo erectus*. *Journal of Human Evolution*, 51, 350–359.

Balzeau, A., & Radovčić, J. (2008). Variation and modalities of growth and development of the temporal bone pneumatization in Neandertals. *Journal of Human Evolution*, 54, 546–567.

Bastir, M., & Rosas, A. (2009). Mosaic evolution of the basicranium in *Homo* and its relation to modular development. *Evolutionary Biology*, 36(1), 57–70.

Beals, M. E., Frayer, D. W., Radovčić, J., & Hill, C. A. (2016). Cochlear labyrinth volume in Krapina Neandertals. *Journal of Human Evolution*, 90, 176–182.

Berger, L., Ruiter, D., Churchill, S., Schmid, P., et al. (2010). *Australopithecus sediba*: A new species of *Homo*-like australopith from South Africa. *Science*, 328, 195–204.

Berger, L. R., Hawks, J., de Ruiter, D. J., Churchill, S. E., et al. (2015). *Homo naledi*, a new species of the genus *Homo* from the Dinaledi Chamber, South Africa. *eLife*, 4, e09560.

Bermúdez de Castro, J. M., Arsuaga, J. L., Carbonell, E., Rosas, A., et al. (1997). A hominid from the Lower Pleistocene of Atapuerca, Spain: Possible ancestor to Neandertals and modern humans. *Science*, 276, 1392–1395.

Boë, L.-J., Heim, J. L., Honda, K., Maeda, S., et al. (2007). The vocal tract of newborn humans and Neanderthals: Acoustic capabilities and consequences for the debate on the origin of language. A reply to Lieberman (2007). *Journal of Phonetics*, 35(4), 564–581.

Boë, L., Heim, J., Honda, K., & Maeda, S. (2002). The potential Neandertal vowel space was as large as that of modern humans. *Journal of Phonetics*, 30, 465–484.

Bouzouggar, A., Barton, N., Vanhaeren, M., d'Errico, F., et al. (2007). 82,000-year-old shell beads from North Africa and implications for the origins of modern human behavior. *Proceedings of the National Academy of Sciences of the United States of America*, 104(24), 9964–9969.

Braga, J., Loubes, J., Descouens, D., Dumoncel, J., et al. (2015). Disproportionate cochlear length in genus *Homo* shows a high phylogenetic signal during apes' hearing evolution. *PLoS One*, 10(6), e0127780.

Brown, C., & Waser, P. (1984). Hearing and communication in blue monkeys (*Cercopithecus mitis*). *Animal Behaviour*, 32, 66–75.

Brown, C., & Waser, P. (1988). Environmental influences on the structure of primate vocalizations. In D. Todt, P. Goedeking, & D. Symmes (Eds.), *Primate vocal communication* (pp. 51–66). Berlin: Springer.

Brunet, M., Guy, F., Pilbeam, D., Mackaye, H., et al. (2002). A new hominid from the Upper Miocene of Chad, Central Africa. *Nature*, 418, 145–151.

Caramazza, A., Chialant, D., Capasso, R., & Miceli, G. (2000). Separable processing of consonants and vowels. *Nature*, 403, 428–430.

Carbonell, E., Bermúdez de Castro, J., Parés, J., Pérez-González, A., et al. (2008). The first hominin of Europe. *Nature*, 452, 465–469.

Caron, F., d'Errico, F., Del Moral, P., Santos, F., et al. (2011). The reality of Neandertal symbolic behavior at the Grotte du Renne, Arcy-sur-Cure, France. *PLoS One*, 6(6), e21545.

Clark, A., Glanowski, S., Nielsen, R., Thomas, P., et al. (2003). Inferring nonneutral evolution from human-chimp-mouse orthologous gene trios. *Science*, 302, 1960–1963.
Coleman, M. N. (2009). What do primates hear? A meta-analysis of all known nonhuman primate behavioral audiograms. *International Journal of Primatology*, 30, 55–91.
Coleman, M. N., & Colbert, M. W. (2010). Correlations between auditory structures and hearing sensitivity in non-human primates. *Journal of Morphology*, 271, 511–532.
Coleman, M. N., & Ross, C. (2004). Primate auditory diversity and its influence on hearing performance. *Anatomical Record Part A: Discoveries in Molecular, Cellular, and Evolutionary Biology*, 281, 1123–1137.
Coleman, M. N., Kay, R., & Colbert, M. (2010). Auditory morphology and hearing sensitivity in fossil New World Monkeys. *Anatomical Record*, 293, 1711–1721.
Constantino, P., & Wood, B. (2004). *Paranthropus* paleobiology. In E. Baquedano & S. Rubio Jara (Eds.), *Miscelanea en homenaje a Emiliano Aguirre, Paleoantropologia* (Vol. 3, pp. 137–151). Madrid: Museo Arqueologico Regional.
Dean, M., & Wood, B. (1981). Metrical analysis of the basicranium of extant hominoids and *Australopithecus*. *American Journal of Physical Anthropology*, 54, 63–71.
Dean, M., & Wood, B. (1982). Basicranial anatomy of Plio-Pleistocene hominids from East and South Africa. *American Journal of Physical Anthropology*, 59, 157–174.
de Boer, B. (2012). Loss of air sacs improved hominin speech abilities. *Journal of Human Evolution*, 62(1), 1–6.
d'Errico, F., Zilhao, J., Julien, M., Baffier, D., et al. (1998). Neandertal acculturation in Western Europe? A critical review of the evidence and its interpretation. *Current Anthropology*, 39(Supp.)), S1–S44.
Elder, J. (1934). Auditory acuity of the chimpanzee. *Journal of Comparative Psychology*, 17, 157–183.
Elder, J. (1935). The upper limit of hearing in chimpanzee. *American Journal of Physiology*, 112, 109–115.
Fant, C. G. M. (1973). *Speech sounds and features*. Cambridge, MA: MIT Press.
Fitch, W. T., & Reby, D. (2001). The descended larynx is not uniquely human. *Proceedings of the Royal Society of London B: Biological Sciences*, 268(1477), 1669–1675.
Frenz, D., McPhee, J., & van de Water, T. (2001). Structural and functional development of the ear. In A. Jahn & J. Santos-Sacchi (Eds.), *Physiology of the ear* (2nd ed., pp. 191–214). San Diego: Singular.
Fu, Q., Hajdinjak, M., Moldovan, O. T., Constantin, S., et al. (2015). An early modern human from Romania with a recent Neanderthal ancestor. *Nature*, 524(7564), 216–219.
Grodzinksy, Y. (2000). The neurology of syntax: Language use without Broca's area. *Behavioral and Brain Sciences*, 23, 1–71.
Gunz, P., Stoessel, A., Neubauer, S., Kuhrig, M., et al. (2013). Morphological integration of the bony labyrinth and the cranial base in modern humans and Neandertals. *Proceedings of the European Society for the Study of Human Evolution, 2013*, 104. (Abstract).
Harmand, S., Lewis, J. E., Feibel, C. S., Lepre, C. J., et al. (2015). 3.3-million-year-old stone tools from Lomekwi 3, West Turkana, Kenya. *Nature*, 521(7552), 310–315.
Harvati, K. (2007). Neanderthals and their contemporaries. In H. Winfried & I. Tattersall (Eds.), *Handbook of paleoanthropology* (pp. 1717–1748). Berlin: Springer.
Heffner, R. S. (2004). Primate hearing from a mammalian perspective. *Anatomical Record Part A: Discoveries in Molecular, Cellular, and Evolutionary Biology*, 281, 1111–1122.
Henshilwood, C. S., D'Errico, F., Marean, C. W., Milo, R. G., et al. (2001). An early bone tool industry from the Middle Stone Age at Blombos Cave, South Africa: Implications for the origins of modern human behaviour, symbolism and language. *Journal of Human Evolution*, 41(6), 631–678.
Henshilwood, C., d'Errico, F., Yates, R., Jacobs, Z., et al. (2002). Emergence of modern human behavior: Middle Stone Age engravings from South Africa. *Science*, 295, 1278–1280.
Henshilwood, C., d'Errico, F., Vanhaeren, M., van Niekerk, K., et al. (2004). Middle Stone Age shell beads from South Africa. *Science*, 304, 404.

Holloway, R., Broadfield, D., & Yuan, M. (2004). *The human fossil record. Vol. 3, Brain endocasts. The paleoneurological evidence*. New York: Wiley-Liss.
House, E. L. (1972). The ear. In G. Bourne (Ed.), *The chimpanzee. Histology, reproduction and restraint* (Vol. 5, pp. 153–192). Baltimore, MD: University Park Press.
Hughes, G., & Halle, M. (1956). Spectral properties of fricative consonants. *The Journal of the Acoustical Society of America*, 28(2), 303–310.
ITUR. (2000). Radiocommunication vocabulary: International Telecommunication Union recommendations. ITU-R V.573-5. https://www.itu.int/dms_pubrec/itu-r/rec/v/R-REC-V.573-5-200709-S!!PDF-E.pdf
Jaubert, J., Verheyden, S., Genty, D., Soulier, M., et al. (2016). Early Neanderthal constructions deep in Bruniquel Cave in southwestern France. *Nature*, 534(7605), 111–114.
Kay, R., Cartmill, M., & Balow, M. (1998). The hypoglossal canal and the origin of human vocal behavior. *Proceedings of the National Academy of Sciences of the United States of America*, 95, 5417–5419.
Kimbel, W., Rak, Y., & Johanson, D. (2004). *The skull of Australopithecus afarensis*. Oxford: Oxford University Press.
Kojima, S. (1990). Comparison of auditory functions in the chimpanzee and human. *Folia Primatologica*, 55, 62–72.
Krause, J., Lalueza-Fox, C., Orlando, L., Enard, W., et al. (2007). The derived FOXP2 variant of modern humans was shared with Neandertals. *Current Biology*, 17(21), 1908–1912.
Ladefoged, P., & Maddieson, I. (1996). *The sounds of the world's languages*. Oxford: Blackwell.
Lai, C. S., Fisher, S. E., Hurst, J. A., Vargha-Khadem, F., et al. (2001). A forkhead-domain gene is mutated in a severe speech and language disorder. *Nature*, 413(6855), 519–523.
Laitman, J., Heimbuch, R., & Crelin, E. (1979). The basicranium of fossil hominids as an indicator of their upper respiratory system. *American Journal of Physical Anthropology*, 51, 15–34.
Lameira, A. R., Maddieson, I., & Zuberbühler, K. (2014). Primate feedstock for the evolution of consonants. *Trends in Cognitive Sciences*, 18(2), 60–62.
Lieberman, P. (1975). *On the origin of language*. New York: Macmillan.
Lieberman, P. (2007a). Current views on Neanderthal speech capabilities: A reply to Boe et al. (2002). *Journal of Phonetics*, 35, 552–563.
Lieberman, P. (2007b). The evolution of human speech: Its anatomical and neural bases. *Current Anthropology*, 48(1), 39–66.
Lieberman, P., Laitman, J. T., Reidenberg, J. S., & Gannon, P. J. (1992). The anatomy, physiology, acoustics and perception of speech: Essential elements in analysis of the evolution of human speech. *Journal of Human Evolution*, 23(6), 447–467.
Lindblom, B., & Maddieson, I. (1988). Phonetic universals in consonant systems. In L. Hyman & C. Li (Eds.), *Language, speech, and mind: Studies in honour of Victoria A. Fromkin* (pp. 62–78). London: Routledge.
MacNeilage, P. F. (1998). The frame/content theory of evolution of speech production. *Behavioral and Brain Sciences*, 214, 499–511.
MacNeilage, P. F. (2008). *The origin of speech*. Oxford: Oxford University Press.
Maddieson, I. (1984). *Patterns of sounds*. Cambridge: Cambridge University Press.
Mallo, M. (2003). Formation of the outer and middle ear, molecular mechanisms. *Current Topics in Developmental Biology*, 57, 85–113.
Mallo, M., Schrewe, H., Martin, J., Olson, E., et al. (2000). Assembling a functional tympanic membrane: Signals from the external acoustic meatus coordinate development of the malleal manubrium. *Development*, 127, 4127–4136.
Martínez, I., & Arsuaga, J. L. (1997). The temporal bones from Sima de los Huesos Middle Pleistocene site (Sierra de Atapuerca, Spain). A phylogenetic approach. *Journal of Human Evolution*, 33(2/3), 283–318.
Martínez, I., Rosa, M., Arsuaga, J., Jarabo, P., et al. (2004). Auditory capacities in Middle Pleistocene humans from the Sierra de Atapuerca in Spain. *Proceedings of the National Academy of Sciences of the United States of America*, 101(27), 9976–9981.

Martínez, I., Arsuaga, J. L., Quam, R., Carretero, J. M., et al. (2008). Human hyoid bones from the middle Pleistocene site of the Sima de los Huesos (Sierra de Atapuerca, Spain). *Journal of Human Evolution*, 54(1), 118–124.

Martínez, I., Rosa, M., Quam, R., Jarabo, P., et al. (2013). Communicative capacities in Middle Pleistocene humans from the Sierra de Atapuerca in Spain. *Quaternary International*, 295, 94–101.

Masali, M., Maffei, M., & Borgognini Tarli, S. M. (1991). Application of a morphometric model for the reconstruction of some functional characteristics of the external and middle ear in Circeo 1. In M. Piperno & G. Scichilone (Eds.), *The Circeo 1 Neandertal skull: Studies and documentation* (pp. 321–338). Rome: Instituto Poligrafico e Zecca Dello Stato.

Masterton, B., Heffner, H., & Ravizza, R. (1969). The evolution of human hearing. *The Journal of the Acoustical Society of America*, 45(4), 966–985.

McComb, K., & Semple, S. (2005). Coevolution of vocal communication and sociality in primates. *Biology Letters*, 1(4), 381–385.

McDougall, I., Brown, F., & Fleagle, J. (2005). Stratigraphic placement and age of modern humans from Kibish, Ethiopia. *Nature*, 433, 733–736.

McPherron, S., Alemseged, Z., Marean, C., Wynn, J., et al. (2010). Evidence for stone-tool-assisted consumption of animal tissues before 3.39 million years ago at Dikika, Ethiopia. *Nature* 466, 857–860.

Mellars, P. (2010). Neanderthal symbolism and ornament manufacture: The bursting of a bubble? *Proceedings of the National Academy of Sciences of the United States of America*, 107(47), 20147–20148.

Meyer, M., Arsuaga, J.-L., de Filippo, C., Nagel, S., et al. (2016). Nuclear DNA sequences from the Middle Pleistocene Sima de los Huesos hominins. *Nature*, 531(7595), 504–507.

Mitani, J., Hunley, K., & Murdoch, M. (1999). Geographic variation in the calls of wild chimpanzees: A reassessment. *American Journal of Primatology*, 47, 133–151.

Moggi-Cecchi, J., & Collard, M. (2002). A fossil stapes from Sterkfontein, South Africa, and the hearing capabilities of early hominids. *Journal of Human Evolution*, 42, 259–265.

Moore, J. K., & Linthicum Jr., F. H. (2007). The human auditory system: A timeline of development. *International Journal of Audiology*, 46(9), 460–478.

Morgan, T., Uomini, N., Rendell, L., Chouinard-Thuly, L., et al. (2015). Experimental evidence for the co-evolution of hominin tool-making teaching and language. *Nature Communications*, *6*.

Nishimura, T., Mikami, A., Suzuki, J., & Matsuzawa, T. (2006). Descent of the hyoid in chimpanzees: Evolution of face flattening and speech. *Journal of Human Evolution*, 51, 244–254.

Nowak, M. A. (2000a). Evolutionary biology of language. *Philosophical Transactions of the Royal Society of London B: Biological Sciences*, 355(1403), 1615–1622.

Nowak, M. A. (2000b). The basic reproductive ratio of a word, the maximum size of a lexicon. *Journal of Theoretical Biology*, 204(2), 179–189.

Nowak, M. A., & Krakauer, D. C. (1999). The evolution of language. *Proceedings of the National Academy of Sciences of the United States of America*, 96(14), 8028–8033.

Nowak, M. A., Krakauer, D. C., & Dress, A. (1999). An error limit for the evolution of language. *Proceedings of the Royal Society of London B: Biological Sciences*, 266(1433), 2131–2136.

Nowak, M. A., Plotkin, J. B., & Jansen, V. A. (2000). The evolution of syntactic communication. *Nature*, 404(6777), 495–498.

O'Connor, K. N., & Puria, S. (2008). Middle-ear circuit model parameters based on a population of human ears. *The Journal of the Acoustical Society of America*, 123(1), 197–211.

Owren, M., & Bernacki, R. (1988). The acoustic features of vervet monkey alarm calls. *The Journal of the Acoustical Society of America*, 83(5), 1927–1935.

Owren, M., & Cardillo, G. (2006). The relative roles of vowels and consonants in discriminating talker identity versus word meaning. *The Journal of the Acoustical Society of America*, 119(3), 1727–1739.

Owren, M., Hopp, S., Sinnott, J., & Petersen, M. (1988). Absolute auditory thresholds in three Old World Monkey species (*Cercopithecus aethiops, C. neglectus, Macaca fuscata*) and humans (*Homo sapiens*). *Journal of Comparative Psychology*, 102, 99–107.

Patterson, N., Richter, D. J., Gnerre, S., Lander, E. S., et al. (2006). Genetic evidence for complex speciation of humans and chimpanzees. *Nature*, 441(7097), 1103–1108.
Pike, A. W., Hoffmann, D., García-Diez, M., Pettitt, P. B., et al. (2012). U-series dating of Paleolithic art in 11 caves in Spain. *Science*, 336(6087), 1409–1413.
Poza-Rey, E. M., Lozano, M., & Arsuaga, J. L. (2015). Brain asymmetries and handedness in the specimens from the Sima de los Huesos site (Atapuerca, Spain). *Quaternary International*, 433, 32–44. http://dx.doi.org/10.1016/j.quaint.2015.10.004.
Quam, R. (2006). *Temporal bone anatomy and the evolution of acoustic capacities in fossil humans*. Ph.D. dissertation, Binghamton University (SUNY), Binghamton, NY.
Quam, R., & Rak, Y. (2008). Auditory ossicles from southwest Asian Mousterian sites. *Journal of Human Evolution*, 54, 414–433.
Quam, R., Martinez, I., & Arsuaga, J. L. (2006). Middle Pleistocene auditory ossicles from the Sierra de Atapuerca (Spain). *American Journal of Physical Anthropology*, 42(Suppl), 149. (Abstract).
Quam, R., Martínez, I., Lorenzo, C., Bonmatí, A., et al. (2012). Studying audition in fossil hominins: A new approach to the evolution of language? In M. Jackson (Ed.), *Psychology of language* (pp. 47–95). Hauppage, NY: Nova Science Publishers, Inc.
Quam, R., Martínez, I., & Arsuaga, J. L. (2013a). Reassessment of the La Ferrassie 3 Neandertal ossicular chain. *Journal of Human Evolution*, 64(4), 250–262.
Quam, R., De Ruiter, D., Masali, M., Arsuaga, J., et al. (2013b). Early hominin auditory ossicles from South Africa. *Proceedings of the National Academy of Sciences of the United States of America*, 110(22), 8847–8851.
Quam, R., Martínez, I., Rosa, M., Bonmatí, A., et al. (2015). Early hominin auditory capacities. *Science Advances*, 1(8), e1500355.
Quam, R., Lorenzo, C., Martínez, I., Gracia-Téllez, A., et al. (2016). The bony labyrinth of the middle Pleistocene Sima de los Huesos hominins (Sierra de Atapuerca, Spain). *Journal of Human Evolution*, 90, 1–15.
Rak, Y., & Clarke, R. J. (1979). Aspects of the middle and external ear of early South African hominids. *American Journal of Physical Anthropology*, 51, 471–474.
Rodríguez-Vidal, J., d'Errico, F., Pacheco, F. G., Blasco, R., et al. (2014). A rock engraving made by Neanderthals in Gibraltar. *Proceedings of the National Academy of Sciences of the United States of America*, 111(37), 13301–13306.
Rosowski, J. (1991). The effects of external and middle ear filtering on auditory threshold and noise-induced hearing loss. *The Journal of the Acoustical Society of America*, 90, 124–135.
Rosowski, J. (1994). Outer and middle ears. In R. R. Fay & A. N. Popper (Eds.), *Comparative hearing: Mammals* (pp. 172–247). New York: Springer.
Rosowski, J. (1996). Models of external- and middle-ear function. In H. L. Hawkins, T. A. McMullen, A. N. Popper, & R. R. Fay (Eds.), *Auditory computation* (pp. 15–61). New York: Springer.
Rosowski, J., & Graybeal, A. (1991). What did *Morganucodon* hear? *Zoological Journal of the Linnean Society*, 101, 131–168.
Ruff, C., Trinkaus, E., & Holliday, T. (1997). Body mass and encephalization in Pleistocene *Homo*. *Nature*, 387, 173–176.
Scally, A., Dutheil, J. Y., Hillier, L. W., Jordan, G. E., et al. (2012). Insights into hominid evolution from the gorilla genome sequence. *Nature*, 483(7388), 169–175.
Senut, B., Pickford, M., Gommery, D., Mein, P., et al. (2001). First hominid from the Miocene (Lukeino Formation, Kenya). *Comptes Rendus de l'Académie des Sciences de Paris*, 332, 137–144.
Seyfarth, R., & Cheney, D. (1984). The acoustic features of vervet monkey grunts. *The Journal of the Acoustical Society of America*, 75, 1623–1628.
Seyfarth, R., & Cheney, D. (2003). Signalers and receivers in animal communication. *Annual Review of Psychology*, 54, 145–173.
Seyfarth, R. M., Cheney, D. L., & Marler, P. (1980). Monkey responses to three different alarm calls: Evidence for semantic communication and predator classification. *Science*, 210, 801–803.

Shannon, C. (1948). A mathematical theory of communication. *The Bell System Technical Journal*, 27, 379–423.

Shannon, C. E., & Weaver, W. (1949). *The mathematical theory of communication*. Champaign-Urbana, IL: University of Illinois Press.

Sherwood, C., Broadfield, D., Holloway, R., Gannon, P., et al. (2003). Variability of Broca's area homologue in African great apes: Implications for language evolution. *Anatomical Record Part A: Discoveries in Molecular, Cellular, and Evolutionary Biology*, 271(2), 276–285.

Sherwood, R. (1999). Pneumatic processes in the temporal bone of chimpanzee (*Pan troglodytes*) and gorilla (*Gorilla gorilla*). *Journal of Morphology*, 241, 127–137.

Sivian, L., & White, S. (1933). On minimum audible sound fields. *The Journal of the Acoustical Society of America*, 4, 288–321.

Sponheimer, M., Passey, B., de Ruiter, D., Guatelli-Steinberg, D., et al. (2006). Isotopic evidence for dietary variability in the early hominin *Paranthropus robustus*. *Science*, 314, 980–982.

Spoor, F., Wood, B., & Zonneveld, F. (1994). Implications of early hominid labyrinthine morphology for the evolution of human bipedal locomotion. *Nature*, 369, 645–648.

Tattersall, I. (2004). What happened in the origin of human consciousness? *Anatomical Record Part B: The New Anatomist*, 276B, 19–26.

Tattersall, I. (2009). Human origins: Out of Africa. *Proceedings of the National Academy of Sciences of the United States of America*, 106(38), 16018–16021.

Tobias, P. V. (1987). The brain of *Homo habilis*: A new level of organization in cerebral evolution. *Journal of Human Evolution*, 16, 741–761.

van der Merwe, N., Thackeray, J., Lee-Thorp, J., & Luyt, J. (2003). The carbon isotope ecology and diet of *Australopithecus africanus* at Sterkfontein, South Africa. *Journal of Human Evolution*, 44, 581–597.

Villmoare, B., Kimbel, W. H., Seyoum, C., Campisano, C. J., et al. (2015). Early *Homo* at 2 8 Ma from Ledi-Geraru, Afar, Ethiopia. *Science*, 347(6228), 1352–1355.

Waser, P., & Brown, C. (1986). Habitat acoustics and primate communication. *American Journal of Primatology*, 10, 135–154.

West, C. (1985). The relationship of the spiral turns of the cochlea and the length of the basilar membrane to the range of audible frequencies in ground dwelling mammals. *The Journal of the Acoustical Society of America*, 77(3), 1091–1101.

White, T., Asfaw, B., DeGusta, D., Gilbert, H., et al. (2003). Pleistocene *Homo sapiens* from Middle Awash, Ethiopia. *Nature*, 423, 742–747.

White, T. D., WoldeGabriel, G., Asfaw, B., Ambrose, S., et al. (2006). Asa Issie, Aramis and the origin of Australopithecus. *Nature*, 440(7086), 883–889.

Wood, B., & Lonergan, N. (2008). The hominin fossil record: Taxa, grades and clades. *Journal of Anatomy*, 212(4), 354–376.

Zuberbühler, K. (2000a). Referential labelling in Diana monkeys. *Animal Behaviour*, 59(5), 917–927.

Zuberbühler, K. (2000b). Interspecies semantic communication in two forest primates. *Proceedings of the Royal Society of London B: Biological Sciences*, 267(1444), 713–718.

Zuberbühler, K. (2002). A syntactic rule in forest monkey communication. *Animal Behaviour*, 63(2), 293–299.

Zuberbühler, K., & Lemasson, A. (2014). Primate communication: Meaning from strings of calls. In F. Lowenthal & L. Lefebvre (Eds.), *Language and recursion* (pp. 115–125). New York: Springer Science+Business Media.

MIX
Papier aus verantwortungsvollen Quellen
Paper from responsible sources
FSC® C105338

If you have any concerns about our products,
you can contact us on
ProductSafety@springernature.com

In case Publisher is established outside the EU,
the EU authorized representative is:
Springer Nature Customer Service Center GmbH
Europaplatz 3, 69115 Heidelberg, Germany

Printed by Libri Plureos GmbH
in Hamburg, Germany